GREEN CONSTRUCTION PROJECT MANAGEMENT AND COST OVERSIGHT

GREEN CONSTRUCTION PROJECT MANAGEMENT AND COST OVERSIGHT

Sam Kubba Ph.D., LEED AP

AMSTERDAM • BOSTON • HEIDELBERG • LONDON
NEW YORK • OXFORD • PARIS • SAN DIEGO
SAN FRANCISCO • SINGAPORE • SYDNEY • TOKYO
Architectural Press is an imprint of Elsevier

ELSEVIER

Architectural
Press

Architectural Press is an imprint of Elsevier
The Boulevard, Langford Lane, Kidlington, Oxford OX5 1GB, UK
30 Corporate Drive, Suite 400, Burlington, MA 01803, USA

Notices
Knowledge and best practice in this field are constantly changing. As new research and experience broaden our understanding, changes in research methods, professional practices, or medical treatment may become necessary.

Practitioners and researchers must always rely on their own experience and knowledge in evaluating and using any information, methods, compounds, or experiments described herein. In using such information or methods they should be mindful of their own safety and the safety of others, including parties for whom they have a professional responsibility.

To the fullest extent of the law, neither the Publisher nor the authors, contributors, or editors, assume any liability for any injury and/or damage to persons or property as a matter of products liability, negligence or otherwise, or from any use or operation of any methods, products, instructions, or ideas contained in the material herein.

Library of Congress Cataloging-in-Publication Data
Kubba, Sam.
  Green construction project management and cost oversight/Sam Kubba.
    p. cm.
  Includes bibliographical references and index.
  ISBN 978-1-85617-676-7 (alk. paper)
  1. Sustainable construction. 2. Building–Cost control. I. Title.
  TH880.K8393 2010
  690.068′4–dc22

                                                                        2009052215

British Library Cataloguing-in-Publication Data
A catalogue record for this book is available from the British Library.

For information on all Architectural Press publications
visit our Web site at www.elsevierdirect.com

Printed in the United States of America

10 11 12 13 14   10 9 8 7 6 5 4 3 2 1

**Working together to grow
libraries in developing countries**

www.elsevier.com | www.bookaid.org | www.sabre.org

ELSEVIER   BOOK AID
            International   Sabre Foundation

To my mother and father,
Who bestowed on me the gift of life…
And to my wife and four children,
Whose love and affection inspired me on…

Contents

1

"GREEN" AND "SUSTAINABILITY" DEFINED

2

ELEMENTS OF GREEN DESIGN AND CONSTRUCTION

3

GREEN PROJECT REQUIREMENTS AND STRATEGIES

4

GREEN CONSTRUCTION COST MONITORING

5

HOW THE CONSULTANT FUNCTIONS IN THE REQUISITION PROCESS

6

CHOOSING MATERIALS AND PRODUCTS

7

PROJECT COST BREAKDOWN

8

GREEN DESIGN AND CONSTRUCTION ECONOMICS

9

UNDERSTANDING SPECIFICATIONS

10

LITIGATION AND LIABILITY ISSUES

11

GREEN PROJECT COMMISSIONING

12

GREEN BUSINESS DEVELOPMENT

Foreword

"Building green" and "sustainability" are words that permeate just about every publication these days dealing with construction, leasing, asset management, financing, and development. In much of the same way that non-animal based food products are often labeled "cholesterol free" so as to cash-in on the health craze, so too are the words "green" and "sustainability" with respect to buildings.

Dr. Kubba recognizes that "green" and "sustainability" are becoming the new paradigm, especially for end users under a corporate directive to lease or acquire LEED buildings. This book cuts through the hype with pragmatic information and procedures providing the user with sufficient information to make the necessary decisions with confidence when deciding to go "green," and the steps to implement a successful green building program. Besides the societal and health benefits that accrue to the occupants, energy efficient buildings are a necessity. Although at the time of this writing crude oil is priced at $70/barrel, it was only 14 months ago when it reached its peak of $147. Given that most buildings have an expected useful life of about 50 years, peak oil will be realized during this term, and the ensuing competition for this resource will manifest itself in an increase in all operating costs. This pertains not only to building energy consumed, but to embedded energy as well, which is incorporated in all building materials and services. Just as we as a nation must re-think our over consumption, so too must we re-think our built environment to take into consideration energy efficiency and increases in productivity.

Regardless of corporate directives, going green is simply good business. End users will be seeking buildings that are cost efficient to operate, and such buildings will have a competitive advantage over those that are not.

Sustainability is becoming standard practice—not only because it corresponds to best practices, but out of necessity.

Carl de Stefanis, PE, CEM, LEED AP
President
Energy Reduction Solutions, Inc.
White Plains, New York 10604
www.e-reduction.com

Acknowledgments

A book of this scope would not have been possible without the active and passive support of many friends and colleagues who have contributed greatly to my thinking and insights during the writing of this book and who were in many ways instrumental in the crystallization and formulation of my thoughts on the subjects and issues discussed within. To them I am heavily indebted, as I am to the innumerable people and organizations that have contributed ideas, comments, photographs, illustrations, and other items that have helped make this book a reality.

I must also unequivocally mention that without the unfailing fervor, encouragement, and wisdom of Mr. Roger Woodson, president of Lone Wolf Enterprises Ltd., and Mr. Kenneth McCombs, Senior Acquisitions Editor, Elsevier Science and Technology Books, this book would still be on the drawing board. It is always a great pleasure working with them. I must likewise acknowledge the wonderful work of Ms. Maria Alonso, Assistant Editor, Elsevier Science & Technology Books, and thank her for her unwavering commitment and support. I also wish to thank Mrs. Jacqueline Wallace, a highly valued and dedicated member of the Lone Wolf team, for copyediting the first drafts. I would especially like to salute and express my deepest appreciation to Sarah Binns, Associate Project Manager, who saw the book through production, Macmillan India Ltd., who laid out the manuscript, and Eric DeCicco for the excellent cover design.

I am particularly indebted to the U.S. Green Building Council (USGBC) and its staff for their assistance, continuous updates, and support on the new LEED™ 2009 version 3 Rating System.

I would also be amiss if I failed to acknowledge my wife Ibtesam, for her loving companionship and continuous support and for helping me prepare some of the line illustrations. And last but not least, I wish to record my gratitude to all those who came to my rescue during the final stretch of this work-the many nameless colleagues-architects, engineers, and contractors who kept me motivated with their ardent enthusiasm, support, and technical expertise. To these wonderful professionals, I can only say, "Thank you." I relied upon them in so many ways, and while no words can reflect the depth of my gratitude to all of the above for their assistance and advice, in the final analysis, I alone must bear responsibility for any mistakes, omissions or errors that may have found their way into the text.

Introduction

THE GREEN MOVEMENT—OVERVIEW

The Green Movement—Myths and Realities

The public's perception of the green movement has evolved considerably since its early formative days. So much so that Jerry Yudelson, author of *Green Building Revolution*, says, "The green building revolution is sweeping across not only the United States but most of the world. It's a revolution inspired by an awakened understanding of how buildings use resources, affect people, and harm the environment."

Sustainable development principles are increasingly being applied in real estate—today voluntarily, tomorrow by regulation. Perhaps the primary reason most developers have not already jumped on the environment friendly or "green" building bandwagon is the misplaced notions of cost overruns or impracticality of construction.

The myth that sustainability costs more ignores recent research as well as the reality that for any society to thrive and prosper, it must seek to create a healthy balance between its environmental, social, and economic dimensions. Sustainability is not just about building green but about building a community and sustaining a quality way of life. As championed by President Obama and his cabinet, as a community we can no longer afford to delay pursuing new sources of energy such as wind, solar, and geothermal. These efforts could help create new jobs, attract new businesses, and most importantly reduce our energy costs and create a healthy environment.

But why shouldn't we build green? Perhaps because some are still unconvinced of its benefits due to the many myths that float around the construction and real estate industries regarding green building. And although green building has made tremendous strides in the past few years, there remain many misconceptions and myths floating in the mainstream such as:

Myth 1: Green buildings are too expensive.
Reality: This is a very common misconception which has been debunked many times and yet continues to linger on, largely because many of the high-profile green projects that get builders' attention are at the very

Doi: 10.1016/B978-1-85617-676-7.00014-2

high-end. Green building practices are incorporated in all levels of building. And while the initial cost of green buildings may sometimes be slightly higher, the operational cost is lower when compared with conventional buildings. Green buildings have proven themselves to be very cost effective in the longer term. Furthermore, there are numerous strategies for inexpensive green building, from right-sizing the structure to optimal value engineering to reducing waste, among many others.

Myth 2: Green buildings are usually ugly and lack the aesthetic quality of traditional buildings.
Reality: The exterior of a green/sustainable building and a conventional building look very much alike. In fact, many of today's green buildings are virtually indistinguishable from traditional buildings. A vegetated roof for example would not normally be visible from ground level. Moreover you don't have to mount rows and rows of solar panels to be green. And if you do want to go with solar power, there are numerous ways to creatively integrate PV (photovoltaic) panels into a project that are both attractive and effective.

Likewise, with respect to affordable and mixed income housing, it has been shown that there is minimal overall impact on a neighborhood. Buildings in well-designed rural settings are often clustered to leave areas of open space that utilize efficient, attractive design.

Myth 3: Green buildings do not fetch higher rental rates or capitals compared with conventional buildings because consumers only care about the bottom line.
Reality: Recent studies show that green buildings achieved much higher rentals and capitals as a result of reduced operational costs and higher productivity of employees. Tenants and homebuyers do care about green and healthier environments and are willing to pay for it!

Myth 4: Green building is a fad, and therefore not very important.
Reality: There is increasing interest and continuous growth in green building and green building certification—so much so that it has become part of the mainstream in the building industry.

Myth 5: Green buildings are unable to provide the comfort levels that many of today's occupants demand.
Reality: Green buildings are typically more comfortable and healthier than conventional buildings because they are designed to blend in with nature and reduce adverse impacts on the environment, which is not the case with traditional buildings.

Myth 6: Green building is primarily about material selection.
Reality: Many people previously equated green building with the use of "green materials" such as those possessing high recycled content, low

embodied energy, minimal VOCs, etc. And while green building materials are an important aspect of green building, they remain only a small part of the overall equation. Other important factors include site selection, water conservation, and energy performance.

Myth 7: Green building products are hard to find.
Reality: While some green building products may not always be readily accessible—either because they are not manufactured nationwide or they are hard to find in certain parts of the country—it is often possible to find satisfactory alternatives. Moreover, the number of green products and systems that are now available on the market has increased considerably over the past few years and continues to grow. Today green building products are in the hundreds, if not thousands, and are considered part of the mainstream. Building Green, Inc. currently publishes two comprehensive directories (*GreenSpec* and *Green Building Products*) with performance data and contact information on the many green products on the market.

Myth 8: Green buildings utilize traditional tools and traditional techniques.
Reality: Successful green building design often utilizes an integrated design approach where a number of consultants are involved as a team and the architect takes on the role of team leader rather than sole decision maker. Furthermore, locally available materials and techniques are used in addition to the latest technology.

Myth 9: Green building products don't function as well as conventional ones.
Reality: Typical examples of products that often get a bad rap include double-flush and low-flow toilets and fiberglass insulation. Many people still think that 1.6 gallon-per-flush toilets don't work, even though these fixtures were mandated for all new construction over a decade ago. Moreover, when customers are asked to comment on their satisfaction with their new 1.6-gal., high-efficiency toilet fixtures, the majority say they double-flush the same number of times or fewer with their new efficient fixture than with their old water waster. Regarding the myth that inhaling fiberglass fibers can lead to cancer, this is also false. It is therefore important to research a product and seek a scientific basis behind any efficiency claims prior to formulating a final opinion regarding its suitability. However, generally speaking, most new modern green products work as well, if not better than traditional products.

Myth 10: It is difficult and complicated to build green.
Reality: This statement is false; in fact many builders consider green building to be very easy and it compares favorably with conventional building.

Building green is a business that uses common sense, is fundamental, and does not require rocket science to implement.

Myth 11: High rise green construction is not possible.
Reality: All modern techniques that apply to conventional building can be employed in building green. Green concepts therefore will not inhibit or restrict building design or space usability.

Myth 12: Existing conventional buildings cannot be converted into energy efficient buildings.
Reality: Existing buildings can be converted into green/sustainable buildings. In fact, there are various scientific ratings and checklists that builders can use to redesign and realign traditional buildings to modern green standards. Moreover, President Obama, upon becoming president, committed the administration to retrofitting 75 percent of all existing federal buildings.

Ratings such as LEED™ for existing buildings, Canada's Go Green Plus and the Japanese CASBEE certification system all encourage such conversions. It is imperative, therefore, to ensure that the general public is aware of how baseless such myths are and to do everything possible to stamp them out.

Myth 13: To build green, it is necessary to sign up for a green program or third-party certification.
Reality: While certification programs such as Green Globes™ and the U.S. Green Building Council's LEED are terrific at increasing exposure and furthering the green movement, builders don't necessarily have to get involved with them to build green. However, there are many financial and other benefits to attaining certification.

Myth 14: It's an all-or-nothing proposition.
Reality: Developers and construction professionals are frequently categorized into two separate groups: those who mainly build green and those who don't build green at all. Very often, builders utilize green concepts and green products without being aware of it. However, with the increased awareness and demand for green products, many builders and manufacturers are moving toward green building.

The Green Movement—Historic Overview

For a better understanding of the modern green movement, it helps to trace its origins back to the beginning. It is rarely easy to determine exactly when a movement may have started; some associate its beginning with Rachel Carson's (1907–1964) book *Silent Spring* and the legislative fervor of the 1970s or with Henry David Thoreau, where in his book,

Marine Woods, he advocated for an awakening to the need for conservation and federal preservation of virgin forests as well as for the respecting nature. Still others believe it had its roots in the energy crises of the 1970s and the creative approaches to saving energy that transpired from it, including tighter building envelopes and the use of active and passive solar design. This was further inspired by a small group of enlightened architects and environmentalists who began to question the wisdom of the conventional method of building.

This nascent "environmental movement" captured the attention and imagination of the general public, particularly after the imposition of the 1973 OPEC oil embargo, which caused an upward spike of gasoline prices and for the first time, long lines of vehicles at gas stations around the country. The energy crisis caused many to clamor for a reexamination of the wisdom of our reliance on fossil fuels for transportation and buildings. Indeed, the 1970s saw numerous legislative steps to clean up the environment, such as the National Environmental Policy Act, the Clean Air Act, the banning of DDT, the Water Pollution Control Act, the Endangered Species Act, and the founding of Earth Day.

The American Institute of Architects responded to this crisis by forming an energy task force that was later followed by an AIA Committee on Energy. According to committee member Dan Williams, two groups were formed—one looked mainly at passive systems, such as reflective roofing materials and environmentally appropriate siting of buildings to achieve energy savings, while the other looked primarily into solutions that involved the use of new technologies. This was transformed into a more broadly scaled AIA Committee on the Environment (COTE) in 1989, and the following year, the AIA (through COTE) and the AIA Scientific Advisory Committee on the Environment managed to obtain funding from the Environmental Protection Agency (EPA) to embark on the development of a building products guide based on life-cycle analysis, which was published in 1992. But even after the dangers of the energy crisis began to subside, the pioneering efforts in energy conservation for buildings continued its advance forward.

During the 1980s and early 1990s we also witnessed global conservation efforts by sustainability proponents such as Robert Berkebile and Sandra Mendler in the United States; Thomas Herzog of Germany; Norman Foster and Richard Rogers of England; and Kenneth Yeang of Malaysia. In 1987 the UN World Commission on Environment and Development, under Norwegian prime minister Gro Harlem Bruntland, put forward a definition for the term "sustainable development," as that which "meets the needs of the present without compromising the ability of future generations to meet their own needs."

A UN Conference on Environment and Development was held in Rio de Janeiro in 1992, which proved to be a spectacular success and which drew delegations from 172 governments and 2400 representatives of non-governmental organizations. This momentous event witnessed the passage of the Rio Declaration on Environment and Development, a blueprint for achieving global sustainability. Following on the heels of the Rio de Janeiro Summit, the AIA chose sustainability as its theme for the June 1993 UIA/AIA World Congress of Architects. An estimated 6000 architects from around the world attended the Chicago event.

Bill Clinton's election to the presidency in November 1992 encouraged a number of proponents of sustainability to circulate the idea of "greening" the White House itself. And on Earth Day April 21, 1993 President Bill Clinton launched his ambitious "greening the White House" project—an effort that saved more than $1.4 million in its first six years, primarily from improvements made to the lighting, heating, air conditioning, water sprinklers, insulation, and energy and water consumption reduction.

The "greening of the White House" also included the 600,000 sq ft Old Executive Office Building located across from the White House, as well as an energy audit by the Department of Energy (DOE), an environmental audit led by the Environmental Protection Agency (EPA), and a series of well attended design charettes with the aim of formulating energy-conservation strategies using off-the-shelf technologies. Nearly a hundred design professionals, engineers, environmentalists, and government officials took part in these charettes. The results of these energy-conservation strategies were numerous improvements to the nearly 200-year-old mansion, such as reducing atmospheric emissions by an estimated 845 tons of carbon per annum and an estimated $300,000 in annual energy and water savings.

The flood of federal greening projects was but one of several forces propelling the sustainability movement in the 1990s. For example, President Clinton issued several executive orders, the first being on September 14, 1998, which called upon the federal government to improve its use of recycled and "environmentally preferred" products (including building products). This was followed by an executive order in June 1999 to encourage government agencies to improve energy management and reduce emissions in federal buildings through improved design, construction, and operation. A third executive order was issued in April 2000 requiring federal agencies to integrate environmental accountability into their daily decision-making and long-term planning.

During the eight years of George W. Bush's presidency, it was taken a little further by installing three solar systems, including a thermal setup on the pool cabana to heat water for the pool and showers, and photovoltaic panels to supplement the mansion's electrical supply.

The greening of the White House proved to be a great success and created a demand to green other properties in the massive federal portfolio, like the Pentagon, the Presidio, and the U.S. Department of Energy Headquarters, as well as three national parks: Grand Canyon, Yellowstone, and Alaska's Denali. The AIA/COTE and the U.S. Department of Energy signed a memorandum of understanding in 1996 to work together on research and development with the aim of formulating a program containing a series of roadmaps for the construction of sustainable buildings in the 21st Century.

During this frenzy of green activity, individual federal departments were also making significant headway. For example, the Navy undertook eight pilot projects, including the Naval Facilities Engineering Command (NAVFAC) headquarters at the Washington Navy Yard. And in 1997, the Navy initiated development of an online resource, the *Whole Building Design Guide* (WBDG) whose main mission is to incorporate sustainability requirements into mainstream specifications and guidelines. Seven other federal agencies have now joined this project, which is now managed by the National Institute of Building Sciences (NIBS).

The emergence of the green movement as a major force was due to the influence of many people from all walks of life. Visionaries and innovative thinkers have for many years recognized the need for serious changes in how we treat our environment. The championing of green issues by forward thinking politicians, celebrities and visionaries went a long way to addressing some of the environmental concerns that captivated the public's imagination during the early years of this century. This was further helped by Vice-President Al Gore's release in May 2006 of his acclaimed documentary film *An Inconvenient Truth*, which projected global warming into the popular consciousness and raised public awareness of many issues including that our quality of life is endangered, that our water is contaminated with toxic chemicals, and our natural resources are running out.

President-elect Barack Obama was also an outspoken vocal advocate on the campaign trail for sustainability, with regard to both the environment and the economic stimulus. Upon taking office, President Obama put green building at the forefront of his sustainability agenda; he proposed expanding federal grants that assist states and municipalities to build LEED-certified public buildings. According to Jerry Yudelson, "The impact of the Obama administration on green building is going to be to make it a permanent part of the economic, cultural, and financial landscape." The Obama administration's ambitious economic recovery plan is also designed to create new green jobs. President Obama has frequently stressed the need to build a green economy as a means to keep America competitive in the global labor market, while reducing our impact on the environment.

A new report by ICF International (ICFI), concludes that the proposed "Green New Deal" environmental measures included in President

FIGURE I Transportation accounts for about 28 percent of all greenhouse gas emissions. A recent report by ICF International estimates that President Obama's $800 billion economic stimulus package will save a minimum of 61 million tonnes of greenhouse gas emissions annually, which is roughly the equivalent of taking 13 million cars off the road.

Obama's $800 billion economic stimulus package is estimated to save a minimum 61 tonnes a year in greenhouse gas emissions, which if accurate is quite substantial, the equivalent of taking roughly 13 million cars off the road. All of the green measures included in the Senate version of the economic stimulus package have survived, even though the controversial bill was cut from $838 billion to $789 billion.

A green movement activist, Lindsay McDuff says, "When politicians create or formulate policies, the business industries are consequently affected. With the rise in green policy, business executives from every arena are jumping on the green movement bandwagon, basically out of the growing market demand. Being green has become a selling advantage in the business world, and eager companies are starting to jump at the chance to get ahead." The green movement has been transformed into a global hodge-podge of activists and diverse organizations seeking eco-friendly solutions to environmental concerns around the world.

Many scholars consider the green building movement to be mainly a reaction to crises fostered by efforts to make buildings more efficient and revamp the way energy, water, and materials are used. Perhaps more importantly, it is about enhancing our communities and minimizing the impact of building on the environment through better site location, better design, construction, operation, maintenance, and removal.

Defining Green Building and Sustainability

"Green building" and "sustainable architecture" are relatively new terms in our vocabulary; their core message is essentially to improve conventional design and construction practices and standards so that the buildings we build today will last longer, be more efficient, cost less to operate, increase productivity, and contribute to healthier living and working environments for their occupants. But more than that, green building and sustainability are also about increasing the efficiency with which buildings and their sites utilize and conserve energy, water, and materials, about protecting our natural resources, and improving the built environment.

Green building denotes a fundamental change in our understanding of how we design and construct buildings today. It is clear that the green building phenomenon has, over the last decade or more, significantly impacted both the U.S. and international construction markets. Recent studies have shown that buildings in the United States consume roughly one third of all primary energy produced and almost two thirds of electricity produced. Research also shows that roughly 30 percent of all new and renovated buildings in the U.S. have poor indoor environmental quality caused by noxious emissions, pathogens, and emittance of harmful substances present in building materials. There are continuing efforts to address these environmental impacts; one of the proposed solutions was implementing sustainability practices in construction project objectives.

But incorporating sustainable practices into conventional design and construction procedures is an approach that would require redefining and reassessing the current roles played by project participants in the design/construction process to ensure effective contribution to sustainable project objectives. A primary characteristic of successful sustainable design is applying a multidisciplinary and integrated "total" team approach of the various project members and stakeholders, particularly during the early design phases. This integrated team approach helps ensure that the project will be a more energy efficient and healthier building for both its occupants and its owner.

As international awareness of green issues increased, various international conferences were taking place such as the Green Building Challenge (October 1998), held in Vancouver, B.C., which was a well-attended affair with representatives from 14 nations. The goal of these conferences was to create an international assessment tool that takes into account regional and national environmental, economic, and social equity conditions. The green building movement encouraged other parallel efforts to take shape. In the United States, for example, we saw the founding of the U.S. Green Building Council (USGBC), and in the

United Kingdom, the Building Research Establishment was working on its own building assessment method known as BREEAM. The USGBC developed the Leadership in Energy and Environmental Design (LEED™) Green Building Rating System which has become the leading and most widely accepted green building rating system in the United States, as witnessed by its dramatic growth over the past few years.

There has been a striking increase in recent years in the number of projects seeking LEED™ certification from the U.S. Green Building Council (USGBC). This confirms the substantial inroads green building is making into the mainstream design and construction industry. During the early stages of the green building movement, many builders were reluctant to join or encourage this movement. However, this reluctance has diminished dramatically over the last decade and the construction industry is now making serious efforts to embrace this initiative.

The LEED NC rating system now has an international forum having been introduced into countries like Canada, India, China, the United Arab Emirates, and Israel. One of the Indian Green Building Council's stated objectives, for example, is to achieve 1 sq ft of green building for every Indian by 2012. Prem C. Jain, the council's chairman, says that India currently has an estimated 240 million sq ft of green buildings in place.

While there has been tremendous interest regarding green building issues, the amount of money allocated to research has been minimal and presently constitutes a mere 0.2 percent of all federally funded research, which comes to about $193 million annually. This corresponds approximately to a bare 0.02 percent of the estimated $1 trillion value of buildings constructed in the U.S. annually, even though the building construction industry represents over 10 percent of the U.S. GDP. The relative insignificance of funding allocated for green building compared to funding for other research topics needs to be corrected. The federal government and other relevant funding sources should be encouraged to provide appropriate financial support to these research programs and readily achievable strategies. We cannot move forward unless we significantly improve our green building practices. The alternative is to experience a dramatic increase in the negative impacts of the built environment on human and environmental health in the coming decades.

The challenges we face of critical issues like global warming, water shortages, indoor environmental quality problems, and destruction of our ecosystem are grave. It has been clearly documented that conventionally constructed buildings contribute substantially to the environmental problems emerging in the U.S. For example, it has been estimated that current building operations account for about 38 percent of U.S. carbon dioxide emissions, 71 percent of electricity use, and according to the Environmental Information Administration (2008), almost 40 percent of total energy use; the latter number increases to an estimated 48 percent if the energy required to make building materials and constructing buildings are included. It is further estimated that buildings consume roughly 13.6 percent of the country's potable water and the EPA estimates that waste from demolition, construction, and remodeling amounts to 136 million tons of landfill additions annually. Construction and remodeling of buildings accounts for 3 billion tons, or roughly 40 percent, of raw material used globally each year.

One example of the significant impact that green building research is having can be seen by the impact of carbon emissions on global warming, which has recently received national attention. This has resulted in various organizations (like the AIA, ASHRAE, USGBC, and the Construction Specifications Institute) collectively adopting the 2030 Challenge, which consists of a series of goals—the main one being that all new construction will have net zero carbon emissions by the year 2030, and that an equivalent amount of existing square footage will be renovated to use half of its previous energy use.

In March 2007 the UN came out with a report that clearly reaffirms buildings' role in global warming, and according to Achim Steiner, UN under-secretary general and UNEP executive director, "Energy efficiency, along with cleaner and renewable forms of energy generation, is one of the pillars upon which a de-carbonized world will stand or fall." He goes on to say, "This report focuses on the building sector. By some conservative estimates, the building sector world-wide could deliver emission reductions of 1.8 billion tonnes (1 tonne = 1000 kilograms = 2025 pounds) of CO_2. A more aggressive energy efficiency policy might deliver over 2 billion tonnes or close to three times the amount scheduled to be reduced under the Kyoto Protocol." But to meet the 2030 Challenge, a fundamental change in our present approach and knowledge of building energy issues is needed.

The construction industry today is facing unprecedented and growing pressures, originating from a global economic crisis, rising material costs, an increase in natural disasters, and the dramatic impact of the green consumer, among other things. Together these trends have motivated the industry to increasingly reevaluate its position and adopt sustainable design and construction methods in an effort to construct more efficient buildings designed to conserve energy and water, improve building operations, and enhance the health and well-being of the general population.

The market share of green building will continue to develop and rise, partly because of the unprecedented level of state and local government interest and initiatives, such as the use of various incentive-based techniques to encourage green building practices. Unfortunately these efforts have encountered some obstacles and challenges along the way, including the high cost of these new incentive programs, lack of available resources, and implementation difficulties. In an effort to assist communities in overcoming these obstacles, the American Institute of Architects (AIA) commissioned a report, *Local Leaders in Sustainability—Green Incentives*, which defines and explains these various programs, examines the main barriers to success, and highlights best practice examples.

The U.S. Built Environment

The Department of Commerce (2008) statistics show that the construction market accounts for 13.4 percent of the $13.2 trillion U.S. GDP, of which the value of green building construction is projected to increase to $60 billion by 2010 (Source: McGraw-Hill Construction, 2008. Key Trends in the European and U.S. Construction Marketplace: SmartMarket Report). It is further estimated that 82 percent of corporate America will be greening at least 16 percent of their real estate portfolios in 2009, and 18 percent of these corporations are expected to be greening more than 60 percent of their real estate portfolios.

Several sectors in the green building industry such as office, education, healthcare, government, hospitality, and industrial are expected to grow in the coming years. Of these, the three largest segments for non-residential green building construction—office, education, and health care—is expected in 2008 to account for more than 80 percent of total non-residential green construction (Source: FMI—2008; U.S. Construction Overview).

Green-registered and certified projects reflect a diverse cross-section of the industry. As an example, many roofing companies have made the determination to focus on green technologies that allow their customers to harness the energy a rooftop solution is capable of providing. However, these companies recognize that to be successful and meet the challenges that green technology brings requires a reevaluation of how their company is to function, with a serious financial commitment in terms of manpower, technology, equipment, and training.

New green technologies are also having a positive economic impact on the plumbing industry and are spurring economic growth for plumbing contractors around the country. Plumbing contractors are taking an active role in pushing for the installation of water and energy efficient systems, and through the installation and use of green technologies, are promoting energy efficiency and water conservation.

Project designers and property owners as stakeholders, both public and private, play a critical role in pursuing sustainable design and construction practices, since from the beginning they were the creators and driving force of the built environment concept. With both the source (designer/consultant) and the end user (owner) readily adopting sustainable design practices, it became necessary for the contractor/builder to become an active team member along with the architect, mechanical/electrical/civil/structural engineer, landscape designer, etc., if green building projects were to be successfully executed. An experienced builder has much to offer in terms of input on aspects like material selection, specification, system performance, minimizing construction waste, etc. The contractor can also assist in streamlining construction and value engineering methods to achieve the green project's overall objectives.

Most of the research conducted on the costs and benefits of green buildings come to the conclusion that energy and water savings on their own outweigh the initial cost premium in most green buildings, and the median increase that green buildings may incur is less than 2 percent when compared with constructing conventional non-green buildings. This is at direct variance with the public perception and common held belief that green buildings are much more expensive than conventional buildings. In a 2007 survey by the World Business Council for Sustainable Development, it was found that business leaders believe green buildings to be on average 17 percent more expensive than conventionally designed buildings. Another recently published international study, *Greening Buildings and Communities: Costs and Benefits,* also concludes "Most green buildings cost 0–4 percent more than conventional buildings, with the largest concentration of reported "green premiums" between 0–1 percent. Green premiums increase with the level of greenness but most LEED buildings, up through gold level, can be built for the same cost as conventional buildings."

Finally, Greg Kats, the above study's lead author and a managing director of Good Energies, one of the study's main supporters, says, "The deep downturn in real estate has not reduced the rapid growth in demand for and construction of green buildings," and "This suggests a flight to quality as buyers express a market preference for buildings that are more energy efficient, more comfortable, and healthier."

CHAPTER

1

"Green" and "Sustainability" Defined

1.1 GENERAL—OVERVIEW OF THE GREEN BUILDING MOVEMENT

Enormous changes have taken place in the construction industry and the architectural/engineering professions over the last decade in the promotion of environmentally responsible buildings. Indeed, the green movement has invaded almost all areas of our society, including the construction and home-building industries. But "green" means more than just recycling empty bottles and cans and taking public transportation to work. With respect to building green and sustainability, the emphasis should be on designing and erecting buildings that are more energy efficient, use natural or reclaimed materials in their construction, and are more in harmony with the environments in which they exist. Sustainable buildings are more efficient in their use of valuable resources such as energy, water, materials, and land than conventional buildings or buildings that are simply built to code. Green buildings are therefore kinder to the environment, and provide indoor spaces that are typically more healthy, comfortable, and productive.

Buildings have become an area of focus for green investment dollars largely because they are primary contributors to impacting the environment. Studies show that buildings are the world's heaviest consumers of natural resources, which is why many architects, engineers, contractors, and builders today have started to reevaluate how residential and commercial buildings are being built. Moreover, various incentive programs are now in place that encourage and sometimes stipulate that developers and federal agencies go green—both nationally and internationally. But while sustainable or green building is a strategy for creating healthier and more energy efficient eco-friendly buildings, recent research and experience clearly shows that buildings designed and operated with their life-cycle impacts in mind most often provide substantially greater environmental, economic, and social benefits. Additionally, incorporating green strategies and materials during the early design phase is an outstanding approach to increase a project's potential market value. This strongly suggests that to succeed, the green-building process requires the implementation of an integrated approach to building design and construction.

In fact, an integrated approach to green building is pivotal to a project's success, which means that all aspects of a project, from the site selection to the structure, to interior finishes, are all carefully considered. Focusing on only one aspect of a building can have a profoundly adverse impact on the project as a whole. For example, continuing research has shown that designing an inefficient building envelope can adversely impact indoor environmental quality as well as increase energy costs, whereas a sustainable development can help lower operating costs over the life of a building by increasing productivity and utilizing less energy and water. Sustainable developments can also provide tenants and occupants with a healthier and more productive working environment as a result of improved indoor air quality. Similarly, exposure to materials like asbestos, lead, and formaldehydes which may have high volatile organic compound (VOC) emissions in a building can precipitate significant health problems because of poor indoor air quality and may create what is known as Sick Building Syndrome (SBS).

The objectives of those who engage in green building are usually to achieve both ecological and aesthetic harmony between a structure and its surrounding environment. The exterior appearance and style of sustainable buildings is not always immediately distinguishable from their more conventional counterparts that are built to code. As mentioned earlier, buildings can have an enormous impact on the environment— both during the construction phase and through their operation and maintenance. Moreover, according to Rob Watson, author of the Green Building Impact Report issued in November 2008, "The construction and operation of buildings require more energy than any other human activity. The International Energy Agency (IEA) estimated in 2006 that buildings used 40 percent of primary energy consumed globally, accounting for roughly a quarter of the world's greenhouse gas emissions (Figure 1.1). Commercial buildings comprise one-third of this total. Urbanization trends in developing countries are accelerating the growth of this sector relative to residential buildings, according to the World Business Council on Sustainable Development (WBCSD)." Buildings also account for an estimated 71 percent of all electricity consumed in America and 40 percent of global carbon dioxide emissions.

The utilization of sustainable/green building strategies and practices offer a unique opportunity to create environmentally sound and resource-efficient buildings. Employing an integrated design approach can achieve this by having the stakeholders—architects, engineers, land planners, building owners and operators, as well as members of the construction industry collaborate as a team to design the project.

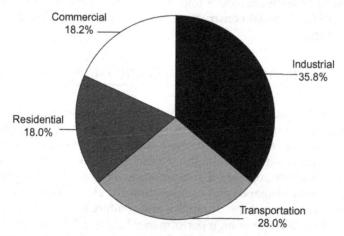

FIGURE 1.1 A pie chart showing U.S. total greenhouse gas emissions in 2005. The Energy Information Administration (EIA) typically breaks down U.S. energy consumption into four end-use categories: industry, transportation, residential, and commercial. Almost all residential greenhouse emissions are CO_2, and are strongly related to energy consumption. The chart shows that the residential sector generates very little greenhouse gases other than CO_2, and so accounts for only 18 percent of total greenhouse gas emissions measured in MMT CO_2 equivalents. (Source: National Association of Home Builders, Paul Emrath and Helen Fei Liu.)

Leading the green building assault is the federal government which is the nation's largest single landlord. In this regard the General Services Administration recently announced that it will be applying stringent green-building standards to its $12 billion construction portfolio of post offices, courthouses, border stations, and other buildings.

The CEO and founding chairman of the U.S. Green Building Council, Richard Fedrizzi, clearly echoed this when he said, "The federal government has been at the forefront of the sustainable building movement since its inception, providing resources, pioneering best practices, and engaging multiple federal agencies in the mission of transforming the built environment." The first-ever White House Summit on Federal Sustainable Buildings conference was held January 24–25, 2006, in which over 150 federal facility managers and decision makers attended and 21 government agencies participated in formulating and witnessed the signing of the Federal Leadership in High Performance and Sustainable Buildings Memorandum of Understanding (MOU). Signatory agencies to this MOU commit to federal leadership in the design, construction, and operation of high-performance and sustainable buildings. Moreover, the MOU represents a significant achievement by the federal government through its cumulative effort to define common strategies and guiding principles of green building. The signatory agencies will need to coordinate their efforts with others in the private and public sectors if these goals are to be achieved. More importantly, the narrowing gap between green and conventional construction is a sign that green construction is coming of age.

1.2 WHAT IS GREEN DESIGN? BASIC GREEN CONCEPTS

The term "green building" is relatively new to our language and a precise definition is elusive. The EPA for example defines it as, "the practice of creating structures and using processes that are environmentally responsible and resource-efficient throughout a building's life-cycle from siting to design, construction, operation, maintenance, renovation and deconstruction. This practice expands and complements the classical building design concerns of economy, utility, durability, and comfort. Green building is also known as a sustainable or high performance building." In the Gothenburg European Council meeting of June 2001, sustainable development was defined as a means of meeting the needs of the present generation without compromising those of the future. But however one wishes to define the term, green building or sustainable development has had a prolific impact on the U.S. construction market in the last decade, although its complete impact on the building construction industry and its suppliers is still being evaluated and may not be completely known for at least another decade.

Although the United States is the undisputed global leader in the construction of green buildings, other countries around the world are increasingly investing in sustainability. The European Union (EU) agreed on a new sustainable development strategy that has the potential to determine how the EU economy evolves in the coming decades. There are various green building assessment systems used around the world, such as the Building Research Establishment's Environmental Assessment Method (BREEAM), the Comprehensive Assessment System for Building Environmental Efficiency (CASBEE), Green Globes™ U.S., and Green Building (GB) Tool (Mago and Syal 2007).

Green building strategies mainly relate to land-use, building design, construction, and operation that in aggregate help minimize or mitigate a building's overall impact on the environment. The primary objective of green buildings is therefore to improve the efficiency with which buildings use available natural resources such as energy, water, and materials, while simultaneously minimizing a building's adverse impact on human health and the environment. There are many ways that green construction methods can be employed to build a new building designed for long-term operations and maintenance savings. Likewise, our nation has a vast existing building stock that can be made greener—and studies show that many building owners are interested in doing just that.

Once the myths and misinformation that surround sustainability and green design are set aside, a number of pertinent strategies become apparent and that will help achieve the objectives of building green. "Myth and misinformation surround the topic of sustainability, clouding its definition and purpose, and blurring the lines between green fact and fiction," says Leah B. Garris, senior associate editor at Buildings magazine. Another green building proponent, Alan Scott, principal, Green Building Services in Portland, OR states, "You can have a green building that doesn't really 'look' any different than any other building." Ralph DiNola, also a principal with Green Building Services echoes this statement, believing that a level of sustainability can easily be achieved by designing a green building that looks "normal." DiNola goes on to say that "People don't really talk about the value of aesthetics in terms of the longevity of a building. A beautiful building will be preserved by a culture for a greater length of time than an ugly building." Thus, for a building to be sustainable, it is important that it has the potential for a long-term, useful life. Aesthetics is a pivotal factor to achieving longevity, and longevity is pivotal to achieving sustainability.

But sustainability is more about conscious choices than about spending on superfluous options in hopes of earning an increased return on investment. Sustainability is about understanding nature and working with it and not against it. Finally, sustainability and building green is not about constructing structures that purport to be environmentally responsible but in reality sacrifice tenant/occupant comfort. This does not suggest that purchasing green products or recycling assets at the end of their useful lives is not sustainable, because it is. It is also appropriate for both the environment and for the health of a building's tenants and employees. However, a developer or building owner should not make a final determination without first taking the time to research the various options that will best work for your project and offer the best possible return on investment. When little or no time is set aside to sorting through the many possible sustainable options that are available, a decision may be made at the last minute, which often turns out to be the wrong one.

According to some, sustainability starts with the climate, and the primary reason that green strategies are considered green is because they work in harmony with the surrounding climatic and geographic conditions and not against them. This however requires a thorough understanding of the environment in which a project is being designed in order to fully utilize them to a project's advantage. Architects and designers that specialize in green building are fully aware of the need to be familiar with year-round weather conditions such as temperature, rainfall, humidity, site topography, prevailing winds, indigenous plants, etc., in order to succeed in sustainable design. But while climate impacts sustainability in one form or another, affected to some extent by a project's location, a building's degree of successful sustainable design can be measured by comparing it to a baseline condition. That baseline condition relates to the microclimate and environmental conditions of a building's location.

Furthermore, to successfully achieve sustainability, it is crucial to identify and reduce a building's need for resources that are scarce or locally unavailable, such as water and energy, and encourage the use of readily available resources such as the sun, rainwater, wind, etc. A full understanding of the microclimate is imperative because it reflects a comprehension to what is readily available and at a project's disposal such as the sun for heating and lighting, the wind for ventilation, and rainwater for irrigation and other water requirements. When considering sustainability, there are five principal categories that come to mind:

- Sustainable sites
- Water conservation
- Energy efficiency
- Materials and resource conservation
- Indoor environmental quality.

(This list is modeled after the Washington, D.C.-based U.S. Green Building Council's LEED™ rating system.)

Water conservation and energy efficiency both rely significantly on the climate, whereas indoor environmental quality and materials and resource conservation are almost entirely independent of climate. Site sustainability on the other hand depends on climate to some degree, but more specifically on the specifications and microelements that are particular to a specific site. It is important to note that different regions or locations may encounter different climates—from hot, arid to humid, freezing, and windy. Thus, understanding a region's climate and readily available resources can help avoid applying inappropriate techniques to a project which will have an adverse impact and invariably increase costs.

1.3 ESTABLISHING MEASURABLE GREEN CRITERIA

In December 1983 the United Nations gave birth to the World Commission on Environment and Development (WCED) with the main intention of addressing growing concerns "about the accelerating deterioration of the human environment and natural resources and the consequences of that deterioration for economic and social development." The establishment of this commission is recognition by the UN General Assembly that the environmental problems we face are global in nature. The UN determined that it was in the best interest of all nations to establish common policies for sustainable development (Report of the World Commission on Environment and Development: Our Common Future, http://www.un-documents.net/wced-ocf.htm). Following the formation of the WCED came the Brundtland Commission in 1987 which produced the Brundtland Report in August of the same year. Many found the findings of this report troubling; it stated, among other things:

"The 'greenhouse effect,' one such threat to life support systems, springs directly from increased resource use (Figure 1.2). The burning of fossil fuels and the cutting and burning of forests release carbon dioxide (CO_2). The accumulation in the atmosphere of CO_2 and certain other gases traps solar radiation near the earth's surface, causing global warming. This could cause sea level rises over the next 45 years large enough to inundate many low lying coastal cities and river deltas. It could also drastically upset national and international agricultural production and trade systems.

Another threat arises from the depletion of the atmospheric ozone layer by gases released during the production of foam and the use of refrigerants and aerosols. A substantial loss of such ozone could have catastrophic effects on human and livestock health and on some life forms at the base of the marine food chain. The 1986 discovery of a hole in the ozone layer above the Antarctic suggests the possibility of a more rapid depletion than previously suspected."

The report goes on to say, "A variety of air pollutants are killing trees and lakes and damaging buildings and cultural treasures, close to and

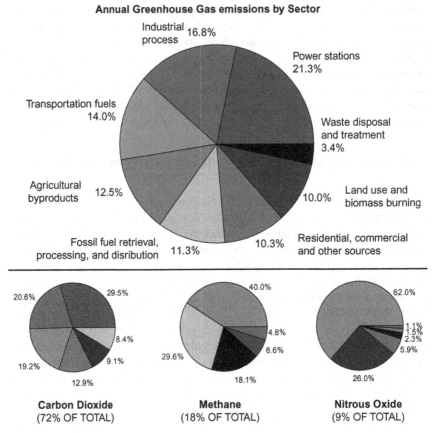

FIGURE 1.2   A pie chart showing global anthropogenic greenhouse gas emissions broken down into eight different sectors for the year 2000. Concentrations of several greenhouse gases have increased over time and human activity may be increasing the greenhouse effect through release of carbon dioxide. (Source: Wikipedia for Schools.)

sometimes thousands of miles from points of emission. The acidification of the environment threatens large areas of Europe and North America. Central Europe is currently receiving more than one gram of sulfur on every square meter of ground each year. The loss of forests could bring in its wake disastrous erosion, siltation, floods, and local climatic change. Air pollution damage is also becoming evident in some newly industrialized countries." Figure 1.3 is a graphic illustration of the greenhouse effect.

It is unfortunate that the substantial building boom we experienced in recent years was often underpinned by inferior design and construction strategies and highly inefficient HVAC systems, placing buildings at the top of the list of contributors to global warming. There have been many attempts by federal and private organizations to address these problems, and due partly to these efforts we are now witnessing a surge of interest

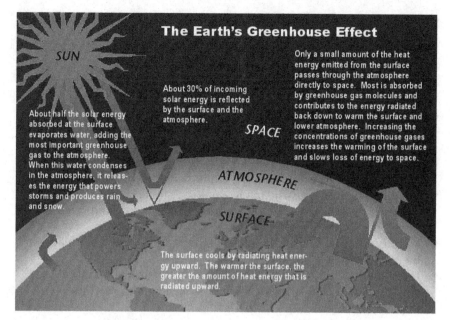

The Earth's Greenhouse Effect

SUN

About half the solar energy absorbed at the surface evaporates water, adding the most important greenhouse gas to the atmosphere. When this water condenses in the atmosphere, it releases the energy that powers storms and produces rain and snow.

About 30% of incoming solar energy is reflected by the surface and the atmosphere.

SPACE

Only a small amount of the heat energy emitted from the surface passes through the atmosphere directly to space. Most is absorbed by greenhouse gas molecules and contributes to the energy radiated back down to warm the surface and lower atmosphere. Increasing the concentrations of greenhouse gases increases the warming of the surface and slows loss of energy to space.

ATMOSPHERE

SURFACE

The surface cools by radiating heat energy upward. The warmer the surface, the greater the amount of heat energy that is radiated upward.

FIGURE 1.3   The Earth's greenhouse effect. Greenhouse gases included in the order of relative abundance are water vapor, carbon dioxide, methane, nitrous oxide, and ozone. Greenhouse gases come from natural sources and human activity. (Source: Darren Samuelsohn, www.earthportal.org.)

in green concepts and sustainability. Many developers and project owners have become aware of the numerous benefits of incorporating green strategies and are increasingly aspiring to achieve LEED™ certification for their buildings. Green building rating systems generally have two main objectives—creating incentives to produce high performance buildings and building demand for sustainable construction. Green buildings have shown to be economically viable, ecologically benign, and sustainable over the long term.

This encouraged the collaboration of the Partnership for Achieving Construction Excellence and the Pentagon Renovation and Construction Program Office, who together recently issued a Field Guide for Sustainable Construction with the goal of assisting and educating field workers, as well as supervisors and managers and other stakeholders, in making appropriate decisions that can assist the project team to meet sustainable project goals. To do this the guide discusses various topics, most of which are incorporated in the LEED and other rating systems, including:

- Ensure that sustainable construction requirements are met by addressing specific procurement strategies.
- Identify methods and means to reduce the environmental impact of construction on the project site and surrounding environment.

- Identify and employ environmentally friendly building materials and avoid use of harmful and toxic materials.
- Identify ways and means to minimize or eliminate waste on construction projects.
- Identify materials that can be recycled at each phase of construction as well as approaches to support onsite recycling efforts.
- Employ techniques that improve a building's energy performance and reduce energy consumed during construction, as well as determine available opportunities to use renewable energy sources.
- Identify reusable materials and methods that will facilitate potential future reuse of a facility, systems, equipment, products, and materials.
- Identify the latest construction technologies that can be utilized during the construction process to promote efficiency and reduce waste (e.g., paper).
- Identify ways and means to improve the health, safety, and quality of life for construction workers.
- Delineate methods that ensure that during construction indoor environmental quality measures are managed and correctly implemented.

Various programs exist that outline important green criteria. The Department of Energy (DOE) for example has an Environmental Protection Program, the goals and objectives of which are "to implement sound stewardship practices that are protective of the air, water, land, and other natural and cultural resources impacted by DOE operations and by which DOE cost effectively meets or exceeds compliance with applicable environmental, public health, and resource protection laws, regulations, and DOE requirements. This objective must be accomplished by implementing environmental management systems (EMSs) at DOE sites. An EMS is a continuing cycle of planning, implementing, evaluating, and improving processes and actions undertaken to achieve environmental goals."

It is important for architects, designers, property developers, contractors, and other stakeholders to have a clear understanding of the various certification programs available, and why certification will help property owners remain competitive in an increasingly green market. Certification implies independent verification that a building has met accepted guidelines in these areas, as outlined for example, in the LEED Green Building Rating System. LEED certification of a project is a recognized testimonial to its quality and environmental stewardship, particularly since the LEED green building rating system has become widely accepted by public and private owners.

The significant inroads into the mainstream design and construction industry by the LEED rating system has embolden contractors and property developers into realizing that it is in their interest to contribute to a project's success in achieving green objectives. This can be accomplished by first understanding the LEED process and its specific role in achieving LEED credits, and second, through early involvement and participation throughout the various project phases by implementing a team approach in an integrated design process. Moreover, measureable benchmarks are necessary to enable verification and confirm a building's acceptable performance. It should be pointed out that ASHRAE puts the responsibility of defining design intent requirements squarely on the shoulders of the project owner. But it would be difficult, if not impossible, to correctly evaluate a building or project's performance unless certain information is made available regarding the criteria upon which the project's design and execution was based. This requires that a project's plans and specifications, etc. are prepared in a manner that can provide measurable results, without which it would not be possible to make a meaningful evaluation of a project to see if it has met the specified objectives and original design intent. Furthermore, before measureable sustainable criteria can be established, it is necessary to agree first on a finite definition of green construction and to articulate specifically what is required.

In promoting LEED, the USGBC emphasizes the simplicity of the system. But the uniqueness of the LEED certification system is that it typically mandates performance over process. Moreover, the USGBC through the application of its widely circulated and recently updated LEED V3 scoring system and other efforts has dramatically changed the way many contractors and their subcontractors operate. LEED V3 has also taken into account the impact of microclimate and has now added regional priority to its rating system.

The National Association of Home Builders (NAHB) has also put forward a set of green home building guidelines that "should be viewed as a dynamic document that will change and evolve as new information becomes available, improvements are made to existing techniques and technologies, and new research tools are developed." The NAHB states that its Model Green Home Building Guidelines were written to help move environmentally-friendly home building concepts further into the mainstream marketplace; it is one of two rating systems that make up NAHBGreen, the National Green Building Program.

The NAHB point program consists of three different levels of green building—bronze, silver, and gold. The system is available to builders wishing to implement these guidelines to rate their projects. "At all levels, there are a minimum number of points required for each of the seven guiding principles to assure that all aspects of green building are addressed and that there is a balanced, whole-systems approach. After reaching the thresholds, an additional 100 points must be achieved by implementing any of the remaining line items." The table below (Figure 1.4) outlines the necessary points needed for the three different threshold levels.

While the general appearance of a green building may be similar to other building forms, the conceptual approach in their design is

	Bronze	Silver	Gold
Lot Design, Preparation, and Development	8	10	12
Resource Efficiency	44	60	77
Energy Efficiency	37	62	100
Water Efficiency	6	13	19
Indoor Environmental Quality	32	54	72
Operation, Maintenance, and Homeowner Education	7	7	9
Global Impact	3	5	6
Additional Points from Sections of Your Choice	100	100	100

FIGURE 1.4   The NAHB three-tier point system is available to builders wishing to use these guidelines to rate their projects. Points required for the three different levels of green building are: bronze, silver, and gold.

fundamentally different because it revolves around a concern for the potential impact on the environment and it seeks to extend the life span of natural resources, provide human comfort and well-being, security, productivity, and energy efficiency. Designing for green buildings will also result in reduced operating costs including energy, water, and other intangible benefits. The Indian Green Building Council (IGBC), which administers the LEED India rating system for example, points to a number of salient green building attributes as outlined below:

- Minimal disturbance to landscapes and site condition
- Use of recycled and environmentally-friendly building materials
- Use of nontoxic and recycled/recyclable materials
- Efficient use of water and water recycling
- Use of energy efficient and eco-friendly equipment
- Use of renewable energy
- Indoor air quality for human safety and comfort
- Effective controls and building management systems.

Other organizations, like the Whole Building Design Guide (WBDG), also have established certain objectives and principles relating to sustainable design, including:

Objectives:
- Avoid resource depletion of energy, water, and raw material
- Prevent environmental degradation caused by facilities and infrastructure throughout their life-cycles
- Create built environments that are livable, comfortable, safe, and productive.

Principles:

- Optimize site potential
- Optimize energy use
- Protect and conserve water
- Use environmentally-preferred products
- Enhance indoor environmental quality (IEQ)
- Optimize operations and maintenance procedures.

James Woods, executive director of The Building Diagnostics Research Institute notes, "Building performance is a set of facts and not just promises. If the promises are achieved and verified through measurement, beneficial consequences will result and risks will be managed. However, if the promises are not achieved, adverse consequences are likely to lead to increased risks to the occupants and tenants, building owners, designers and contractors, and to the larger interests of national security and climate change."

Another sustainable design expert, Alan Bilka, who is with ICC Technical Services, quite rightly points out that "Over time, more and more 'green' materials and methods will appear in the coders and/or have an effect on current code text. But the implications of green and sustainable building are so wide and far reaching that their effects will most certainly not be limited to one single code or standard. On the contrary, they will affect virtually all codes and will spill beyond the codes. Some green building concepts may become hotly contested political issues in the future, possibly requiring the creation of new legislation and/or entirely new government agencies."

1.4 BENEFITS OF GOING GREEN—INCENTIVES AND BARRIERS

Over the last decade or so, architects, designers, constructors, and building owners have increasingly taken an interest in green building. Encouraged by national and local programs, green building is flourishing throughout the nation, as well as globally. Thousands of projects have been constructed over recent years that provide tangible evidence of what green building can accomplish in terms of improved comfort levels, aesthetics, and energy and resource efficiency.

Some of the primary benefits of building green include:

- Reducing energy consumption
- Protection of ecosystems
- Improved health of occupants
- Increased productivity.

Studies have shown that building operating expenses often represent less than 10 percent of an organization's cost structure, whereas personnel usually comprises the remaining 90 percent. This strongly suggests that even minor improvements in worker comfort can result in substantial dividends in performance and productivity. There is also a growing body of evidence linking more efficient buildings with improved working conditions, leading to reduced turnover and absenteeism, increased productivity, and other benefits. This, in turn, has become a major contributing factor to the growth of building efficiency, particularly with respect to the occupants and tenants who live and work inside these structures.

With regard to existing buildings, Incisive Media's 2008 Green Survey: Existing Buildings concluded that nearly 70 percent of commercial building owners have already implemented some form of energy monitoring system. The survey also confirmed that energy conservation is the most widely employed green program in commercial buildings, followed by recycling and water conservation. And 84 percent of respondents to Turner's "Green Building Barometer" confirm that their green buildings have resulted in lower energy costs, and 68 percent report lower overall operating costs. Furthermore, nearly 65 percent of building owners who have built green buildings claim that their investments have already produced a positive return on investment.

Likewise, developers and property owners now generally agree that among the more tangible benefits of attaining LEED certification for a building is the ability to use that achievement as a marketing tool. Tenants and employees have shown a clear preference for living and working in certifiably green buildings, resulting in greater demand and therefore the capability to attract quality tenants and get higher rents. Likewise, designers and contractors with LEED-certified buildings in their portfolios often find that they have a greater competitive advantage.

1.4.1 Incentives and Tax Deductions

In addition to the health and environmental benefits of living and working in a green building, many local and state governments, utility companies, and other entities nationwide are now offering rebates, tax breaks, and other incentives to encourage the incorporation of eco-friendly elements in proposed building projects. In fact, the majority of large cities in the United States now provide financial and other incentives for building green. To date, there are more than 65 local governments throughout the nation that have already made a commitment to LEED standards in building construction, with some reducing the entitlement process by up to a year in addition to various tax credits.

Energy costs have become a major office building expense, although this can be reduced by as much as 30 percent, and savings continue to increase with the development of new technologies.

The California Building Industry Association (CBIA) for example claims that there is considerable evidence that the state's new-homebuyer tax credit enacted at the beginning of 2009 is helping to generate new-home sales, and in turn, job-generating home construction. Furthermore, the U.S. government now provides an Energy Efficient Commercial Building Tax Deduction and other possible tax breaks through the Energy Policy Act of 2005. This has created an important incentive for commercial developers to construct energy-efficient buildings as well as incentives for efficiency upgrades to residences.

Links to the various funding sources for green building that are available to homeowners, industry, government organizations, and nonprofits in the form of grants, tax credits, loans, and other sources can be found on the U.S. Environmental Protection Agency Website (http://www.epa.gov). Likewise, the Database of State Incentives for Renewables & Efficiency (DSIRE) which is a nonprofit project funded by the U.S. Department of Energy through the North Carolina Solar Center and the Interstate Renewable Energy Council, also has information regarding local, state, federal, and utility incentives available for switching to renewable or efficient energy use. The DSIRE website is: http://www.dsireusa.org. Another way to achieve federal tax credits is by using energy-efficient products such as those proposed by the U.S. government's ENERGY STAR® program. The website is: www.energystar.gov.

1.4.2 Green Building Programs

Numerous cities throughout the United States are now promoting the use of various external green building programs. The City of Seattle is one such example that is promoting a number of green building programs such as:

- Built Green: This is a nonprofit residential green building program developed by the Master Builders Association of King and Snohomish Counties in partnership with Seattle, King County, and a number of local environmental groups.
- Energy Star Homes: A program for new homes that was created by the U.S. EPA and U.S. Department of Energy.
- LEED™ for Homes: A newly proposed residential rating system by the U.S. Green Building Council that is currently in the pilot stage, but which is due to be operational shortly.
- Multifamily: Consisting of apartments, townhomes, and condominiums.

Seattle has also commenced the implementation of a Sustainable Building Policy that requires all new city-funded projects and renovations in excess of 5000 square feet of occupied space to attain a LEED^{TM} Silver rating. This policy affects all city departments that are involved with construction, such as the Department of Planning and Development (DPD), which monitors implementation of the policy. Seattle's green building program is now called CITY Green Building and resides within DPD.

1.4.3 Defining Sustainable Communities

The concept of sustainable communities remains somewhat elusive but is evolving with time as our knowledge and understanding of sustainability develops. Some community planners have started to formulate a perception or vision of how such a community will grow to allow the sustainability of its citizens' core values which include: the community, social equity, environmental stewardship, economic prosperity, opportunity, and security.

A number of cities in the United States have already started to adopt a comprehensive plan that includes goals and policies designed to help guide development toward a more sustainable future. This new "green urbanism" seeks to apply leading edge tools, models, strategies, and technologies to motivate cities into meeting sustainability goals and policies. The application of an integrated, whole-systems approach to community or neighborhood planning allows the city to achieve increased environmental protection levels. There are other compelling inducements for building owners and property developers to invest in green buildings, especially in the LEED certification program, which include the financial benefit of operating a more efficient and less expensive facility. Furthermore, adhering to LEED guidelines will go a long way to ensuring that the facilities are designed, constructed, and operated more effectively, largely because LEED prefers project teams to focus on operating life-cycle costs, not initial construction costs. Additionally, many states are now also offering incentives in the form of various tax benefits for green building and LEED compliance. An example of this is the State of New York, where Governor Pataki has created the New York State Green Building Tax Credit Program to promote the funding of concepts and ideas that encourage green building practices.

It is puzzling that until recently no single organization has had the courage to take a strong initiative to bring green construction to the American home market. Earlier residential green building programs that did exist were often sponsored by local homebuilder associations (HBAs), nonprofit organizations, or municipalities. Realizing that this situation was unacceptable and unlikely to last, the National Association of Homebuilders (NAHB) and the NAHB Research Center (NAHB RC) took preemptive action and produced the Model Green Home Building Guidelines and various other utility programs. But while the NAHB program may have provided many of the answers relating to the nation's residential building market, with respect to commercial construction LEED was until recently and for all practical purposes the only game in town.

However, an alternative to LEED called Green Globes has recently arrived from Canada and hopes to provide the U.S. commercial construction industry with a simpler, less expensive method for assessing and rating a building's environmental performance. Green Globes is a web-based auditing tool that was developed by a Toronto-based environmental consultant, Energy and Environment Canada. The system's greatest strength is purported to be its rapid and economical method for assessing and rating the environmental performance of new and existing buildings. Rights to market the program in the United States were recently purchased by the Green Building Initiative (GBI) which has budgeted more than $800,000 as a first step to promote national awareness of Green Globes as a viable alternative to the LEED program throughout the construction and development community and try and capture a significant percentage of its market share.

The building codes of California and New York are among the national leaders in sustainable development. California Governor Arnold Schwarzenegger's solar initiatives were in fact pivotal in motivating and creating a solar industry in the state and which has become the nation's largest market. On December 14, 2004, Governor Schwarzenegger signed Executive Order S-20-04, which requires the design, construction, and operation of all new and renovated state-owned facilities to be LEED Silver-certified. New York City's Local Law 86 (also known as "The LEED Law") took effect in January 2007. It basically requires that many of New York City's new municipal buildings, as well as additions and renovations to its existing municipal buildings, achieve certain standards of sustainability that would meet various LEED criteria. New Mexico followed by passing a major green-building tax credit in 2007, and Oregon passed a 35 percent tax credit for the employment of solar energy systems. These various tax incentives have induced green building tenant attraction and retention to grow stronger, and major tenants are increasingly favoring healthier air quality over luxury amenities in premium properties, thus making a green building a sound investment.

1.5 EMERGING DIRECTIONS—A RECENT UPSURGE IN GREEN BUILDING

The world of building design has over the last decade or so been increasingly going green, so that today it has become an inherent part of our global culture. This has forced the construction industry to undergo a fundamental change in how it does business and executes projects. Most architects, designers, engineers, developers, contractors, manufacturers, and federal, state, and local governments are enthusiastically embracing this emerging green phenomenon. Moreover, according to a 2008 Green Building Market Barometer online survey of commercial real estate executives conducted by New York City-based Turner Construction, even the credit collapse of 2008 has not affected the desire of developers to go green. Green building has not only become global, but in recent years it has been shown that the trend toward sustainable design is indeed much more than a passing trend.

The increasing prominence of the public's environmental awareness has become an integral component of the corporate mainstream and general global awareness of the human impact on the environment as well as increased consumer demand for sustainable goods and services are creating new challenges and opportunities for businesses in all aspects of construction-related industries. Many corporations have responded to these challenges by becoming more aware of these environmental impacts. Several voluntary initiatives and partnerships like the U.S. Green Building Council (USGBC) and Green Globes continue to raise corporate awareness and encourage increased participation in the green movement. Furthermore, in addition to the U.S., LEED and other building rating systems can be found the world over, in countries as diverse as Britain, Mexico, Australia, Spain, Canada, India, China, United Arab Emirates, and Japan, to name but a few. Green building continues to impact and transform the building market, as it fundamentally changes our perception of how we design, inhabit, and operate our buildings. Some of the factors that are accelerating the push toward building green include: An increased demand for green construction, particularly in the residential sector, increasing levels of government initiatives, and improvements in the quality and availability of sustainable materials.

This growing demand for building projects that use environmentally friendly and energy-efficient materials has spurred a strong green movement in the construction industry. Furthermore, some general contracting companies like DPR Construction, Inc. are well prepared and well placed to deliver successful green projects, with 28 percent of their professionals having acquired LEED accreditation, which may be the highest percentage in the nation among general contractors. The increasing demand for sustainability has forced businesses to seek ways to become more sustainable, mainly by focusing their efforts on improving their buildings' energy efficiency. Improving building efficiency will also help reduce fixed operating costs.

Even with the present downturn in the economy and the construction industry, the amount of "green buildings" currently being built in the United States is estimated to be in excess of $10 billion. According to the Department of Commerce (2008), the construction market constitutes about 13.4 percent of the $13.2 trillion U.S. GDP, which includes all commercial, residential, industrial, and infrastructure construction. Commercial and residential building construction on their own account for about 6.1 percent of the GDP (Source: Department of Construction, 2008). Furthermore, as of 2006, the USGBC's LEED® system certified 775 million sq ft of commercial office space as green. This represents a mere 2 percent of U.S. commercial office space. However, this is expected to increase exponentially, with green buildings accounting for 5–10 percent of the U.S. commercial construction market by 2010. These are but a few of the general indicators that point to an accelerating forward movement in the green construction market.

There are also many indicators that show the shift toward green construction is a global trend, with more developed areas putting more resources into improving efficiency and sustainability. In 2008 for example, the McGraw Hill "Global Green Building Trends" stated that 67 percent of global construction firms reported at least 16 percent of their projects as green buildings. They also project that by 2013 that number should climb to 94 percent of construction firms. If these forecasts are correct, it would be a clear signal that green construction has not only become part of the mainstream, but can expect a significant share of the $4.7 trillion global construction industry.

Internationally, Europe has achieved the highest level of sustainable building activity, with 44 percent of European construction firms building green on at least 60 percent of their projects. Slightly behind are North America and Australia, but the gaps are gradually closing. However, the area with the greatest potential for market growth is Asia where the number of firms dedicated to green construction is projected to increase three-fold between 2009 and 2013.

It should be noted that the quantity and quality of much of the data relating to business and the environment remains wanting, to say the least. In this regard, government agencies, corporations, nonprofit groups, and academic institutions have often remained lethargic and produced relatively little to quantify or assess simple measures of business environmental impact. It is interesting to note that although the American Institute of Architects (AIA) has concluded that buildings are the leading source of greenhouse gas emissions in the United States, a recent online survey conducted by Harris Interactive shows that only 4 percent of U.S. adults were aware of this fact.

With this in mind, Frank Hackett, an energy conservation sales consultant for Mayer Electric Supply Co., Inc., says that one of the most basic things a business can do to improve its efficiency is to update or retrofit its lighting system. One example is to modify and update existing lighting fixtures to use the more energy efficient T-5 or T-8 fluorescent lamps as opposed to the T-12 models that are often used. Replacement of the magnetic ballasts in the lighting can also increase the system's energy efficiency. According to Department of Energy estimates, a significant percentage of a business's typical energy bill is made up of lighting costs and being able to reduce this can have a favorable economic impact. And as more local governments and federal agencies hop on the "green" bandwagon, it is very likely that updating and retrofitting existing lighting systems could eventually become mandatory. Other energy-saving options that should be considered include automatic control systems that take advantage of natural light and automatically switch off when no one is around.

The U.S. Department of Energy (DOE) recently issued a ruling that states must now certify that their building codes meet the requirements in ASHRAE/IESNA's 2004 energy efficiency standard. The American Society of Heating, Refrigerating and Air-conditioning Engineers (ASHRAE) is now moving forward in developing the nation's first standard for high-performance, green commercial buildings (Standard 189.1P). This standard would require buildings to be significantly more efficient than required by its Standard 90.1-2007, Energy Standard for Buildings Except Low-Rise Residential Buildings. The U.S. Green Building Council also reaffirmed its commitment to the development of Standard 189.1P, which when completed will be America's first national standard developed to be used as a green building code. Standard 189.1P is being developed as an ANSI standard, created specifically for adoption by states, localities, and other building code jurisdictions that are ready to require a minimum level of green building performance for all commercial buildings.

Green building initiatives were by now taking place on all sides of the equation. President Bush on October 3, 2008 signed into law H.R. 1424 and extended the Energy Efficient Commercial Building Tax Deduction as part of the Emergency Economic Stabilization Act of 2008. This is a tax deduction and not a tax credit, i.e., an amount will be subtracted from the gross taxable income and not directly subtracted from the tax owed. This recently created program can offer benefits to the taxpayer and be used as an incentive to assist in choosing energy-efficient building systems.

California passed its Revised Title 24 Code in October 2005. Additionally, a new law that took effect on January 1, 2009 states that owners of all non-residential properties in California are mandated to make available to tenants, lenders, and potential buyers the energy consumption of their buildings as part of the state's participation in the federal ENERGY STAR program. This data will then be transmitted to the Environmental Protection Agency's ENERGY STAR Portfolio Manager who will benchmark the information under its ENERGY STAR standards. Once this data is assembled, beginning in 2010, building owners will be required to disclose the data and ratings. The majority of major cities within the United States have since initiated some form of energy efficiency standards for new construction and existing buildings.

Another example is Washington State, which in April 2005 began requiring that all state-funded construction projects with more than 5000 square feet be built green. Following this, in May 2006, Seattle approved a plan offering incentives to encourage site-appropriate packages of greening possibilities that would include green roofs, exterior vertical landscaping, interior green walls, air filtration, and storm water runoff management. Seattle also boasts becoming the first municipality in the United States to adopt the USGBC's Leadership in Energy and Environmental Design (LEED) Silver rating for its own major construction projects.

Pennsylvania currently has the second highest number of LEED-certified buildings at 83 and is just behind California. Pennsylvania now has four state funds, including a $20-million Sustainable Energy Fund that offers grants and loans for energy efficiency and renewable energy projects. Philadelphia has also recently enacted a "Green Roofs Tax Credit" to encourage the installation of roofs that supports living vegetation; it has also proposed a "sustainable zoning" ordinance that mandates incorporating green roofs for buildings that occupy a minimum of 90,000 square feet or more.

The Baltimore, Maryland City Planning Commission voted in April 2007 to require developers to incorporate green building standards into their projects by 2010. Boston also amended its zoning code to require that all public and private development projects in excess of 50,000 square feet be constructed to green building standards. When Washington, D.C.'s Green Building Act of 2006 went into effect in March 2007, it became the first major U.S. city to require LEED compliance for private projects. The application of these new green building standards became mandatory in the district in 2009 for privately owned, non-residential construction projects with 50,000 square feet or more; public projects are also required to comply with these new standards.

An increasing number of states including Arizona, Arkansas, Colorado, Connecticut, Florida, Michigan, and Nevada have followed suit, in addition to the nearly 60 cities and counties nationwide. Furthermore, with the increasing onslaught of green building into the mainstream, it seems that soon green or sustainable building will cease to be an option, but rather will become a requirement to be adopted. Stacey Richardson, a product specialist with the Tremco Roofing and Building Maintenance division, says, "It is the way of the future, and industry developments in new green technology will provide building owners increasing access to energy-saving, environmentally-friendly systems and materials. Everything from bio-based adhesives and sealants, low-VOC or recycled-content building products, to the far-reaching capabilities of nanotechnology—the movement of building "renewable" and "energy-efficient" will only continue to strengthen." Even colleges and universities such as Harvard University, Pennsylvania State, the University of Florida, the University of South Carolina, the University of California-Merced, and others have taken steps to go green.

The study, Greening Buildings and Communities: Costs and Benefits, by Landmark International, which is purported to be the largest international study of its kind and based on extensive financial and technical analyses of 150 green buildings in 33 U.S. states and 10 countries worldwide built from 1998 to 2008, provides the most detailed findings to date on the costs and financial benefits of building green. Below are some key findings outlined in the report:

- *Most green buildings cost 0–4 percent more than conventional buildings, with the largest concentration of reported "green premiums" between 0–1 percent. Green premiums increase with the level of greenness but most LEED buildings, up through gold level, can be built for the same cost as conventional buildings. This stands in contrast to a common misperception that green buildings are much more expensive than conventional buildings.*
- *Energy savings alone make green building cost effective. Energy savings alone outweigh the initial cost premium in most green buildings. The present value of 20 years of energy savings in a typical green office ranges from $7/sf (certified) to $14/sf (platinum), more than the average additional cost of $3 to $8 per square foot for building green.*
- *Green building design goals are associated with improved health and with enhanced student and worker performance. Health and productivity benefits remain a major motivating factor for green building owners, but are difficult to quantify. Occupant surveys generally demonstrate greater comfort and productivity in green buildings.*
- *Green buildings create jobs by shifting spending from fossil fuel-based energy to domestic energy efficiency, construction, renewable energy, and other green jobs. A typical green office creates roughly one-third of a permanent job per year, equal to $1/sf of value in increased employment compared to a similar non-green building.*
- *Green buildings are seeing increased market value (higher sales/rental rates, increased occupancy, and lower turnover) compared to comparable conventional buildings. CoStar, for example, reports an average increased sales price from building green of more than $20/sf, providing a strong incentive to build green even for speculative builders.*
- *Roughly 50 percent of green buildings in the study's data set see the initial "green premium" paid back by energy and water savings in five years or less. Significant health and productivity benefits mean that over 90 percent of green buildings pay back an initial investment in five years or less.*

- *Green community design (e.g., LEED-ND) provides a distinct set of benefits to owners, residents, and municipalities, including reduced infrastructure costs, transportation and health savings, and increased property value. Green communities and neighborhoods have a greater diversity of uses, housing types, job types, and transportation options and appear to better retain value in the market downturn than conventional sprawl.*
- *Annual gas savings in walkable communities can be as much as $1,000 per household. Annual health savings (from increased physical activity) can be more than $200 per household. CO_2 emissions can be reduced by 10–25 percent.*
- *Upfront infrastructure development costs in conservation developments can be reduced by 25 percent, approximately $10,000 per home.*
- *Religious and faith groups build green for ethical and moral reasons. Financial benefits are not the main motivating factor for many places of worship, religious educational institutions, and faith-based nonprofits. A survey of faith groups building green found that financial cost effectiveness of green building makes it a practical way to enact the ethical/moral imperative to care for the Earth and communities. Building green has also been found to energize and galvanize faith communities.*

Achieving sustainability on a large scale and attaining a corresponding market share remains elusive, even with the tremendous boom in green construction. LEED-registered projects today still represent only 5 percent of the total square footage in U.S. new construction. The ultimate target according to Environmental Design + Construction magazine is 25 percent of the entire market.

Although the U.S Green Building Council (USGBC), a nonprofit membership organization, was founded in 1993, it was only in the last few years that it has become a potent driving force in the green building construction movement. Its popularity was achieved largely through the development of its commercial building rating system known as Leadership in Energy and Design (LEED). Earning LEED certification typically starts in the early planning stage, where the interested stakeholders make the decision to pursue certification. Once a determination is made, the next step is to register the project and pay the required fee. After the project is completed and commissioned and all the required numbers are in with the supporting documentation, the project is submitted for evaluation, certification, and final determination.

The USGBC has had an enormous impact on green building and has emerged as a clear leader in fostering and furthering green building efforts throughout the world. In the United States, the LEED™ Green Building Rating System is increasingly becoming the national standard for green building; it is also internationally recognized as a major tool for the design and construction of high performance buildings and sustainable projects. The USGBC is clearly looking to the future and has recently issued its updated LEED V3. The USGBC has also formulated a strategic plan for the period 2009–2013 in which it outlines the key strategic issues that face the green building community such as:

- "A shift in emphasis from individual buildings toward the built environment and broader aspects of sustainability, including a more focused approach to social equity;
- Need for strategies to reduce contribution of the built environment to climate change;
- Rapidly increasing activity of government in green building arena;
- Lack of capacity in the building trades to meet the demand for green building;
- Lack of data on green building performance;
- Lack of education about how to manage, operate, and inhabit green buildings; and
- Increasing interest in and need for green building expertise internationally."

One of the primary indicators reflecting international interest in USGBC and LEED is the large annual attendance and the increasing number of countries represented at Greenbuild, USGBC's International Conference and Expo. In 2008, nearly 30,000 people attended the Greenbuild conference held in Boston (source: USGBC, Green Building Facts) as compared to only 4200 attending the same event in Austin, Texas in 2002. The 2008 attendees include representatives from all 50 states, 85 countries and six continents. This highlights the international importance of the annual Greenbuild conferences and expos and strongly suggests that Greenbuild has become an important forum for international leaders in green building to exchange ideas and information.

For example, one of the many international delegations included a recent high-level delegation from China headed by the Vice Minister of the Ministry of Construction, Mr. Qiu Baoxing. This is significant because over the past decade China's economy has been expanding and growing at an alarming rate, and some forecasts expect it to become the largest economy in the world by 2020. With such growth, however, comes a series of potentially severe environmental problems. China has made substantial inroads to addressing these challenges and reversing many of these environmental trends. To facilitate this, China announced a new energy efficiency strategy, of which green building is a primary component. The Chinese Ministry of Construction and the USGBC also signed a Memorandum of Understanding identifying points of duel interest for collaboration in the promotion of environmentally responsible buildings in both China and in the U.S.

Many project teams are today globally applying the LEED rating system as developed in the United States. However, the USGBC recognizes that certain criteria, processes, or technologies may not always be appropriate for all countries, and that successful strategies for encouraging and practicing green building will differ from one country to another, depending on local conditions, traditions, and practice. To address this reality, the USGBC has agreed to sanction other countries to license LEED and allow them to adapt the rating system to their specific needs, providing LEED's high standards are not compromised. Many countries worldwide have expressed an interest to being LEED licensed, and countries like Canada and India have already achieved this.

On the international stage, the USGBC works through the World Green Building Council (WorldGBC), which was formed in 1999 by David Gottfried. One of WorldGBC's goals is to assist other countries in establishing their own councils and find means to work effectively with policy makers and local industry. The WorldGBC is devoted to transforming the global property industry to sustainability and is presently composed of a federation of 10 national green building councils that include the United States and Canada. Its mission is to serve as a forum for knowledge transfer between green building councils and to support and promote the individual green building council's members. Additionally, the WorldGBC has the mission of recognizing global green building leadership and encouraging the development of market-based environmental rating systems. These exciting global developments are a reflection of some of the countless ways the U.S. Green Building Council and its members are contributing to the international green building movement.

Following the terrorist attacks on the World Trade Center in New York and in Mumbai, India, many architects and building owners are now demanding that their facilities be designed to have greater blast resistance and to better withstand the effects of violent tornadoes and hurricanes, for example by the use of blast-resistant windows with protective glazing. Windows of federal buildings now are required to be designed to provide protection against such potential threats. Likewise, there are increasing demands from both governments and the public alike for structures to be sustainable and meet environmental requirements. These reflect emerging trends that when taken together necessitate a greater emphasis on industry expertise, innovation, and solutions capable of addressing a wide range of environmental challenges.

The unprecedented growth of the USGBC is evidenced by the dramatic increase in the number of certified and registered projects since LEED was first launched in 2000—an increase of nearly 700 percent a year. By the end of 2007, the square footage of U.S. office and commercial space registered or certified under LEED reached 2.3 billion, an increase of more than 500 percent from two years earlier. Certified projects are

FIGURE 1.5   The Sydney Opera House, which overlooks the harbor, is considered one of the most recognizable buildings in the world and has become the city's landmark. It consists of 14 freestanding sculptures of spherical roofs and sail-like shells sheathed in white ceramic tiles. The original plan to build the opera house was won in competition in 1957 by the late Danish architect John Utzon, but because his vision and design were too advanced for the architectural and engineering capabilities at the time, it wasn't until 1973 that the Opera House was finally opened. (Source: Best Places of the World.)

projects that have been completed and verified through the USGBC's process, while registered projects are those that are still in the process of design or construction. This astonishing increase in LEED project registration for new construction—roughly 40 percent in 2008—is very significant as it is a clear indicator of future trends.

Bob Schroeder, industry director (Americas) for Dow Corning's construction business, echoes this and says, "Today, sustainable design has been recognized by the industry and the public as a critical factor in achieving high quality architecture and benefiting the building owners, the companies that occupy these structures, and the wider community."

Most scholars consider great architecture to be a delicate balance of form and function. Yet high-rise buildings are being constructed globally with increasing ferocity, often without any particular concern for due diligence, the environment, or aesthetics. With all this evident commotion there are several interesting trends now emerging, such as the building of spectacular landmarks as exemplified by the Sydney Opera House (Figure 1.5), and Burj Dubai (Figure 1.6) which will reportedly be

FIGURE 1.6 An illustration of Burj Dubai—the tallest skyscraper in the world—by SOM for Emaar Properties, UAE. The height of the completed tower is estimated to be 2684 feet with a budget of about $800,000,000. It reportedly contains more than 160 habitable floors, 56 elevators, apartments, shops, swimming pools, spas, corporate suites, Italian fashion designer Giorgio Armani's first hotel, and an observation platform on the 124th floor. (Source: Skidmore, Owings & Merrill.)

the highest building in the world. A driving force behind the creation of national landmarks is basically twofold: the primary desire is that cities are often searching for recognizable symbols to foster local pride; the second is essentially an economic one, e.g., increasing tourism.

Elements of Green Design and Construction

2.1 GENERAL

Green building and contracting has become a hot selling point with home and business customers and can add tangible value to your business for years to come. It is also an excellent way to differentiate your company from the competition. Specializing in green building and sustainability basically means incorporating environmentally friendly techniques and sustainable practices into your operations. Among the many advantages of building green include promotion of resource conservation, such as energy efficiency, renewable energy, use of materials that have a reduced carbon footprint, water conservation; minimizing environmental impacts and generating less waste during construction; creating a healthy and comfortable environment, while at the same time reducing operation and maintenance costs. In fact, the entire life-cycle of the building and its components is considered, as well as the economic and environmental impact and performance. It obviously helps if you have among your employees a LEED™ Accredited Professional.

An increasing number of designers, builders, and building owners are getting involved in green building practices. National and local programs instigating and promoting green building are growing and reporting increasing successes, while hundreds of certified green projects across the nation and internationally provide tangible evidence of what sustainable building design can accomplish in terms of aesthetics, comfort, and energy and resource efficiency.

Roger Woodson, a well-known author and contractor says, "In theory, you don't have to know much about construction to be a builder who subs all the work out to independent contractors, but as the general contractor it is you who will ultimately be responsible for the integrity of the work."

Today more than ever buildings have a tremendous impact on the environment both during construction and through their operation. "Green/sustainable building" is a loosely defined collection of strategies such as land-use, building design, construction, and operation that reduce environmental impacts. Green building practices facilitate the creation of environmentally sound and resource-efficient, high-performance buildings by employing an integrated team approach to design where architects, engineers, land planners, building owners and operators, and constructors pool their resources and design the building.

The Department of Energy (DOE) has also developed an Environmental Protection Program, the goals and objectives of which are "to implement sound stewardship practices that are protective of the air, water, land, and other natural and cultural resources impacted by DOE operations and by which DOE cost effectively meets or exceeds compliance with applicable environmental, public health, and resource protection laws, regulations, and DOE requirements. This objective must be accomplished by implementing environmental management systems (EMSs) at DOE sites. An EMS is a continuing cycle of planning, implementing, evaluating, and improving processes and actions undertaken to achieve environmental goals." Some of these goals and objectives include:

1. Goal: Protect the environment through waste prevention.

   Objective: Minimize environmental hazards, protect environmental resources, minimize life-cycle costs and liability of DOE programs, and maximize operational capability by eliminating or minimizing the generation of wastes that would otherwise require storage, treatment, disposal, and long-term monitoring and surveillance.

2. Goal: Protect the environment through reduction of environmental releases.

   Objective: Minimize environmental hazards, protect environmental resources, minimize life-cycle costs and liability of DOE programs, and maximize operational capability by eliminating or minimizing the use of toxic chemicals and associated releases of pollutants to the environment that would otherwise require control, treatment, monitoring, and reporting.

3. Goal: Protect the environment through environmental preferable purchasing.

   Objective: Minimize environmental hazards, conserve environmental resources, minimize life-cycle costs and liability of DOE programs, and maximize operational capability through the procurement of recycled content, bio-based content, and other environmentally preferable products thereby minimizing the economic and environmental impacts of managing toxic byproducts and hazardous wastes generated in the conduct of site activities.

4. Goal: Protect the environment through incorporation of environmental stewardship in program planning and operational design.

   Objective: Minimize environmental hazards, conserve environmental and energy resources, minimize life-cycle costs and liability of DOE programs, and maximize operational capability by incorporating sustainable environmental stewardship in the commissioning of site operations and facilities.

5. Goal: Protect the environment through post-consumer material recycling.

   Objective: Protect environmental resources, minimize life-cycle costs of DOE programs, and maximize operational capability by diverting materials suitable for reuse and recycling from landfills, thereby minimizing the economic and environmental impacts of waste disposal and long-term monitoring and surveillance.

All project team members need to have a clear understanding of LEED certification and the role it can play in improving property owners' competitive edge in an increasingly green market. Certification also gives independent verification that a building has achieved accepted standards in these areas, as outlined in the LEED Green Building Rating System. LEED certification of a project provides recognition of its quality and environmental stewardship. The LEED green building rating system is widely accepted and recognized by both the public and private sectors, further fueling the demand for green building certification systems.

Since its inception, the LEED rating system has made significant inroads into the mainstream design and construction industry, and contractors and property developers are realizing that they too can contribute toward a project's success in achieving green objectives. This would be accomplished first by understanding the LEED process and the specific role they can play in achieving LEED credits, and then, through early involvement and participation throughout the different project phases by incorporating a team approach in an integrated design process. Needless to say, measurable benchmarks are needed in order to achieve verification and confirm a building's acceptable performance. ASHRAE places this responsibility of defining design intent requirements squarely on the shoulders of the owner. However, it is not possible to correctly evaluate a building or project unless certain information is made available regarding the criteria on which the project's design and execution was based. A project's plans and specifications, etc. must therefore be prepared in a manner that can achieve measurable results. Otherwise, a meaningful evaluation to see if a project has met the required results and original design intent is not possible. Furthermore, before measureable green criteria can be established, it is necessary to first agree on a finite definition of green construction and to specify exactly what is required to be achieved.

2.2 GREEN RATING STANDARDS USED WORLDWIDE

The first environmental certification system was created in the United Kingdom in 1990, called the Building Research Establishment Environmental Assessment Method (BREEAM). This was followed in the United States by USGBC's introduction in 1998 of the LEED green building rating system, based substantially on the BREEAM rating system. The Green Globes rating system is an adaptation of the Canadian version of BREEAM and was released in the United States by the Green Building Initiative (GBI) in 2005. There are numerous other rating systems used in countries around the world, each with its pros and cons depending on the type of certification targeted for a specific building. However, LEED followed by Green Globes are currently the most widely applied systems for commercial construction in the United States.

Today, USGBC has rating systems for new construction, existing buildings, core and shell, commercial interiors, schools, retail, homes, healthcare, and neighborhood development, whereas the GBI has a rating system for commercial buildings, which includes new construction buildings and existing buildings. The GBI also partners with the National Association of Home Builders to promote green homes.

Rating and certification systems are required to verify the sustainability and "greenness" of buildings in the market. They basically inform us how eco-friendly and environmentally sound a building is, delineate to what extent green components have been incorporated, and identify the sustainable principles and practices that have been employed. Moreover, rating or certifying a green building helps remove some of the subjectivity that often surrounds buildings that have not been certified. Moreover, rating a green building makes a property more marketable by informing tenants and the public about the environmental benefits of a property, and also discloses the additional innovation and effort that the owner has invested to achieve a high performance building. Certification also reveals the level of sustainability achieved.

A holistic approach to design translates into a strategic integration of mechanical, electrical, and materials systems. Such an approach often creates substantial efficiencies, the complexities of which are not always apparent. Rating a green building identifies these differences objectively, and quantifies their contribution to energy and resource efficiency. The fact that rating systems typically require independent third-party testing of the various elements means there is less risk that these systems will not perform as predicted. Furthermore, formally rating or certifying a building dramatically reduces the risk of the possibility of falsely marketing a building under the perception that it is green when in fact it is not. Below are examples of some of the more widely used rating systems in the United States:

LEED™

This rating system was developed by the U.S. Green Building Council (USGBC) and continues to be the most widely applied rating system in the United States for commercial buildings. The LEED system consists of several rating categories, applicable to different points in a building's lifecycle and is discussed in other parts of this chapter. The General Services Administration (GSA) and other municipalities and government departments, as well as an increasing number of private investors and owners, have instituted policies requiring LEED certification for new construction projects. The USGBC also holds an annual green building conference—Greenbuild, which helps promote the green building industry.

Green Globes™

This rating system is an interactive, Web-based commercial green building assessment protocol offered by the Green Building Initiative (GBI) and is discussed in the following section of this chapter. It offers immediate feedback on the building's strengths and weaknesses and automatically generates links to engineering, design, and product sources. The system basically evaluates buildings in seven areas (www.thegbi.org/greenglobes). Green Globes continues to gain traction. Indeed, the parent company of Green Globes was recently acquired by a highly respected, established company: Jones Lang LaSalle.

Energy Star®

This is a joint government-backed program of the Environmental Protection Agency (EPA) and the U.S. Department of Energy (www.energystar.gov). Since its inception in 1992, the program has overcome many market barriers and helped revolutionize the marketplace for cost-effective, energy-efficient products and services. It has now transformed into an international standard for energy efficient consumer products and has been adopted by numerous countries including the European Union, Canada, Japan, and Australia. The mission of the program is to help businesses and individuals protect the environment through superior energy efficiency. The program is also designed for existing buildings, consisting of an energy performance rating system that is free with an online tool that focuses on energy performance. With Energy Star, impacts of other factors such as indoor air quality, materials, or water conservation are not taken into consideration. The program has grown to encompass more than 60 product categories for the home and workplace, new homes, and superior energy management within organizations. The system basically compares the energy performance of a specific building to that of a national stock of similar buildings. The data entered into the ENERGY STAR Portfolio Manager tool will model energy consumption based on a building's size, occupancy, climate, and space type. A minimum of one year's utility information input is required before the property can be assigned a rating (from 1–100). In order to apply for and receive the ENERGY STAR label, buildings must acquire a score of 75 or more.

Other Green Building Standards Worldwide

Many countries throughout the world continue to develop their own standards of energy efficiency for buildings. Only a small number of these systems are currently being applied in the U.S., but they may still prove influential in the emerging green building industry. Examples include BREEAM in the UK, Green Star in Australia, and BOMA Go Green Plus in Canada. Below are examples of building environmental assessment tools currently being used by different countries around the world.

Australia

The Green Building Council of Australia (GBCA) developed a green building standard known as the Green Star environmental rating system, which is accepted as the Australian industry standard for green buildings. The system is essentially a comprehensive, national, voluntary environmental rating system that evaluates the environmental design and construction of buildings in Australia. It has now been mandated in three states as a minimum for government office accommodation. The Green Star environmental rating tools for buildings benchmark the potential of buildings based on nine environmental impact categories. There are other standards that are used such as EER (Energy Efficiency Rating) and NABERS (National Australian Built Environment Rating System), which is a government initiative to measure and compare the environmental performance of Australian buildings. NABERS is a performance-based rating system for existing buildings. It now incorporates the Australian Building Greenhouse Rating (ABGR), and has been re-named NABERS Energy for Offices.

Canada

LEEDS™ and Green Globes™ are the two most widely applied green rating systems in Canada. The Canada Green Building Council, which was established in December 2002, acquired an exclusive license in 2003 from the USGBC to adapt the LEED rating system to Canadian circumstances. The Canadian LEED for Homes rating system was released on March 3, 2009. Canada has also implemented the R-2000 program which is a made-in-Canada home building technology to promote construction that goes beyond its building code to increase energy efficiency and promote sustainability. An optional feature of the R-2000 home program is the EnerGuide rating service, which is available across Canada and which allows homebuilders and homebuyers to measure and rate the performance of their homes, and confirm that those specifications and standards have been met.

The R-2000 Program is a collaboration between the Canadian Home Builders' Association (CHBA) and the Office of Energy Efficiency (OEE) of Natural Resources Canada (NRCan). For close to three decades, the two partners have worked together on the components that make up the R-2000 Program. Regional initiatives based on R-2000 include Energy Star for New Homes, Built Green, Novoclimat, GreenHome, Power Smart for New Homes, and GreenHouse. In March 2006, Canada's first green building point of service, Light House Sustainable Building Centre, opened in Vancouver, BC which is funded by Canadian government departments and businesses to help implement green building practices.

China

China reportedly has the biggest construction volume in the world; nearly half of the world's new building construction is now estimated to be in China. Yet the green building industry is still in its infancy in China, and green building demand continues to be driven mainly by

multinational companies. There are currently two sets of national building energy standards being applied in China (one for public buildings
and another for residential buildings). China has in place mandatory
building energy standards that are narrow in scope and currently lacks
a strong regulatory framework to incorporate energy efficient standards
in construction. Moreover, Ministry of Construction (MoC) enforcement remains problematic, and the central government established a
building inspection program to monitor the implementation of building
energy efficiency. Design institutions, developers, and construction companies will lose their licenses or certificates under this program if they
do not comply with the regulations. China also recently launched a new
green building standard meant to complement better known labels like
BREEAM (UK) and LEED, which are presently only used in office buildings for multinationals or upscale apartments.

Standards and ratings: Multinational companies have taken the lead
to promote green building construction in China by pursuing more
stringent LEED certification. The trend was initiated by Plantronics, a
California-based electronics company, when it achieved a LEED gold certification for its new manufacturing and design centre in Suzhou. This
was followed by Nokia who received its first LEED gold certification
globally for the Nokia China Campus in Beijing. Other big names followed, including Siemens and BHP Billiton.

A Memorandum of Understanding (MOU) on building energy efficiency was signed between the U.S. and China in which the United States
pledged $15 million for a joint U.S.-China clean energy research center.
The MoC also recently introduced the "Evaluation Standard for Green
Building" (GB/T 50378-2006), which resembles in structure and rating
process of the USGBC's LEED (which is also being used). The building
energy consumption data will be collected by MoC, and will be used to
assess building performance; a three-star green building certificate will be
awarded to qualified buildings. The Green Olympic Building Assessment
System (GOBAS) is another green building rating system that was initially based on Japan's Comprehensive Assessment System for Building
Environment Efficiency (CASBEE). High-performance building projects
are being supported both by the government and the private business sector. The World Green Building Council (WBGC), a Sustainable
Buildings and Climate Initiative (SBCI) member, has assisted the ministry
of construction in China to establish the China Green Building Council.
This Council is also supported by the USGBC.

France

French President Nicolas Sarkozy instigated a "Le Grenelle de
l'Environnement" ("Grenelle Environment Round Table"), to define the
key points of public policy on ecological and sustainable development

issues for the coming five years and to find ways to redefine France's environment policy. The process led to a set of recommendations released at the end of October 2007 which were presented to the French parliament in early 2008. Six working groups were formed composed of representatives of the central government, local governments, employer organizations and trade unions, and NGOs to debate and address various themes such as climate change, energy (the building sector consumes 42.5 percent and transports 31 percent of total French energy) biodiversity, natural resources, health and the environment, production and consumption, democracy and governance, and competitiveness, development patterns, and environmental employment. Within the framework of the "Grenelle de l'Envronnement," the performance acceleration is designed to meet with the following objectives for tertiary buildings:

- Low consumption buildings (BBC) by 2010 with minimum requirements concerning the levels of renewable energy and CO_2 absorption materials by 2012
- Passive new buildings (BEPAS) or positive buildings (BEPOS) by 2020
- Labels for refurbishment of existing BBC buildings.

All these developments match the European and international regulations and frameworks.

Additionally, two property companies, AXA Real Estate Investment Management and ING Real Estate, recently set up a rival rating system in the UK called Green Rating which allows owners to compare properties' sustainability across Europe. It was recently launched in France, Spain, Italy, the Netherlands, and Germany. In 2010, it is expected to be launched in the U.S. and Japan. The advantage of Green Rating is that it is an assessment of the energy efficiency of a building and looks at both the building materials and the waste generated by the building. The audit process has been tested on 50 sites in Europe, and is broken into four stages:

- Data collection: The site technical manager collects data in six areas: energy use, carbon emissions, water use, waste, proximity to transport links, and the health of the employees.
- Onsite survey and interview with the site manager or owner: A Green Rating assessor inspects the building and the equipment.
- Energy modeling: A model is created to see how energy is used across areas of the building.
- Recommendations: A Green Rating advises how improvements could be made. An additional audit could then be carried out a year or two after the first.

Germany

In January 2009 the first German standard was developed for sustainable buildings by the DGNB (Deutsche Gesellschaft für nachhaltiges

Bauen e.V.—German Sustainable Building Council) and the BMVBS (Bundesministeriums für Verkehr, Bau und Stadtentwicklung—Federal Ministry of Transport, Building, and Urban Affairs) to be used as a tool for the planning and evaluation of buildings. The DGNB is a clear and easy-to-understand rating system covering all relevant topics of sustainable construction. There are six subjects covered in the evaluation: ecology, economy, social-cultural and functional topics, techniques, processes, and location. Outstanding buildings can achieve awards in the categories of bronze, silver, and gold.

There are a number of German organizations that employ green building techniques such as:

- The Solarsiedlung (Solar Village) in Freiburg, Germany, which features energy-plus houses
- The Vauban development, also in Freiburg
- Houses designed by Baufritz, incorporating passive solar design, heavily insulated walls, triple-glaze doors and windows, non-toxic paints and finishes, summer shading, heat recovery ventilation, and greywater treatment systems
- The new Reichstag building in Berlin, which produces its own energy.

India

GRIHA is a rating system for green buildings developed by The Energy and Resources Institute (TERI) of India, which plays a key role in developing green building awareness and strategies in the country. GRIHA was adopted and endorsed by the government of India as the national green building rating system for the country and measures are being taken to spread awareness about this rating system, and TERI signed a memorandum of understanding with the Union Ministry of New and Renewable Energy to this effect. The aim of GRIHA is to ensure that all types of buildings become green buildings. One of the strengths of GRIHA is that it puts great emphasis on local and traditional construction knowledge and even rates non-air conditioned buildings as green.

GRIHA uses 32 criteria to evaluate and rate buildings, totaling a maximum of 100 points. A building must score at least 50 to apply for certification. Preserving landscape during construction, soil conservation after construction, and reducing air pollution are some of the qualifying criteria. Buildings also need to quantify energy consumption in absolute terms and not percentages alone.

The Confederation of Indian Industry (CII) is also playing an active role in promoting sustainability in the Indian construction sector. The CII is the central pillar of the Indian Green Building Council or IGBC. The IGBC in turn is licensed by the LEED Green Building Standard from the U.S. Green Building Council and is currently responsible for certifying

LEED New Construction and LEED Core and Shell buildings in India. All other projects are certified through the U.S. Green Building Council. There are many energy efficient buildings in India, situated in a variety of climatic zones (Figure 2.1).

In June 2007, the Indian Bureau of Energy Efficiency (BEE) launched the Energy Conservation Building Code (ECBC), which specifies the energy performance requirements for all commercial buildings that are to be constructed in India. The ECBC has set energy efficiency standards for design and construction of any building of minimum conditioned area of 1000 square meters (10,764 square feet) and a connected power demand of 500 KW or 600 KVA. On February 25, 2009, the BEE launched a five-star rating scheme for office buildings operated only in the daytime in three climatic zones—composite, hot and dry, and warm and humid.

Israel

A voluntary new standard for "Buildings with Reduced Environmental Impact" SI-5281 was approved in November 2005 and is awarded to new or renovated residential and office buildings that comply with the requisite requirements and criteria. The standard is comprised of four sections covering energy, land, water, wastewater, drainage, and other environment-related elements. A building that meets the prerequisites in each section and accumulates the minimum number of credit points in every environment-related sphere is eligible for "green building" certification.

FIGURE 2.1   The Sohrabji Godrej Green Business Centre in Hyderabad, India. This was the first Platinum rated green building under the LEED rating system, outside the U.S., boasting energy savings of 63 percent. (Source: Confederation of Indian Industry.)

The standard is based on a point rating system. Thus, a cumulative score of 55–75 points entitles a building to a "green building" label, while a cumulative score of more than 75 points allows it to be certified as an "outstanding green building." Together with complementary standards 5282-1, 5282-2 for energy analysis and 1738 for sustainable products, it provides a system for evaluating environmental sustainability of buildings.

United States Green Building Council's LEED rating system has also been implemented on several building projects in Israel and there is a strong industry drive to introduce an Israeli version of LEED in the near future. Because of different climatic conditions and building construction methods in Israel, the LEED rating system cannot be adopted as is.

Japan

A joint industrial/government/academic project was created in 2001 with the support of the Housing Bureau, Ministry of Land, Infrastructure, Transport and Tourism (MLIT). This led to the creation of the Japan GreenBuild Council (JaGBC)/Japan Sustainable Building Consortium (JSBC), which in turn created the Comprehensive Assessment System for Building Environmental Efficiency (CASBEE) system. The CASBEE system was developed according to the following principals:

- Structured in a manner to award high assessments to superior buildings, thereby enhancing incentives to designers and other stakeholders
- To be applicable to buildings in a wide range of applications
- To be as simple as possible
- To take into consideration issues and problems peculiar to Japan and Asia.

CASBEE certification is currently available for new construction, existing buildings, renovations, urban development, heat island, urban areas, plus buildings and detailed homes.

The CASBEE system is composed of four assessment tools that are intended to correspond to a building's lifecycle. The collective name for these four tools is the "CASBEE Family," and its expanded tools are designed to serve at each stage of the design process. Each tool is designed for a separate purpose and target user, with the purpose of accommodating a wide range of uses (offices, schools, apartments, etc.) in the buildings evaluation process. Furthermore, the process of obtaining CASBEE certification differs from LEED in that the LEED certification process starts at the beginning of the design process, with review and comments taking place throughout the design and construction of a project. Although CASBEE's latest version for new construction ranking uses pre-design tools, certification consists primarily of site visits once the building is completed.

Malaysia

The main organization promoting green practices and building techniques is the Standards and Industrial Research Institute of Malaysia (SIRIM). However, Malaysia has now put in place a new rating system called the Green Building Index for commercial and residential properties. The GBI was developed by Pertubuhan Akitek Malaysia (PAM) and the Association of Consulting Engineers Malaysia (ACEM). It is a profession-driven initiative to lead the Malaysian property industry toward becoming more environmentally friendly (Figure 2.2). The GBI rating system provides an opportunity for developers to design and construct sustainable buildings that can provide increased energy savings, water savings, a healthier indoor environment, better connectivity to public transport, and the adoption of material recycling and greenery for their projects.

The Green Building Index (GBI) has six key criteria, which are: energy efficiency, indoor environmental quality, sustainable site and management, and materials, resources, and water efficiency. Based on the scores achieved, commercial buildings will be rated and then certified as silver, gold, or platinum. The final award is presented one year after the building is first occupied. Buildings are also required to be re-assessed every three years in order to maintain their GBI rating by ensuring that the buildings are well maintained.

FIGURE 2.2 The PTM Green Energy Office Building (GEO Building) is Malaysia's first GBI certified green building, and is designed as an administration-cum-research office for Pusat Tenaga Malaysia (Malaysia Energy Centre). The GEO building is built on a five-acre site in Seksyen 9, Bandar Baru Bangi, Selangor, Malaysia. It is among the first Malaysian government office buildings whose design is based on green concepts and is environmentally friendly. (Source: PTM-GreenBuildingIndex.)

Internationally there are other green rating systems such as LEED (the United States and Canada), Energy Star (U.S.), BREEAM (Britain), CASBEE (Japan), Green Star and NABERS (Australia).

Mexico

The Mexico Green Building Council (CMES) is the principal organization dedicated to promotion of best practices that improve the environmental performance of buildings and fostering sustainable building technology and policy in Mexico. It is an independent nonprofit, nongovernmental organization that works from within the construction industry in order to promote a broad-based transition toward sustainability. Its stated mission is to promote sustainable development through the realization and construction of a superior built environment.

Under a new partnership agreement, ASHRAE and the Mexico Green Building Council (CMES) will work together to promote buildings that are healthy, environmentally responsible, comfortable and productive, and profitable. The agreement is part of ASHRAE's new strategy for a global environment, committing the society to working with organizations with shared interests and values.

Mexico City's government is promoting the creation of a standard certification program for green buildings, and the city's minister of the environment announced that certified green buildings will be eligible for up to a 25 percent discount on property taxes. The Green Building Certification Program will have three levels of certification: lowest rating from 21 to 50 points; an efficiency level from 51 to 80 points; and an excellence level from 81 to 100 points. The higher the level, the greater the level of property tax discounts. Initially the tax discount will be a voluntary scheme, but is intended to become mandatory in the future.

New Zealand

In July 2005, the New Zealand Green Building Council (NZGBC) was formed; this is a not-for-profit, industry organization dedicated to accelerating the development and adoption of market-based green building practices.

In 2006/2007, several major milestones were achieved, including the NZGBC becoming a member of the World GBC and launching of the Green Star NZ (an office design tool). Green Star is a comprehensive, national, voluntary environmental rating scheme that evaluates the environmental attributes and performance of New Zealand's buildings using a suite of rating toolkits developed to be applicable to each building type and function. Green Star was developed by the New Zealand Green Building Council (NZGBC) in partnership with the building industry.

South Africa

The Green Building Council (GBC) of South Africa was launched in 2007; its stated mission is "To promote, encourage and facilitate green building in the South African property and construction industry through market-based solutions, by:

- Promoting the practice of green building in the commercial property industry
- Facilitating the implementation of green building practice by acting as a resource centre
- Enabling the objective measurement of green building practices by developing and operating a green building rating system
- Improving the knowledge and skills base of green building in the industry by enabling and offering training and education."

The South African GBC has since developed Green Star SA rating tools, based on the Green Building Council of Australia tools, to provide the property industry with an objective measurement tool for green/sustainable buildings and to recognize and reward environmental leadership. Each Green Star SA rating tool reflects a different market sector (e.g., office, retail, multi-unit residential, etc.). Green Star SA-Office was the first tool developed and was released in final form (version 1) at the Green Building Council of South Africa Convention & Exhibition '08 in November 2008. South Africa is in the process of incorporating an energy standard SANS 204 which aims to provide energy-saving practices as a basic standard in the South African context. Green Building Media, which was launched 2007, has also played an instrumental role in green building in South Africa.

United Kingdom

The Association for Environment Conscious Building (AECB) was founded in 1989 and incorporated in January 2005 to increase awareness within the construction industry of the necessity to respect the environment and to promote sustainable building in the UK. The AECB is now under the Energy Performance of Building Directive (EPBD), which Europe has made a mandatory energy certification since January 4, 2009. A key part of this legislation is that the EPBD requires all EU countries to enhance their building regulations and to introduce energy certification schemes for buildings. All countries are also required to have inspections of boilers and air-conditioners.

A mandatory certificate called the Building Energy Rating system (BER) and a certification Energy Performance Certificate (EPC) is needed by all buildings that measure more than 1000 square meters (approximately 10,765 square feet) in all the European nations. Furthermore, when buying or selling a home it is now law to have a certificate. Certificates are also required on construction of new homes and for rented homes the first time the property is let after October 1, 2008. The certificate records how energy efficient a property is as a building and provides A–G ratings. These are similar to the labels now provided with domestic appliances such as refrigerators and washing machines.

The UK Green Building Council (UK-GBC) also called for the introduction of a Code for Sustainable Buildings in March 2009 to cover all non-domestic buildings, both new and existing. Although the Code for Sustainable Buildings is owned by the government, it is developed, managed, and implemented by industry and covers refurbishment as well as new construction. Furthermore, on September 1, 2009, the Welsh Assembly Government planning policy established a national standard for sustainability for most new buildings proposed in Wales.

According to the BREEAM (BRE Environmental Assessment Method) Website, the BREEAM assessment process was created in 1990 as a tool to measure the sustainability of new non-domestic buildings in the UK with the first two versions covering offices and homes. BREEAM is the leading and widely used environmental assessment method in the UK for buildings, setting the standard for best practice in sustainable design and a measure used to describe a building's environmental performance. It has been updated regularly in line with UK building regulations and underwent a significant facelift on August 1, 2008, called BREEAM 2008. Credits are awarded in each of the following areas according to specific performance:

- Management
- Health and well-being
- Energy
- Transport
- Water
- Material and waste
- Land use and ecology
- Pollution.

A set of environmental weightings then enables the credits to be added together to produce a single overall score. This allows a building to be rated on a scale of pass, good, very good, excellent, or outstanding; this is followed by a certificate being awarded to the development.

Some of the dramatic changes to BREEAM 2008 were in response to an evolving and changing construction industry and public agenda, and include:

- Introduction of mandatory credits
- Two-stage assessment process introduced (design stage and post construction stage)
- Additional rating level added (BREEAM outstanding)
- Environmental weightings modified
- CO_2 emissions benchmarks set to align with the new EPC (Environmental Performance Certificate)
- Changes to certain specific credits
- Updated Green Guide Ratings to be available online

- Introduction of BREEAM Healthcare and BREEAM Further Education
- Shell only assessments.

United States

There are numerous sustainable design organizations and programs in place within the United States. The most widely used rating system is the U.S. Green Building Council (USGBC) which promotes sustainability in how buildings are designed, built, and operated, and is best known for the development of the LEED rating system and Greenbuild, a green building conference well known for its promotion of the green building industry and environmental issues.

The USGBC had over 17,000 member organizations by September 2008 from all sectors of the building industry; its mission is to promote buildings that are environmentally responsible, profitable, and healthy places to live and work. The USGBC through its Green Building Certification Institute (GBCI) offers industry professionals an opportunity to receive accreditation as green building professionals. In June 2009, LEED had a complete overhaul of its rating system and introduced a new version (LEED V3), with a two-tier system.

The National Association of Home Builders is a trade association representing homebuilders, remodelers, and suppliers to the industry; it has formed a voluntary residential green building program called NAHBGreen (www.nahbgreen.org). This program incorporates an online scoring tool, national certification, industry education, and training for local verifiers. The online scoring tool is free to both builders and homeowners. The NAHB announced in August 2009 that the number of homebuilders, remodelers, and other members of the real estate and construction industry who hold the Certified Green Professional (CGP) educational designation now tops 4000.

The Green Building Initiative (GBI) is a nonprofit network of building industry leaders working to mainstream building approaches that are environmentally progressive, but also practical and affordable for builders to implement. The GBI has introduced a Web-based rating tool called Green Globes, which is a green management tool that includes an assessment protocol and a rating system and guide for integrating environmentally friendly design into both new and existing commercial buildings (and which is discussed later in the chapter).

Energy Star® is a program established by the U.S. Environmental Protection Agency (EPA) and U.S. Department of Energy that focuses on creating energy efficient homes and buildings designed to protect the environment while at the same time saving money for homeowners and businesses. The system also rates commercial buildings for energy efficiency and provides Energy Star® qualifications for new homes that must meet a series of energy efficiency guidelines established by the EPA.

2.3 LEED™ CERTIFICATION AND RATING SYSTEM

2.3.1 General

Comprehensive documentation can be found on the USGBC and GBCI Websites (www.leedbuilding.org; www.gbci.org), from LEED accreditation requirements to reference guides, careers, and e-newsletters. The most appropriate manner to be able to contribute to the success of a LEED project is to become familiar with the many requirements and opportunities offered by the new program. To succeed in earning LEED certification for a project, the process must start in the initial planning stage, where the stakeholders involved make a commitment to pursue certification. Once this is done, the next step is to register the project and pay an initial flat fee. As part of the newly launched LEED v3, the Green Building Certification Institute (GBCI) has assumed responsibility for administrating LEED certification for all commercial and institutional projects registered under a LEED Rating System.

When the project is completed, and all the numbers are in including preparation of all supporting documentation, the project is submitted for evaluation and certification. Once this has been determined, the project is listed on the LEED project list. The summary sheet showing the tally of credits earned becomes available for most certified projects. To assist in the certification process there is a policy manual that can be accessed online that gives an overview of the program requirements pertaining to the LEED Green Building Rating System, and identifies the policies put in place by the GBCI for the purposes of administering the LEED certification process.

2.3.2 LEED Process Overview

The USGBC and GBCI Websites should always be checked for the latest updates. Basically, the LEED 2009 Green Building Rating System consists of a set of performance standards used in the certification of commercial, institutional, and other building types in both the public and private sectors with the intention of promoting healthy, durable, and environmentally sound practices. A LEED certification is indisputable evidence of independent, third-party verification that a building project has achieved the highest green building and performance measures according to the level of certification achieved. Setting up an integrated project team to include the major stakeholders of the project such as the developer/owner, architect, engineer, landscape architect, contractor, and asset and property management staff is helpful to jumpstart the process. This implementation of an integrated, systems-oriented approach to green project design, development, and operations can yield significant synergies while enhancing the overall performance of a building. During the initial project team meetings, the project's goals will be clarified and delineated and the LEED certification level sought will be established.

Projects must adhere to the LEED Minimum Program Requirements (MPRs) in order to achieve LEED certification. MPRs describe the eligibility for each system and are intended to "evolve over time in tandem with the LEED rating systems." Though there are eight requirements that are standardized for all systems, the thresholds and levels apply differently for each system. Nevertheless, LEED projects must comply with all the applicable MPRs outlined below. To clarify the minimum program requirements, one of the categories will be used as an example—New Construction and Major Renovations:

1. Must comply with all applicable federal, state, and local building-related environmental laws and regulations where the project is located.
2. A LEED project must consist of a complete, permanent building or space. It must be designed for, constructed on, and operated on already existing land. LEED projects are required to include new, ground-up design and construction or major renovation of at least one complete building. Moreover, construction prerequisites and credits may not be submitted for review until substantial completion of construction has been achieved.
3. Must employ a reasonable site boundary: A. The LEED project boundary is to include all contiguous land that is associated with and supports normal building operations for the LEED project. B. The LEED project boundary must normally only include land that is owned by the party that owns the LEED project. C. LEED projects located on a campus must contain project boundaries so that if all the campus buildings become LEED certified, then 100 percent of the gross campus land area would be included within a LEED boundary. D. Any given parcel of real property may only be attributed to a single LEED project building. E. Any tampering with a LEED project boundary is completely prohibited.
4. Project must comply with minimum floor area requirements by incorporating a minimum of 1000 square feet (93 square meters) of gross floor area.
5. Projects must comply with minimum full-time equivalent occupancy rates (FTE). One or more FTE must be served, calculated as an annual average in order to use LEED in its entirety.
6. Project owners must consent to sharing whole-building energy and water usage data with USGBC and/or GBCI for a period of at least five years.

7. The gross floor area of the LEED project must conform to a minimum building area to site area ratio—building must not be less than 2 percent of the gross land area within the LEED project boundary.
8. Registration and certification activity must comply with reasonable timetables and rating system sunset dates, which basically means that if a LEED 2009 project is inactive for four years, the GBCI reserves the right to cancel the registration.

2.3.3 How LEED Works

LEED is a point-based system where building projects earn LEED points for satisfying specific green building criteria. The awarding of points relative to performance is covered under five environmental categories: Sustainable Sites (SS), Water Efficiency (WE), Energy and Atmosphere (EA), Materials and Resources (MR), and Indoor Environmental Quality (IEQ). Additionally there is Innovation in Design (ID), which addresses sustainable building expertise as well as design measures not covered under the five environmental categories, and Regional Priority (RP). Designers can select the points that are most appropriate to their projects to achieve a LEED rating. A total of 100 base points + 10 points (six possible Innovation in Design and four Regional Priority points) is possible. The number of points the project earns determines the level of LEED Certification the project receives, i.e., platinum, gold, silver, or certified ratings are awarded. Moreover, LEED 2009 alignment provides a continuous improvement structure that will enable USGBC to develop LEED in a more predictable manner.

When the USGBC first introduced the Leadership in Energy and Environmental Design (LEED™) green building rating system, Version 1.0 in December 1998, it was considered by all to be a pioneering effort. Since then the LEED Green Building Rating System™ has inspired and instigated global adoption of sustainable green building practices through the adoption and execution of universally understood and accepted tools and performance criteria. And today LEED has become the leading means for certifying green buildings in the United States, and has recently released a new version, LEED 2009, formerly known as LEED V3, and which is the first major LEED overhaul since Version 2.2 came out in 2005. LEED 2009 has been significantly transformed by the many changes, both major and minor, to the rating system and its priorities.

Many of the changes in LEED 2009 are designed to address much of the criticism levied against the LEED system, including an entirely new weighting system which refers to the process of redistributing the available points in LEED in a manner that a credit's point value more accurately reflects its potential to either mitigate the negative or promote positive environmental impacts of a building. Thus, in LEED 2009, credits that most directly address the most significant impacts are given the greatest weight, subject to the system design parameters described above. This has resulted in a significant change in allocation of points compared with earlier LEED rating systems. Generally speaking, the modifications reflect a greater relative emphasis on the reduction of energy consumption and greenhouse gas emissions associated with building systems, transportation, the embodied energy of water and materials, and where applicable, solid waste (e.g., for Existing Buildings: Operations & Maintenance).

Additional improvements include an increased opportunity for innovation credits, and a new opportunity for achieving bonus points for regional priority credits. A less obvious revision in LEED™ 2009 is the reduction of possible exemplary performance credits from a maximum of four to a maximum of three. The intention here was to return to the original intent of the credit, which is to encourage projects to pursue innovation in green building. There are numerous other important modifications and improvements that are discussed below and in the following chapters.

2.3.4 The LEED Points Rating System

LEED™ is a continually evolving basic point-based system that has set the green building standard and has made it the most widely accepted green program in the United States.

The various LEED™ categories differ in their scoring systems based on a set of required "prerequisites" and a variety of "credits" in seven major categories as outlined above. In LEED™ v2.2 for new construction and major renovations for commercial buildings there were 69 possible points and buildings were able to qualify for four levels of certification.

LEED™ 2009 is a significant improvement over earlier LEED™ versions and has become much less complicated. The new USGBG LEED™ Green Building certification levels for all systems are also more consistent and are shown below:

Certified:	40–49 points
Silver:	50–59 points
Gold:	60–79 points
Platinum:	80 plus points

The number of points available per LEED™ system has been increased so that all LEED™ systems have 100 base points as well as 10 possible innovation and regional bonus points which brings the possible total points achievable for each category to 110. Figure 2.3 includes pie charts showing the new LEED™ v2009 for new construction and commercial interiors. Figures 2.4 and 2.5 are examples of buildings that have received various levels of LEED™ certifications.

LEED 2009 Point Distribution – NC

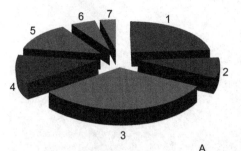

1. Sustainable sites (26 pts)
2. Water efficiency (10 pts)
3. Energy & Atmosphere (35 pts)
4. Materials & Resources (14 pts)
5. Indoor environmental quality (15 pts)
6. Innovations in design (6 pts)
7. Regional priority (4 pts)

A

LEED 2009 Point Distribution – CI

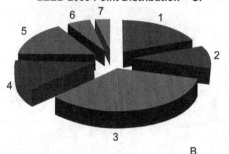

1. Sustainable sites (21 pts)
2. Water efficiency (11 pts)
3. Energy & Atmosphere (37 pts)
4. Materials & Resources (14 pts)
5. Indoor environmental quality (17 pts)
6. Innovations in design (6 pts)
7. Regional priority (4 pts)

B

FIGURE 2.3 A,B Pie charts showing the new LEED™ NC v2009 and LEED™ CI v2009 point-distribution system, which incorporates a number of major technical advancements focused on improving energy efficiency, reducing carbon emissions, and addressing additional environmental and human health concerns.

While the previous maximum achievement in earlier versions of LEED–NC was 69 points, which in LEED 2009 has been increased to 100, it remains unclear sometimes how the added 31 points are distributed. Aurora Sharrard, research manager at Green Building Alliance (GBA), says, "The determination of which credits achieve more than one point (and how many points they achieve) is actually the most complex part of LEED 2009. LEED has always implicitly weighted buildings' impacts by offering more credits in certain sections. However, in an effort to drive greater (and more focused) reduction of building impact, the USGBC is now applying explicit weightings to all LEED credits. The existing weighting scheme was developed by the National Institute of Standards and Technology (NIST). The USGBC hopes to have its own weighting system for future LEED revisions, but currently LEED credits are proposed to be weighted based on the following categories, which are in order of weighted importance:

• Greenhouse gas emissions
• Indoor environmental quality fossil fuel depletion

FIGURE 2.4   The interior of BP America's new government affairs office in Washington, D.C. designed by Fox Architects. The 22,000 square foot building achieved a LEED Platinum level certification. (Source: Fox Architects.)

- Particulates
- Water use
- Human health (cancer)
- Ecotoxicity
- Land use
- Eutrophication
- Smog formation
- Human health (non-cancer)
- Acidification
- Ozone depletion."

The new weighting preferences in the LEED 2009 system put much greater emphasis on energy which is very appropriate as this addresses some of the criticism levied against earlier versions of the LEED Rating

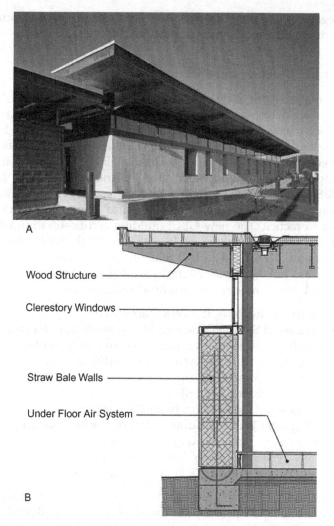

Wood Structure

Clerestory Windows

Straw Bale Walls

Under Floor Air System

FIGURE 2.5 **A.** The Santa Clarita Transit Maintenance is one of the first LEED Gold-certified straw-bale buildings in the world. The resource- and energy-efficient transit facility was designed by HOK and exceeds California Energy Efficiency Standards by more than 40 percent, securing a new standard for straw-bale in high performance building design. **B.** A diagram showing a section taken through an exterior wall of transit facility. The designers reportedly opted for a solar photovoltaic canopy to shade buses and provide nearly half of the building's annual energy needs. An electronic monitoring system is in place to track thermal comfort, energy efficiency, and moisture levels. (Source: HOK Architects.)

System. There has also been an increase in the Innovation and Design (ID) credits from four points to five. An additional point can be achieved for having a LEED Accredited Professional (LEED AP) on the project team, which brings the total ID points to six. The introduction of the new category of Regional Priority also adds another potential four bonus points, bringing the total points possible to 110.

2.3.5 LEED Building Certification Model

Beginning in April 2009, the Green Building Certification Institute (GBCI) assumed responsibility to manage the review and verification process for projects seeking certification under the LEED Green Building Rating System. For LEED v3, the project certification process has moved to the Green Building Certification Institute (GBCI), which is an independent nonprofit organization established in January 2008 with the support of USGBC. The GBCI provides third-party project certification and professional credentials recognizing excellence in green building performance and practice. The new GBCI building certification infrastructure has recently added a network of 10 well-respected certification organizations that are accredited to ISO Standard 17021. These organizations are recognized for their roles in certifying organizations, processes, and products to ISO and other standards and are listed below:

- ABS Quality Evaluations, Inc. (www.abs-qe.com)
- BSI Management Systems America, Inc. (www.bsigroup.com)
- Bureau Veritas North America, Inc. (www.us.bureauveritas.com)
- DNV Certification (http://www.dnv.us/certification/ managementsystems/index.asp)
- Intertek (www.intertek-sc.com)
- KEMA-Registered Quality, Inc. (www.kema.com)
- Lloyd's Register Quality Assurance, Inc. (www.lrqausa.com)
- NSF-International Strategic Registrations (www.nsf.org)
- SRI Quality System Registrar, Inc. (www.sri-i.com)
- Underwriters Laboratories-DQS, Inc. (www.ul.com/mss).

The revised LEED v3 process is a significantly improved ISO-compliant certification process that is adaptable and designed to grow with the green building movement. All LEED certification applications will continue to be submitted through LEED Online. Likewise, the USGBC states that it will continue to administer the development and ongoing improvement of the LEED rating system, and will remain the primary source for LEED and green building education.

It should be noted that the U.S. Green Building Council currently has four LEED programs that are in the process of being launched using the online v3 platform for submittals. These programs are: LEED for Neighborhood Development, LEED Portfolio Program, LEED for Healthcare, and LEED for Retail. Additional information with respect to these programs can be obtained from the GBCI website at: www.gbci.org/DisplayPage.aspx?CMSPageID=132.

2.3.6 LEED v3: What's New?

With the latest overhauling of the LEED Rating System, a new greatly improved system has emerged—LEED Online v3, which is more appropriate and provides enhanced functionality to improve efficiency and productivity. According to the GBCI, the new version is "faster, smarter, and a better user experience. It is designed to be scalable and more robust, through improved design, a more intuitive user interface, better communication between project teams and certifying bodies, and upgrades that respond to the changes in the LEED 2009 rating system." The GBCI also cites on its Website some of the new project management improvement tools incorporated into version 3 such as:

- Project organization—The ability to sort, view, and group LEED projects according to a number of project traits, such as location, design, or management firm, etc.
- Team member administration—Increased functionality and flexibility in making credit assignments, adding team roles, and assigning them to team members. For example, credits are now assigned by team member name rather than by project role.
- Status indicators and timeline—A clearer explanation of the review and certification process and steps are highlighted as they are completed in specific projects. The system now displays specific dates related to each phase and step, including target dates that each review is to be returned to the customer.
- LEED Support for Certification Review and Submittals.
- LEED Online v3 offers many other enabling features to support the LEED certification review process, as well as enhancements to the functionality of submittal documentation and certification forms:
  - End-to-end process support—The new system will guide project teams through the certification process, from initial project registration through the various review phases. Furthermore, it will provide assistance to beginners during the registration phase to help them determine the type of LEED rating system that is best suited for their project.
  - Improved mid-stream communication—A mid-review clarification page allows a LEED reviewer to contact the project team through the system when minor clarifications are required to complete the review.
  - Data linkages—LEED Online v3 automatically fills in fields on all appropriate forms after the user inputs data the first time, which saves time and helps ensure project-wide consistency. Override options are available when required.

- Automatic data checks—New system alerts users when incomplete or required data is missing, thus allowing the user to correct the error before application submission, thus avoiding delays.
- Progressive, context-based disclosure of relevant content—Upon selection of an option, the new system will simplify the process of completing forms by only showing data fields that are relevant to the customer's situation, and hiding all extraneous content.

2.4 THE GREEN GLOBES™ RATING SYSTEM

The Green Globes Website (www.greenglobes.com/) describes the system as "the practical building rating system" and says that, "The Green Globes system is a revolutionary building environmental design and management tool. It delivers an online assessment protocol, rating system, and guidance for green building design, operation, and management. It is interactive, flexible, and affordable, and provides market recognition of a building's environmental attributes through third-party verification." Green Globes is certainly less complicated than USGBC's LEED rating system. It employs a straightforward questionnaire-based format, which is written in lay terms, and is fairly easy to complete even if you lack environmental design experience. The questions are typically a yes/no style and are grouped broadly under seven modules of building environmental performance (management, site, energy, water, resources, emissions, indoor environment). Upon completing the questionnaire, a printable report is automatically generated. Figure 2.6 show a photo of Blakely Hall, the first Green Globes rated building in the United States, which is community center and town hall for Issaquah Highlands, a planned community near Seattle, Washington. Blakely Hall earned two globes (out of a possible four) in the Green Building Initiative's Green Globes. The building incorporates a variety of green attributes such as high energy and water efficiency, integration of daylighting, and the use of locally sourced materials. The implementation of a construction waste management plan also helped divert more than 97 percent of waste from a landfill. Blakely Hall is an example of a "green" building that has achieved various awards, including a LEED Silver award.

The idea and market for green buildings has been growing rapidly, and although there are a number of green building rating systems available in the United States, the two systems most widely used for commercial structures are LEED (Leadership in Energy and Environmental Design) and Green Globes (Go Green Plus). The LEED Green Building Rating System® is focused largely on assessing new construction sustainable high-performance buildings, although existing buildings are also

FIGURE 2.6 Blakely Hall is a community center and town hall for Issaquah Highlands, a planned community near Seattle, Washington. It consists of 7000 square feet and was built to a budget of U.S. $1.5 million. It is used mainly as a meeting place for numerous clubs and groups at Issaquah Highlands. Blakely Hall is the first building in the United States to earn a Green Globes Certification (earned two Globes out of a possible four) as well as a LEED Silver award.

included in the LEED Rating System (Figure 2.7). Go Green mainly targets existing building owners who want to have a more environmentally friendly building. In this section we will analyze and compare Green Globes™ with the LEED Rating System. While Green Globes currently has a minute share of certified buildings in the United States (about 55 buildings), which pales with LEED's™ certified buildings market share, it is making a determined effort to rectify this situation.

2.4.1 An Overview of the Green Building Initiative (GBI) and Green Globes™

The Green Building Initiative (GBI) is a 501(c)(3) nonprofit education organization based in Portland, Oregon. Its mission is to accelerate the adoption of sustainable design and construction practices that result in energy-efficient, healthier, and environmentally sustainable buildings by promoting credible and practical green building approaches for residential and commercial construction. Ward Hubbell serves as president of GBI at the discretion of an independent, multi-stakeholder board of

FIGURE 2.7   The 80,000 ft^2 (7400 m^2) Integrated Learning Centre at Queen's University in Kingston, Ontario, received a four-leaf rating through the BREEAM/Green Leaf program, which is now accessible online as Green Globes. Designed by B+H Architects of Toronto, the project was completed in 2004. The Ottawa-based firm Green & Gold, Inc., implemented the BREEAM/Green Leaf program for the ILC and helped integrate the building analysis tool into the design process. The lighting, ventilation, and water distribution systems, in particular, contributed to the building's high rating. (Source: interiorimages.ca.)

directors that is comprised of construction professionals, product manufacturers, nonprofit organizations, university officials, and other interested parties.

History and Background

The birth of the Green Globes™ system lies in the Building Research Establishment's Environmental Assessment Method (BREEAM); this was exported to Canada from the United Kingdom in 1996 in cooperation with ECD Energy and Environment. Green Globes™ was initially developed as a rating and assessment system to monitor and assess green buildings in Canada. Canada's federal government has been using the Green Globes rating system for several years under the Green Globes name and it has been the basis for the Building Owners and Manufacturer's Association of Canada's Go Green Plus program. Go Green was adopted by BOMA Canada in 2004, and was chosen by Canada's Department of Public Works and Government Services. Green Globes™ has also been adopted by the Continental Association for Building Automation (CABA) to power a building intelligence tool called Building Intelligence Quotient (BiQ).

The Green Globes environmental assessment and rating system represents more than a decade of research and refinement by a wide range of prominent international organizations and experts. The Canadian Standards Association first published BREEAM Canada for Existing Buildings in 1996, with more than 35 individuals participating in its development. In 1999, ECD Energy and Environment collaborated with Terra Choice, the agency that administers the Government of Canada's Environmental Choice program, to develop a more efficient and stream-lined, question-based tool that was later introduced as the BREEAM Green Leaf eco-rating program. Later that year the program led to the formation of Green Leaf for Municipal Buildings with the Federation of Canadian Municipalities. In 2000, BREEAM Green Leaf's development took another step forward by becoming an online assessment and rating tool under the name Green Globes for Existing Buildings. That same year, BREEAM Green Leaf for the Design of New Buildings was adapted for the Canadian Department of National Defense and Public Works and Government Services Canada. The program underwent a further iteration in 2002 by a panel of experts including representatives from Arizona State University, the Athena Institute, BOMA, and a number of Canadian federal departments.

In 2002, Green Globes for Existing Buildings went online in the United Kingdom as the Global Environmental Method (GEM), and endeavors were made to incorporate BREEAM Green Leaf for the Design of New Buildings into the online Green Globes for New Buildings. Green Globes for Existing Buildings was adopted and operated by BOMA Canada in 2004 under the name Go Green Comprehensive (now known as Go Green Plus). The Canadian federal government also later announced plans to adopt Go Green Plus for its entire real estate portfolio. All other Green Globes products in Canada are owned and operated by ECD Energy and Environment Canada.

Additionally in 2004, the Green Building Initiative (GBI) purchased the rights to develop, promote, and distribute Green Globes for New Construction (Green Globes-NC) in the United States. In adapting the system, minor changes were instituted to make the system appropriate for the U.S. market (e.g., converting units of measurement and integration with the ENERGY STAR program). The GBI also committed itself to ensuring that Green Globes continues to reflect best practices and changing opinions and ongoing advances in research and technology. To that end, GBI in 2005 became the first green building organization to be accredited as a standards developer by the American National Standards Institute (ANSI), and Green Globes® rating system is also on track to become the first American National Standard for commercial green buildings. As part of this process, GBI set up a technical committee and subcommittees of more than 75 building science experts, including representatives from several federal agencies, states, municipalities, universities, and leading construction firms, in addition to building developers.

In March 2009 the GBI and the American Institute of Architects (AIA) signed a memorandum of understanding (MOU), which states that the GBI and AIA pledge to work in concert to promote the design and construction of energy efficient and environmentally responsible buildings. An MOU was also signed between the GBI and ASHRAE to collaborate to facilitate the adoption of sustainability principles in the built environment.

2.4.2 Defining the Green Globes™ Rating System

The Green Globes™ v1 assessment protocol covers seven different areas with each area having an assigned number of points that are utilized to quantify overall building performance. These are shown in the table below (Figure 2.8):

The Process

The scoring for the seven Green Globe categories is based on a series of questions that are completed via the online questionnaire within the Green Globes Tool. Normally, there are pop-up "tool tips" embedded within the questionnaire to address frequently asked questions and add clarifications regarding the input data requirements that will appear during the survey. Amy Stodghill, a sustainability consultant who by using a free 30-day trial accessed the online Environmental Assessment for Existing Commercial Buildings, comments, "It is essentially a 22-page

ASSESSMENT CATEGORY	POINTS	PERCENTAGE
Project management	50	5
Site	115	11.5
Energy	360	36
Water	100	10
Resources	100	10
Emissions, effluents, and other impacts	75	7.5
Indoor environment	200	20
TOTAL	**1000**	**100**

FIGURE 2.8 A table depicting the seven different assessment categories of Green Globes. The table shows a clear emphasis on energy, which takes up more than a third of the total points.

questionnaire/survey covering energy, transportation, water, waste reduction and recycling, site management, air and water emissions, indoor air quality (IAQ), purchasing, and communication. It is completed online only and is very user friendly." The time normally required to input data and complete the survey is roughly two to three hours per building; this doesn't include time required to research and gather required information for the survey.

Stodghill also noted that "Each question is weighted with points (in all totaling up to 1000). The overall rating is tracked as questions are answered. The overall rating however is based on a percentage, not on total points. This way there are no penalties for questions that are not applicable (i.e., answering 'no' on water efficient irrigation questions will not be counted against you if you do not have any landscaping)."

Upon completion, the Green Globes system automatically generates a report that is based on answers given. The report lists where the building stands in each major category and lists suggestions for improvement in order to gain a better score.

In order to earn a formal Green Globes rating/certification, a building has to be evaluated by an independent third party that is recognized, trained, and affiliated with GBI. Both new construction and existing buildings can be formally rated or certified within the Green Globes system. Projects that achieve a score of 35 percent or more of the 1000 points possible in the rating system become eligible for a Green Globes rating of one, two, three, or four globes, as follows:

- One globe: 35–54 percent
- Two globes: 55–69 percent
- Three globes: 70–84 percent
- Four globes: 85–100 percent

A summary of rating levels and how they relate to environmental achievement can be seen in Figure 2.9. However, buildings cannot be promoted as having achieved a Green Globes rating until the information submitted has been assessed by a qualified third party.

According to Green Globes-NC, projects are awarded up to 1000 points based on their performance in seven areas of assessment as follows:

1. *Project Management—50 Points* The Green Globes system places an emphasis on integrated design, an approach that encourages multi-disciplinary collaboration from the earliest stages of a project, while also considering the interaction between elements related to sustainability. Most decisions that influence a building's performance (such as siting, orientation, form, construction, and building services) are made at the start of the project and yet it's common, even for experienced designers, to focus on environmental performance late in the process, adding

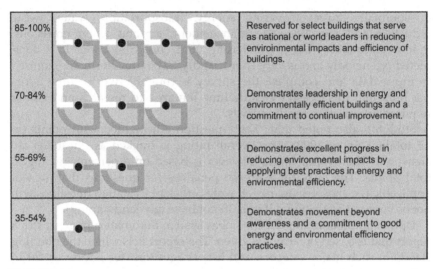

85-100%		Reserved for select buildings that serve as national or world leaders in reducing enviroinmental impacts and efficiency of buildings.
70-84%		Demonstrates leadership in energy and environmentally efficient buildings and a commitment to continual improvement.
55-69%		Demonstrates excellent progress in reducing environmental impacts by appplying best practices in energy and environmental efficiency.
35-54%		Demonstrates movement beyond awareness and a commitment to good energy and environmental efficiency practices.

FIGURE 2.9   Green Globes rating levels in the United States. (Source: Green Building Initiative.)

expensive technologies after key decisions have been made. This is costly as well as ineffective.

To ensure that all of the relevant players are involved, the system tailors questionnaires so that input from team members is captured in an interactive manner, even on those issues which may at first appear to fall outside their mandate. For example, while site design and landscaping may come under the purview of the landscape designers, the questionnaire prompts the electrical engineer to get involved with design issues such as outdoor lighting or security. Thus, the Green Globes format promotes design teamwork and prevents a situation where, despite strong individual resources, the combined effort falls short.

Also included under project management are environmental purchasing, commissioning, and emergency response.

2. Site—115 Points   Building sites are evaluated based on the development area (including site selection, development density, and site remediation), ecological impacts (ecological integrity, biodiversity, air and water quality, microclimate, habitat, and fauna and flora), watershed features (such as site grading, storm water management, pervious cover, and rainwater capture), and site ecology enhancement.

3. Energy—360 Points   To simplify the process of energy performance targeting, Green Globes-NC directs users to the Web interface used for the ENERGY STAR Target Finder software, which helps to generate a realistic energy consumption target. As a result, an aggressive energy performance goal can be set—with points awarded for design and operations strategies that result in a significant reduction in energy consumption—as compared to actual performance data from real buildings.

As previously stated, Green Globes is the only green rating system to use energy data generated through the U.S. Department of Energy's Commercial Buildings Energy Consumption Survey (CBECS), which is widely considered to be the most accurate and reliable source of energy benchmarking information.

In addition to overall consumption, projects are evaluated based on the objectives of reduced energy demand (through space optimization, microclimatic response to site, daylighting, envelope design, and metering), integration of "right sized" energy-efficient systems, on-site renewable energy sources, and access to energy efficient transportation.

4. *Water—100 Points* Projects receive points for overall water efficiency as well as specific water conservation features (such as sub-metering, efficiency of cooling towers, and irrigation strategies), and on-site treatment (of grey water and wastewater).

5. *Resources—100 Points* The resources section covers building materials and solid waste. It includes points for materials with low environmental impact (based on life-cycle assessment), minimal consumption and depletion of resources (with an emphasis on materials that are reused, recycled, bio-based, and, in the case of wood products, certified as having come from sustainable sources), the reuse of existing structures, building durability, adaptability and disassembly, and the reduction, reuse, and recycling of waste.

6. *Emissions, Effluents, and Other Impacts—75 Points* Points in this section are awarded in six categories, including air emissions, ozone depletion and global warming, protection of waterways and impact on municipal waste water treatment facilities, minimization of land and water pollution (and the associated risk to occupants' health and the local environment), integrated pest management, and the storage of hazardous materials.

7. *Indoor Environment—200 Points* According to the U.S. EPA, indoor air can be up to 10 times more polluted than outdoor air, even in cities where the quality of outdoor air is poor. This has obvious health implications, but the consequences are also economic. A study by Lawrence Berkeley National Laboratory found that improving indoor air at work could save U.S. businesses up to $58 billion in lost sick time each year, with another $200 billion earned in increased worker performance.

This section evaluates the quality of the indoor environment based on the effectiveness of the ventilation system, the source control of indoor pollutants, lighting design and the integration of lighting systems, thermal comfort, and acoustic comfort.

GBI states that the process for obtaining formal Green Globes rating/ certification is quite straightforward and consists of implementing the following steps:

Step 1: Purchase a subscription to either Green Globes NC or CIEB

Step 2: Login to Green Globes at the GBI Website with your username and password

Step 3: Select the tool you have purchased (NC or CIEB) to go to Green Globes

Step 4: Add a building and enter basic building information

Step 5: Use step-through navigation and building dashboard to complete the survey

Step 6: Print your report to see your building projected rating and get feedback using automatic reports

Step 7: Order a third-party assessment and Green Globes rating/ certification (if automated report indicates a predicted rating of at least 35 percent of 1000 points)

Step 8: Schedule and complete a third-party building assessment. Third-party assessment for Green Globes-NC occurs in two comprehensive stages: The first stage includes a review of the construction documents developed through the design and delivery process. The second stage includes a walk-through of the building post-construction.

Step 9: Receive the Green Globes rating and certification.

2.4.3 Green Globes™—An Alternative to LEED?

There has been a great deal of interest and discussion about the Green Globes system as opposed to another building certification program, namely, the LEED™ Green Building Rating System. Firstly, it is important to bear in mind that there are a great deal of similarities between both initiatives, largely because they share common roots, and also because they share common ideas of green buildings. However, there are several significant differences that are highlighted below:

The origin of the Green Globes Canada system lies in the Building Research Establishment's Environmental Assessment Method (BREEAM) which was developed in the U.K. and which was later published in 1996 by the Canadian Standards Association. One of BREEAM's creators, ECD Consultants, Ltd., used it as the basis for a Canadian assessment method called BREEAM Green Leaf. At first, BREEAM Green Leaf was created to allow building owners and managers to self-assess the performance of their existing buildings. Green Globes was then developed into a Web-based application of Green Leaf by ECD.

The Green Globes™ system thus became a Web-based green building performance interactive software tool (from Canada), and today competes with the better known, though more complicated and more expensive LEED system from the U.S. Green Building Council (USGBC)—a nonprofit organization based in Washington, D.C. Green Globes™ was introduced to the U.S. market as a potentially viable alternative to the U.S. Green Building Council's LEED® Rating System. The Green Building Initiative (GBI) was established to promote the use of the National Association of Homebuilders' (NAHB) Model Green Home Building Guidelines, and has recently expanded into the nonresidential building market by licensing Green Globes for use in the U.S. GBI is supported by various industry groups including the Wood Promotion Network that object to some provisions in LEED and, as trade associations, are prohibited from joining the USGBC.

When Green Globes was released in Canada in January 2002 it consisted of a series of questionnaires, customized by project phase and the role of the user in the design team (e.g., architect, mechanical engineer, or landscape architect). A total of eight design phases are supported. A separate Green Globes module (Green Globes-CIEB) is available for assessing the performance of existing buildings. The questionnaires produce design guidance appropriate to each team member and project phase. Green Globes users can order a Green Globes third party assessment at any time—upon purchasing a subscription, during the completion of the questionnaire, or after completion of the questionnaire. After an online self-assessment is completed and payment is made, a GBI representative contacts the project manager or project owner to schedule the assessment and provide the assessor name and contact information. Completion of the pre-assessment checklist, which can be downloaded from the Green Globes customer training area, helps prepare for the assessment process.

Methods to formally rate and certify programs are necessary to provide a mechanism to ensure that new construction project teams or facilities management staff is fully aware of the environmental impact of design and/or operating management decisions. It also offers a visible means to quantify/measure their performance, and allows recognition for their achievements and hard work at the end of the process. Green Globes is designed to be cost effective, and through its value-added online system and a comprehensive yet streamlined in person third-party review process, there are significant savings on consulting fees that are normally associated with green certification. There is an annual per building license fee for use of the online tool as well as a third-party assessment fee. Rates are based on a number of factors including size of project (hectares/acres), number of integrated developments, and location (environmentally sensitive areas). Users can register/subscribe for both the annual license for the online Green Globes tools and choose to purchase a third-party assessment (required for certification). Third-party assessor travel expenses are separately billed (Figure 2.10).

It is estimated that the United States is home to more than 100 million buildings, which adds to the urgency to improve the performance of existing structures as a necessary prerequisite for widespread energy efficiency. The GBI says, "The missing element—until last year when GBI introduced Green Globes-CIEB—was a practical and affordable way to measure and monitor performance on an ongoing basis. Green Globes-CIEB allows users to create a baseline of their building's performance, evaluate interventions, plan for improvements, and monitor success—all within a holistic framework that also addresses physical and human elements such as material use and indoor environment."

FIGURE 2.10 The new 356,000 square foot Newell Rubbermaid corporate headquarters in Atlanta, GA, which achieved a two globe rating using the Green Globes New Construction module. (Source: Green Building Initiative.)

Below are some of the costs currently required with respect to the Green Globes rating system:

Green Globes® Existing Building Rating/Certification Package:	$5270* Per Building
Green Globes® New Construction Rating/Certification Package:	$7270* Per Building
Software Subscriptions Costs	**Price**
Green Globes CIEB Existing Building one-year subscription	$1000
Green Globes New Construction	
One project subscription	$500
Three project subscription	$1500
Ten project subscription	$2500

Note: A Green Globes subscription is required for third party assessment/certification

Third Party Assessments/Green Globes Certification	
Green Globes CIEB assessment/rating	$3500*
Green Globes NC Stage I assessment	$2000
Green Globes NC Stage II assessment/rating	$4000*
Green Globes NC Stage I and II assessment/rating	$6000*

Note: *Pricing for buildings over 250,000 square feet in size or departing significantly from standard commercial building complexity will be custom quoted prior to assessment services being performed.

Travel for the GBI assessor to/from building location: Invoice actual expenses +20 percent after assessment OR pay a flat fee of $1000 upfront.

The Green Globes building rating system provides a LEED alternative assessment tool for characterizing a building's energy efficiency and environmental performance. The Green Globes system also provides guidance for green building design, operation, and management. Compared to LEED, some feel that Green Globes' appeal may be enhanced by the flexibility and affordability the system can provide, while simultaneously providing market recognition of a building's environmental attributes through recognized third-party verification. And from a practical and marketing perspective, it should not be necessary to pursue LEED certification in order to demonstrate to tenants, their customers, clients, and building visitors that a building's owners and management are taking steps to be more environmentally responsible—when they have Green Globes.

According to Christine Ervin, former president and CEO of USGBC, "Green Globes offers several very appealing features. Interactive feedback on strategies, interactions and resources can be tailored to 20 different team roles and eight project stages. Numerical assessments are generated at stages for schematic design and construction, designed to coincide with planning and permit approvals."

Green Globes and LEED have many similarities since they both evolved from the same source—the Building Research Establishment's Environmental Assessment Method (BREEAM). For example, Green Globes and the LEED rating systems are very similar in structure. Both systems have four levels of achievement. LEED projects can achieve the following four certifications A. certified, B. silver, C. gold, or D. platinum. Similarly, Green Globes projects can achieve 1, 2, 3, or 4 globes. Both LEED and Green Globes share a common set of green building design practices. There are six focus areas for LEED and seven for Green Globes, but the focus areas are in many respects also similar.

A University of Minnesota study conducted in 2007 that compared LEED (pre-version3) with Green Globes found that the systems were very similar. For example, the study found that "nearly 80 percent of the categories available for points in Green Globes are also addressed in LEED v2.2 and that over 85 percent of the categories specified in LEED v2.2 are addressed in Green Globes." The study further concluded that LEED was characterized as being more rigorous, rigid, and quantitative, whereas Green Globes, while also rigorous, nevertheless maintained greater flexibility. Also, Green Globes focused primarily on energy efficiency as a goal. Likewise, Green Globes was found to be easier to work with, less costly, and less time-consuming than LEED. The same study concluded that there was only moderate dissimilarity between the two rating standards, but that LEED has a slightly greater emphasis on materials choices and Green Globes has a greater emphasis on energy savings. Green Globes also more heavily weights energy systems, up to 36 percent of the total points needed, whereas LEED in its earlier versions limited the energy category to about 25 percent of the total in the rating system. However, in the new LEED v3 version this has been appropriately addressed.

Furthermore, of the many buildings that have been evaluated with both systems, in all but two of the instances, the systems generated comparable ratings. The other two buildings were only marginally different. It should be noted here that the LEED 2009 has addressed many of these issues. In the final analysis, it appears that the primary differences between the two approaches boil down to cost and ease of use. The University of Minnesota study also concluded that "From a process perspective, Green Globes' simpler methodology, employing a user-friendly interactive guide for assessing and integrating green design principles for buildings, continues to be a point of differentiation to LEED's more complex system. While LEED has introduced an online-based system, it remains more extensive and requires expert knowledge in various areas.

Green Globes' Web-based self-assessment tool can be completed by any team member with general knowledge of the building's parameters."

GBI currently oversees Green Globes in the United States. GBI has also become an accredited standards developer under the American National Standards Institute (ANSI) and is in the process of establishing Green Globes as an official ANSI standard. The ANSI process has always been a consensus-based process, involving a balanced committee of varying interests including users, producers, interested parties, and NGOs who basically conduct a thorough technical review through an ANSI-approved, open, and transparent process. The standard continues to be monitored by this committee and will continue to be updated through ANSI approved rules and procedures.

Neither LEED nor Green Globes (or Energy Star for that matter) provides continuous, longitudinal monitoring of energy efficiency or building performance. This indicates that building measurements and ratings are concluded on a one off basis which must then be re-verified later on. This is a significant shortcoming in terms of the practicality of greening existing real estate, since buildings are dynamic and rarely perform in an identical manner week after week. Green Globes-NC is the only environmental rating system that provides early feedback before critical and final decisions are made. This is a proven method for taking advantage of time and cost savings opportunities through integrated design and delivery, while benefitting from a cost-effective and comprehensive third-party assessment program.

Green Globes generates numerical assessment scores at two of the eight project phases; these are the schematic design phase and construction documents phase. These scores can either be used as self-assessments internally, or they can be verified by third-party certifiers. Projects that have had their scores independently verified can use the Green Globes logo and brand to promote their environmental performance. The Green Globes questionnaire corresponds to a checklist with a total of 1000 points listed in seven categories as opposed to LEED's 100 points in seven categories (Figure 2.11).

One of the differences between Green Globes and LEED is that the former offers protection against "non-applicable criteria." Thus, if a builder marks a criterion as "N/A," then he or she will be excused for not gaining points in those areas, which is why the actual number of points available varies by project. For example, if a building code overrides a criterion, then the criterion can be marked as "N/A." Another example would be if points are available for designing exterior lighting to avoid glare and skyglow, but for a project with no exterior lighting, a user can select "N/A," which removes those points from the total number available so as not to penalize the project.

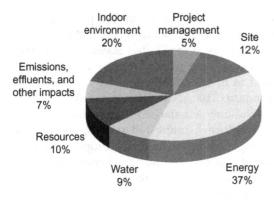

FIGURE 2.11 A pie diagram showing the distribution of points in the Green Globes Rating System. (Source: Building Green LLC.)

A rating of one or more Green Globes is applied to projects based on the percentage of applicable points they have achieved. In Canada the ratings range from one to five Green Globes, while in the U.S. the lowest rating was eliminated and the rest adjusted so that the highest rating is four Globes. Ward Hubbell, executive director of GBI, says that the objective of this was to have something that people are accustomed to—a four-stage system, which is roughly comparable with the four levels of LEED.

It appears that Green Globes is broader than LEED in terms of technical content, including points for topics such as optimized use of space, acoustical comfort, and an integrated design process. It is difficult to compare the levels of achievement required to claim points in the two systems because they are organized differently and also because the precise requirements within Green Globes are not publicly available. Also, unlike LEED, Green Globes encourages energy reduction; it does not require it. LEED also calls for a minimum indoor air quality performance, while Green Globes does not. A recently published AIA report states that LEED makes it mandatory that builders have "some documentation of the initial building energy and operational performance through fundamental commissioning," whereas Green Globes does not.

Recognizing all the mainstream forest certification systems is one of Green Globes' main attractions (and where it differs from LEED) with strong timber-industry lobbying groups supporting GBI in the U.S. Green Globes is more inclusive and opposes favoring FSC over SFI forest certification, and recognizes timber certified through FSC as well as the American Tree Farm System (ATFS), Canadian Standards Association (CSA), and Sustainable Forestry Initiative (SFI), whereas LEED previously referenced only the Forest Stewardship Council's program. Independent research has shown that all of these systems are in fact effective. There are more than 390 million acres of certified forest in North America, but less than one-sixth of that amount is certified by the FSC. The consensus is that legislation to encourage green building in states like Virginia and Arkansas is likely to include Green Globes in addition to LEED. Furthermore, a number of federal agencies such as the Department of the Interior are also reportedly considering an endorsement of Green Globes.

The president of GBI, Ward Hubbell, claims that Green Globes is on par with LEED with respect to overall achievement levels, and says, "We did carry out a harmonization exercise with LEED—not credit-by-credit; we compared objectives." The actual development of the Green Globes system in Canada, as well as its subsequent adaptation for the U.S., has involved many iterations and participation by a wide range of organizations and individuals. Changes originally made to adapt Green Globes for the U.S. market do not appear to be substantive, e.g., converting units of measurement, referencing U.S. rather than Canadian standards and regulations, and incorporation of U.S. programs such as the EPA's Target Finder. Green Globes also awards points for the use of life-cycle assessment methods in product selection, although it doesn't specify how to apply such methods. It is very likely that Green Globes' presence on the American scene has had a beneficial impact on LEED, perhaps prompting it to improve its rating system and release LEED 2009 Version3. It is also important to recognize that Green Globes can attract a significant following that for various reasons is alienated by LEED certifications costs and complexity. This must be good for the green building industry and the environment.

Green Globes supporters tried to block the introduction of LEED into Canada, but lost a close vote in a committee of the Royal Architectural Institute of Canada that in 2003 led to the creation of the Canada Green Building Council (CaGBC). Alex Zimmerman, president of CaGBC, has levied some criticism of Green Globes, noting that in Canada Jiri Skopek, president of ECD Energy and Environment Canada, has been the primary developer of Green Globes and in the past was its sole certifier. Zimmerman also says, "While there are more certifiers now, it is not clear who they are, how they were chosen, or who they are answerable to." GBI responded to this criticism in the U.S. by training a network of independent certifiers to verify Green Globes ratings. These certifiers have access to the report generated by the Green Globes Website, as well as other relevant information such as the project drawings, results of an energy simulation, specifications, and commissioning plan.

Advantages of the Green Globes™ Rating/Certification System is that it is marketed as an economical, practical, and convenient methodology for obtaining comprehensive environmental and sustainability certification for new or existing commercial buildings. It provides a complete, integrated system that has been developed to enable design teams and property managers to focus their resources on the processes of actual environmental improvement of facilities and operations, rather than on costly, cumbersome, and lengthy certification and rating processes.

Among the other advantages to using Green Globes Rating/ Certification Systems are:

- A low registration fee is required to have projects evaluated and informally self-assessed.
- Consultants are not necessary for the certification process—thereby reducing costs.
- Certification requirements are generally less cumbersome and complex than other rating/certification systems.
- On-line Web tools provide a convenient, proven, and effective way to complete the assessment process.
- The entire certification process is fairly rapid, with minimal waiting for final rating/certification.
- The estimated rating number of Green Globes that will be achieved is largely known in advance of the decision to pursue certification because the self assessed score is available to users.
- An upfront commitment to a lengthy and costly rating/certification process is not required.

Finally, to date 19 states have included Green Globes in green building legislation, regulation, or executive order including: Arkansas, Connecticut, Florida, Hawaii, Indiana, Maryland, Minnesota, New Jersey, New York, North Carolina, Oklahoma, South Dakota, Pennsylvania, South Carolina, Kentucky, Illinois, Tennessee, Virginia, and Wisconsin. These states have formally recognized the Green Building Initiative's (GBI) Green Globes environmental assessment and rating system in legislation. This is achieved by passing legislation that is rating system neutral, meaning that it recognizes Green Globes as an equal option alongside LEED and other credible systems.

Green Project Requirements and Strategies

3.1 GENERAL

It is a well-known fact that people spend most of their time inside buildings, and we seem to take for granted the shelter, protection, and comfort that buildings provide. We rarely give much thought to the systems that allow us to enjoy these services unless there is an unfortunate power interruption or some other problem. Moreover, not many Americans fully understand the extent of the environmental consequences that allow us to maintain indoor comfort levels. This may be partly because modern buildings are often deceptively complex. And while buildings provide us with essential shelter in which to live and work, embody our culture, and play a crucial role in our daily lives, they also connect us with our past, while reflecting our greatest legacy for the future. Buildings' functions continue to change as they become increasingly costly to build and maintain, and require constant adjustment to function effectively over their life cycle. But while sustainable design strategies normally cost no more than conventional building techniques, the real goal of interdependence between strategies, known as holistic design, makes determining the true cost difficult to assess. Moreover, we often find that the returns on sustainable design are measured by various intangibles, such as worker productivity, health, and resource economy. But for building owners and designers it is more likely that determinations on sustainable design strategies will be made based first on cost of construction, or by a quick return on investment.

The vast majority of conventionally designed and constructed buildings that prevail today are fortifying their negative impact on the environment as well as on occupant health and productivity. Furthermore, these buildings are becoming increasingly expensive to operate and maintain in an acutely competitive market. Their contribution to excessive resource consumption, waste generation, and pollution is unacceptable and needs to be addressed. Reducing negative impacts on our environment, establishing new environmentally-friendly goals, and adopting guidelines that facilitate the development of green/sustainable buildings such as outlined by the USGBC, Green Globes™, and similar organizations must be a priority for our generation.

The concept of building "green" is no longer in the realm of the theoretical but deep in the mainstream heart of current construction practices, and general acceptance by the industry as well as familiarity with green elements and procedures continue to drive down building costs. Building "green" now offers an opportunity to use our resources more efficiently, while creating healthier buildings and a better environment, in addition to realizing significant cost savings.

Designers, contractors, and other professionals are well aware of the enormous amounts of wood, brick, asphalt, concrete, glass, metal, and plastic that building construction requires. Methods of building construction and materials used also impact the development of the indoor environments that can present an array of health challenges. Green buildings can address many of our environmental problems and have therefore become an essential element of our society. They are designed to meet certain objectives such as protecting occupant health; improving employee productivity; using energy, water, and other resources more efficiently; and reducing the overall impact to the environment. Green buildings are also known as sustainable buildings; they are structures that are designed, built, renovated, operated, or reused in an ecological and resource-efficient manner.

Green or sustainable architecture is particularly relevant in today's world of rapidly dwindling fossil fuel, along with the increasing impact of greenhouse gases on our climate. For this and other reasons, there is a pressing need to find suitable ways to reduce buildings' energy loads, increase building efficiency, and employ renewable energy resources in our facilities. Green construction is environmentally friendly because it uses sustainable, location-appropriate building materials and employs building techniques that reduce energy consumption. Indeed, the primary objectives of sustainable design and construction are to avoid resource depletion of essential resources such as energy, water, and raw materials, and prevent environmental degradation. Sustainability also places a high priority on health, which is partly why green buildings are also comfortable and safe places to live and work (Figure 3.1).

Familiarity with the various "green" certification systems such as LEED and Green Globes is essential for government (federal and state) contractors and certainly is recommended for contractors in the private sector. Various arms of the federal government now require that their public projects meet certain "green" standards, whether it be LEED certification standards, Green Globes, or ENERGY STAR, etc. in addition to various monetary and tax incentives. For example, the General Services Administration requires that all building projects meet the LEED certified level and target the LEED Silver level. The GSA, however, while strongly encouraging projects to apply for certification, does not require it. The U.S. Navy also requires appropriate projects to meet LEED certification requirements, but does not require actual certification. On the other hand, the U.S. Environmental Protection Agency (EPA) requires all new facility construction and acquisition projects consisting of 20,000 square feet or more to achieve a minimum LEED Gold certification. The U.S. Department of Agriculture also now requires all new or major renovation construction to achieve LEED Silver certification.

FIGURE 3.1 The U.S. Green Building Council awarded The Kresge Foundation head-quarters, which is built on a three-acre site in Troy, Michigan, a platinum-level rating, the highest attainable level in the Leadership in Energy and Environmental Design (LEED™) rating system. The state-of-the-art facility was completed in 2006 and serves as a model of sustainable design and an educational resource for the local community. The headquarters integrates a 19th Century farmhouse and barn—part of the offices for many years—with a new contemporary two-level, 19,500-square-foot, glass and steel building. (Source: Kresge Foundation.)

Among the more important reasons why it is worthwhile for owners and developers to consider green building design and construction are:

1. Energy-efficient buildings generally save on operating costs over time, which in a time of sharply rising energy costs can be particularly valuable.
2. Numerous government agencies provide financial incentives for green building projects, which provides a great enticement for building owners and developers to cash in on the "greenness" of their development projects through tax credits, financial incentives, carbon and renewable energy tradable credits, and net metering excess donations.
3. Not many people realize that in some cases it may actually be cheaper to build green. For example, a building that takes advantage of passive solar energy and includes effective insulation may require a smaller, less expensive HVAC system to serve the building. Also, purchasing recycled products can often be cheaper than purchasing

comparable new products, and incorporating a construction plan that minimizes waste will ultimately save on hauling and landfill charges.

4. There is a rising market demand for green buildings, particularly in high-end residential projects and prestigious corporate office projects. A BOMA Seattle survey for example recently found that 61 percent of real estate leaders believe that green buildings enhance their corporate image and 67 percent of these leaders also believe that over the next five years tenants will increasingly make green features of a property an important consideration when choosing space.

5. Green buildings are in a much better position to respond to existing and probable future governmental regulation. For example, in 2007, Washington State Governor Gregoire issued Executive Order 07-02, the Washington Climate Change Challenge, which aimed to establish state goals to reduce greenhouse gas emissions, promote sustainable development, increase clean energy sector jobs, and reduce expenditures on imported fuels. On May 21, 2009, Washington's Governor Chris Gregoire issued an executive order directing state actions to:

   - Reduce greenhouse gas emissions
   - Increase transportation and fuel-saving options for Washington residents
   - Protect Washington's water supplies and coastal areas.

   Building construction and operations are a big factor in nationwide greenhouse gas emissions and energy use, and we can expect future governmental regulations to be directed at the building industry. As in every business, the ability to make accurate predictions regarding the future is pivotal to business success.

6. Green building helps contribute to conservation, and one of the cheapest ways of stretching a limited resource is to conserve it. If new buildings can be built and operated in a way that conserves energy and materials, these limited resources will go farther and minimize the need for capital-intensive projects to increase them.

7. Another reason to support green building practices is that it helps reduce greenhouse gas emissions, which in turn can help prevent climate change. Greenhouse gases are those gases in the atmosphere that are transparent to visible light but which absorb infrared light reflected from the earth, thus trapping heat in the atmosphere. Many naturally occurring gases have this property, including water vapor, carbon dioxide, methane, and nitrous oxide. A number of human-made gases such as some aerosol propellants also have this property. There is a general consensus among climate scientists that greenhouse gases have been increasing in the atmosphere for the past 150 years and that the average temperature of the atmosphere and the oceans has also been increasing during that time.

3.2 GREEN BUILDING PRINCIPALS AND COMPONENTS

The intent of green building is to be sited, designed, constructed, and operated to enhance the well-being of its occupants and to support a healthy community and natural environment with minimal adverse impact on the eco-system (Figure 3.2).

3.2.1 Green Design Principals

The practice of sustainable architecture and construction revolves mainly around innovation and creativity. With green building it is important to employ materials and techniques that do not have an adverse impact on the environment or the building's inhabitants and not to choose materials just for their familiarity. With that in mind, there are numerous recycled products that can be used in the construction of sustainable structures like ceramic floor tiles which can be made from recycled glass. Flooring made from cork oak bark for example is friendly to the environment, since cork harvesting does not harm the trees from

FIGURE 3.2 Bren Hall—faculty and students on the third-floor terrace of Donald Bren School of Environmental Science and Management at UCSB, which is a LEED Gold Pilot project. It utilized silt fencing, straw bale catch basins, and scheduled grading activities in accordance with the project's erosion control plan. (Source: Bren School Website, http://www.esm.ucsb.edu.)

which it is taken. Bamboo flooring is another suitable alternative to wood, and is actually harder and more durable than hardwood floors. Bamboo is also less expensive.

In addition to the traditional building design concerns of economy, utility, durability, and aesthetics, green design strategies underline additional concerns regarding occupant health, the environment, and resource depletion. To address these concerns, there are many green design strategies and measures that can be employed including the following:

- Ensure overall energy efficiency
- Maximize use of renewable energy and materials that are sustainably harvested
- Minimize human exposure to hazardous materials
- Minimize the ecological impact of energy and materials used
- Conserve non-renewable energy and scarce materials
- Use water efficiently and minimize wastewater and run-off
- Encourage occupant bicycle use, mass transit, and other alternatives to fossil-fueled vehicles
- Minimize adverse impacts of materials by employing green products
- Optimize site selection to conserve green space and minimize transportation impacts
- Position buildings to take maximum advantage of sunlight and micro-climate
- Conserve and restore local air, water, soils, flora and fauna.

Implementing these strategies holistically serves to preserve our environment for future generations by conserving natural resources and protecting air and water quality. It also provides critical benefits by increasing comfort and well-being and helping to maintain healthy air quality. Finally, green building strategies are good for the economy by reducing maintenance and replacement requirements, reducing utility bills and lowering the cost of home ownership, and increasing property and resale values. In practical terms, green building is a whole-systems-approach to building design and construction with features that include:

- Using energy-efficient appliances, fixtures, and technologies
- Using water-saving appliances, fixtures, and technologies
- Designing for livable neighborhoods
- Maximizing use of insulation and ventilation
- Building quality, durable structures
- Recycling and minimizing construction and demolition waste
- Using healthy products and building practices
- Incorporating durable, recycled, salvaged, and sustainably harvested materials

- Taking advantage of the sun and site to increase a building's ability for natural heating, cooling, and daylighting
- Landscaping to use native, drought-resistant plants, and water-efficient practices (Figure 3.3).

Perhaps the first and most important step toward sustainability in the area of real estate development is to focus on areas relating to energy efficiency, water efficiency, waste efficiency, and design efficiency on a per building and a whole development basis. The following factors are

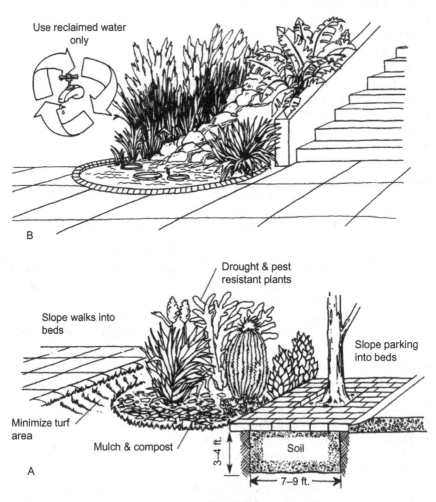

FIGURE 3.3 Two sketches showing the use of native and drought-resistant planting that can significantly enhance the environment in addition to providing opportunities for food and decorative gardens. Sketch B shows the use of reclaimed water. (Source: City of Santa Monica Green Building Design and Construction Guidelines.)

the main components for achieving green building and are rewarded by the majority of green rating systems such as LEED, Green Globes, and BREEM:

1. Integrated Design

This requires an integrated design team, that includes the designers, structural, mechanical, electrical, civil, lighting, plumbing, and landscape engineers, and possibly others, in addition to the contractor, to work with the project owner or developer to find the most effective way to meet the owner's expectations. This is aided by adapting the various systems to each other as an integrated whole and recognizing the interconnectivity of the systems and components that cumulatively make up a building and the disciplines involved in its design. Unlike the traditional approach, integrated design correctly assumes that each system affects the functioning of the other systems, which is why these systems must be harmonized if they are to perform together at maximum efficiency. The ultimate objective must be to optimize the building's performance, thereby reducing the adverse impact on the environment while minimizing its total cost.

2. Site Selection

This is an imperative feature of successful green building; it basically emphasizes the reuse and restoration of existing buildings and sites. Site selection is also concerned with rehabilitating contaminated or brownfield sites (determined by a local, state, or federal agency), as well as preserving natural and agricultural resources. Other features of site selection include promoting biodiversity and maximizing open space by reducing the development footprint; storm water management through supporting natural hydrology and reducing water pollution by increasing pervious area and on-site infiltration; reduction of construction waste; minimizing light trespass by minimizing light pollution associated with interior light in the building and exterior light luminance not to exceed site boundaries; reducing the heat island effect and encouraging use of public or low-environmental-impact transportation options.

3. Energy Efficiency

This is by far the most imperative issue surrounding green building and is also the element of a project that can provide the most significant operating costs reductions. As discussed in Chapter 8, energy efficiency measures may be eligible for federal and state tax credits and other financial incentives as required by the current ASHRAE/IESNA 90.1 standard. The comprising components of this standard are: 1. The building envelope, 2. Heating, ventilation, and air conditioning, 3. Water heating, including swimming pools, 4. Power, including building power distributed generation systems, 5. Lighting, and 6. Other electrical equipment.

While the majority of buildings can reach energy efficiency levels far beyond California Title 24 standards, most generally only strive to meet the standard. It is readily achievable to strive for 40 percent less energy than Title 24 standards. The following strategies contribute to this goal:

- Use an energy-efficient heating/cooling system in conjunction with a thermally efficient building shell. For example, boilers and chillers are major sources of energy usage in buildings and may account for an estimated 20 percent of the total energy usage in a building. Other prudent energy saving opportunities can exist with heat recovery options and thermal energy storage. Install high R-value wall and ceiling insulation; use minimal glass on east and west exposures and light colors for roofing and wall finishes.
- Passive design strategies such as building shape and orientation, passive solar design, and the use of natural lighting can dramatically affect building energy performance.
- Consider incorporating renewable energy sources such as solar, wind, or other alternative energy into the HVAC system to reduce operational costs and minimize the use of fossil fuels. This path can lead the way to emerging future technologies.
- Minimize the electric loads created by lighting, appliances, and systems.
- Use energy management controls, as improperly programmed controls and outdated technology can mislead a building owner to believing that a building is performing more efficiently that it really is. Replacing, upgrading, or reprogramming the temperature controls and the energy management system will ensure equipment operates at maximum efficiency.
- Develop strategies to provide natural lighting where this will improve well-being and productivity. A green building takes advantage of the sun's seasonal position to heat a building's interior in winter and often incorporates design features such as light shelves, overhanging eaves or landscaping to mitigate the sun's heat in summer. Room orientation should be designed to improve ventilation.
- Install high-efficiency lighting systems with advanced lighting control systems and incorporate motion sensors linked to dimmable lighting controls. Inclusion of task lighting can reduce general overhead light levels.
- Use computer modeling when possible to optimize the design of electrical and mechanical systems and the building shell.
- As existing buildings age they require regular maintenance. In this respect, retro-commissioning can be extremely useful. Initiating small changes to a building's operational timing can make a substantial difference in energy usage and savings.

4. Water Efficiency and Conservation

The United States annually draws out an estimated 3700 billion gallons more water from its natural water resources than it returns. Many municipalities have legislation in place requiring storm water and wastewater efficiency measures while the Energy Policy Act of 1992 already requires water conservation for plumbing fixtures. The need to implement water efficiency measures is to conserve our depleting water resources and preserve water for agricultural uses, in addition to reducing the pressure on water-related ecosystems. Efficiency measures that can be implemented to advance water efficiency and conservation include:

- Minimizing wastewater by employing ultra low-flush toilets, low-flow showerheads, and other water conserving fixtures.
- Employ dual plumbing systems that use recycled water for toilet flushing or a gray water system that recovers rainwater or other non-potable water for site irrigation.
- Use a water budget approach that schedules irrigation.
- Use recirculating systems for centralized hot water distribution, and point-of-use hot water heating systems for more distant locations.
- Incorporate self-closing nozzles on hoses and state-of-the-art irrigation controllers.
- Use micro-irrigation to supply water in non-turf areas and landscape to be metered separately from buildings.

5. Materials and Resources

Building material choices can have an enormous impact on the natural environment not only because of the many processes involved such as extraction, production, and transportation, all of which can negatively impact our ecosystem, but also because some materials may release chemicals that are harmful to building occupants. Green building generally avoids using potentially toxic materials such as treated woods, plastics, and petroleum-based adhesives, which can degrade air and water quality and cause health problems. Additionally, building demolition may cause materials to release hazardous or non-biodegradable material pollutants into the natural environment or into drinking water reserves. Sustainable building materials also reduce landfill waste. The following aspects should be considered when choosing a building material for a project:

- Sustainable construction materials and products should be selected. Their sustainability can be measured by several characteristics such as recycled content, reusability, minimum off gassing of harmful chemicals, zero or low toxicity, durability, sustainably harvested materials, high recyclability, and local production. Use of sustainable products increases resource conservation and efficiency and minimizes the adverse impact on the environment.

- Use of dimensional planning and other material efficiency strategies reduces the amount of building materials needed and cuts construction costs. For example, the design of rooms to 4-foot multiples minimizes waste by conforming to standard-sized wallboard and plywood sheets.
- When possible, reuse and recycle construction and demolition materials. For example, using inert demolition materials as a base course for a parking lot keeps materials out of landfills and cuts costs.
- Require waste management plans for managing materials through deconstruction, demolition, and construction.
- Allocate adequate space to facilitate recycling collection and to incorporate a solid waste management program that reduces waste generation.

Employing recycled/reused materials helps to ensure the sustainability of resources. If building projects used only virgin raw materials these materials would gradually be exhausted. As the availability of raw materials become scarce, the prices will rise until the materials are no longer obtainable. This trend has already started, and some raw materials are either no longer available or have become very scarce, and can only be obtained recycled from existing projects. Recycling and reusing materials helps ensure that they will be readily available for future building projects.

6. Indoor Environmental Quality and Safety

Green construction can provide a superior interior environment, which in turn can reduce the rate of respiratory disease, allergy, asthma, sick building symptoms (SBS), and enhance tenant comfort and worker performance. Materials such as carpet, cabinetry adhesives, paint, and other wall coverings with zero or low levels of volatile organic compounds (VOCs) will release less gas and improve the indoor air quality. On the other hand, building materials and cleaning and maintenance products that emit toxic gases, such as volatile organic compounds (VOC) and formaldehyde should be avoided, as these gases can have a detrimental impact on occupants' health and productivity. Daylighting can also improve the interior quality by boosting the occupant's mood with natural light. The potential financial benefits of improving indoor environments are also substantial.

Adequate ventilation and a high-efficiency, in-duct filtration system should be provided. Heating and cooling systems that ensure proper ventilation and filtration can have a dramatic and positive impact on indoor air quality.

Select materials that are resistant to microbial growth to prevent indoor microbial contamination. Provide effective roof drainage, and for the surrounding landscape allow proper drainage of air-conditioning coils, and design other building systems to control humidity.

7. Waste Management

In the United States there is an estimated 31.5 million tons of construction waste produced annually. Furthermore, nearly 40 percent of solid waste in the United States is produced by construction and demolition. Waste management issues are connected to several areas of green building, from waste reduction measures during construction to waste recycling measures.

8. Building Operation and Maintenance

Green-building measures cannot achieve their objectives unless they function as intended. Building commissioning and enhanced commissioning includes testing and adjusting the mechanical, electrical, and plumbing systems to ensure that all equipment meets design intent. It also includes instructing the staff on the operation and maintenance of equipment. As buildings age, their performance generally declines and can only be assured through proper maintenance or retro-commissioning.

Another important aspect of green construction is reduced and easier maintenance. The incorporation of operating and maintenance factors into the design process of a building project can contribute to the creation of healthy working environments, higher productivity, and reduced energy and resource costs. Whenever possible therefore, designers should specify materials and systems that simplify and reduce maintenance and life-cycle costs, use less water, energy, and are cost-effective. Thus for example, by reducing maintenance activities such as painting, the materials needed are saved and the waste and environmental impact of the painting such as VOC gas release and water used in cleanup are reduced.

9. Livable Communities

There are several issues pertaining to community and neighborhood development that should be addressed, such as the application of ecologically appropriate site development practices, the incorporation of high-performance buildings, and the incorporation of renewable energy. In addition, the development of new communities and neighborhoods, and the housing incorporated into such developments, may also involve looking into issues not normally considered in single-structure projects. Such issues may include evaluating the community's location, the proposed structure and density of the community, and the ramifications of the community on transportation requirements. Other issues that should be considered include setting the standards for the community's infrastructure and the standards to be applied to specific development projects within the community, as all these factors influence the environmental impacts of the development, and the ongoing livability of the community as an integrated whole.

Green construction is now entered the mainstream in the United States. Likewise, the escalating costs of energy and building materials, coupled with warnings from the EPA about the toxicity of today's treated and synthetic materials, are prompting architects and engineers to revise their approach to building techniques that employ native resources as construction materials and use nature (daylight, solar, and ventilation) for the heating and cooling process. Green developments are generally more efficient, last longer, and cost less to operate and maintain than conventional buildings. Moreover, green developments generally provide greater occupant comfort and higher productivity than conventional developments, which is why most sophisticated buyers and leasers prefer them, and are usually willing to pay a premium for them.

The vast majority of today's buildings continue to use mechanical equipment powered by electricity or fossil fuels for heating, cooling, lighting, and maintaining indoor air quality. In this regard, the U.S. Department of Energy (DOE) estimates that buildings in the U.S. consume annually more than one-third of the nation's energy and contribute approximately 36 percent of the carbon dioxide (CO_2) emissions released into the atmosphere. This means that the fossil fuels used to generate electricity and condition buildings place a heavy burden on the environment; they emit a plethora of hazardous pollutants such as volatile organic compounds (VOCs) that cost building occupants and insurance companies millions of dollars annually in health care costs. In addition, we have the problem of fossil fuel mining and extraction, which adds to the adverse environmental impacts, while fomenting price instability, which is causing concern among both investors and building owners. Creating buildings that use less energy both reduces and stabilizes costs, and also has a positive impact on the environment.

The U.S. Department of Energy (DOE) early on had the foresight to appreciate the urgent need for buildings that were more energy efficient and in 1998 it took the initiative and decided to collaborate with the commercial building industry to develop a 20-year plan for research and development on energy-efficient commercial buildings. DOE's High-Performance Buildings Program's primary mission is to help create more efficient buildings that save energy and provide a quality, comfortable environment for workers and tenants. The program is targeted mainly toward the building community, particularly building owners/developers, architects, and engineers. Today we have in place the knowledge and technologies required to reduce energy use in our homes and workplaces without having to compromise comfort or aesthetics. Unfortunately however, the building industry remains aloof or uninformed because it is not taking full advantage of these important advances, which is why to this day, many new building projects continue to be designed and operated without properly considering all the environmental impacts.

3.2.2 High Performance and Intelligent Buildings

High performance buildings and building automation have become landmarks in today's society; they typically consist of a programmed, computerized, "intelligent" network of electronic devices that monitor and control the mechanical and lighting systems in a building. Partly because of rising energy costs, an increasing number of new buildings are incorporating central communications systems and the "intelligent" building. Increasing consumer demand for clean renewable energy and the deregulation of the utilities industry have encouraged and energized growth in green power such as solar, wind, geothermal steam, biomass, and small-scale hydroelectric sources of power. In addition, small commercial solar power plants have started to emerge around the country and serve some energy markets, for example in California.

Implementing high performance building projects may be more effective by initially instigating a green design "charrette" or multi-disciplinary kick-off meeting to articulate a clear roadmap for the project team to follow. A crucial advantage of holding a green design charrette during the early stage of the design process is that it offers team professionals (with possible assistance from green design experts and facilitators) the chance to brainstorm on achieving design objectives as well as alternative solutions. This goal-setting approach helps identify green strategies for members of the design team and helps facilitate the group's ability to reach a consensus on performance targets for the project.

Designers of green buildings must pay careful attention to measured performance expectations. Once performance measures are determined, a follow-up is needed to establish performance targets and the metrics to be employed for each measure. Minimum requirements, or baselines, are typically defined by codes and standards, which may differ from one jurisdiction to another. Alternatively, performance baselines can be set to exceed the average performance of a building type, measured against the most recent building built or against the performance of the best documented building of a particular type.

LEED and Green Globes provide various consensus-based criteria to measure performance, along with useful reference to baseline standards and performance criteria. Nevertheless, LEED or Green Globes certification by itself does not ensure high performance in terms of energy efficiency, as certification may have been achieved by acquiring other non-energy related categories such as Materials and Resources or Sustainable Sites. For this reason, specific energy related goals must still be set.

Integrated design is enhanced by the use of computer energy modeling tools such as the Department of Energy's DOE 2.1E and other computer programs. These programs can inform the building team of the impacts of energy-use implications very early in the design process by factoring in relevant information such as climate data, seasonal changes, building massing and orientation, and daylighting. It can also readily prompt investigation and survey of cost-effective design alternatives for the building envelope and mechanical systems by forecasting energy use of various combined alternatives.

Consider the following strategies to ensure success of any green building project:

- Embrace a vision of sustainable principles and a holistic design approach.
- Formulate a clear statement of objectives, design criteria, and priorities for the project.
- Determine a project budget that covers green building measures and which includes a contingency amount for possible additional options. Seek out available tax incentives and grant opportunities.
- Employ a design professional with green building background and experience.
- Determine performance goals for siting, energy, water, materials, and indoor environmental quality along with other sustainable-design goals, and incorporate these goals throughout the design process and the building's life cycle.
- Incorporate building-information modeling (BIM) concepts to enable project team members to create a virtual model of the project and its various systems in 3D in a format that can be shared with the entire project team.
- Select a design and construction team that is committed to green building. Modify the RFQ/RFP selection process to ensure that contractors have appropriate qualifications to identify, select, and implement an integrated system of green building measures.
- Develop a project schedule that allows for systems testing and commissioning (as part of contract documents).
- Ensure the contract documents are complete and that the building design is at an appropriate level of building performance.
- Develop a building management plan that ensures that operating decisions and tenant education are implemented in a satisfactory manner with regard to integrated, sustainable building operations and maintenance.
- All stages of the building's life cycle, including deconstruction, should be considered.
- Create effective incentives and oversight.

But before dwelling too deeply into green design and the integrated design process and to understand its full meaning, it may be prudent to first describe the more conventional design process. The traditional design approach usually starts with the architect and the client agreeing on a budget and design concept, followed by a general massing scheme, typical floor plan, schematic elevations and, usually the general exterior appearance as determined by these design criteria and design intent. The mechanical and electrical engineers are then asked to implement the design and to suggest appropriate systems. Building information modeling (BIM) programs have only recently been introduced and incorporated in the design process.

This conventional design approach, while greatly oversimplified, in fact remains the main method employed by the majority of general-purpose design consultant firms, which unfortunately tends to suppress the achievable performance to conventional levels. Due to the sequential contributions of the members of the design team, the traditional design process consists mainly of a linear structure. The opportunity for optimization is limited during the traditional design process, while optimization in the later stages of the process is usually difficult if at all viable. Thus, while this process may seem adequate and even appropriate, the actual results often tell a different story, with high operating costs and often a sub-standard interior environment. These factors can have a negative impact on a property's ability to attract quality tenants or achieve desirable long-term rentals, in addition to a reduced asset value for the property. Conventional design approaches do not typically take advantage of computer simulations of predicted energy performance, often resulting in dismal performance and steep operating costs to the dismay of owners, operators, and tenants.

3.2.3 Building Information Modeling (BIM)

As building construction continues to grow in complexity and change under the influence of new technologies, there are a number of new programs on the horizon that are having a positive impact on the entire design, planning, and construction community. Among them is the introduction of BIM software, which is the latest trend in computer-aided design and is being touted by many industry professionals as a lifesaver for complicated projects because of its ability to correct errors at the design stage and accurately schedule construction. BIM is comprised of 3D modeling concepts, information database technology, and interoperable software in a computer application environment that design professionals and contractors can use to design a facility and simulate construction.

In this regard, Autodesk says, "building information modeling (BIM) software facilitates a new way of working collaboratively using a model created from consistent, reliable design information—enabling faster decision-making, better documentation, and the ability to evaluate sustainable building and infrastructure design alternatives using analysis to predict performance before breaking ground." In fact, some industry professionals forecast that buildings in the not too distant future will be built directly from the electronic models that BIM and similar programs create, and that the design role of architects and engineers will dramatically change (Figure 3.4). BIM is gradually changing the role of drawings for the construction process, improving architectural productivity, and making it easier to consider and evaluate design alternatives. This modern modeling technology is particularly valuable in sustainable design because it enables project team members to create a virtual model of the structure and all of its systems in 3D in a format that can be shared with the entire project team, thereby facilitating the process of integrating the various design teams' work. This allows team members to identify design issues and construction conflicts and resolve them in a virtual environment before the actual commencement of construction, thus directly promoting the utilization of an integrated team process.

As BIM increases in popularity, it will rapidly become fundamental to building design, visualization studies, cost analysis, contract documents, 3D simulation, and facilities management. Autodesk Revit®, a popular BIM program, is making continuous headway in its market penetration into architectural firms and it is anticipated that in the coming years Revit will have a significant market share of major projects designed in the United States.

FIGURE 3.4 The Highlands Lodge Resort and Spa project, a joint venture of Q&D Construction and Swinerton Builders, Inc. where Vico, a BIM software package, was used. The five-star hotel and high-end luxury condominiums has a total gross floor area of 406,500 sq ft/37,720 sq meters and is built on a roughly 20-acre site at the Northstar-at-Tahoe Ski Resort in Northern California. (Source: Vico Software Inc.)

3.3 IDENTIFYING GREEN PROJECT REQUIREMENTS AND STRATEGIES

3.3.1 Green Design Strategies

There are many recommended practices that can reduce the environmental and resource impacts of buildings, and enhance the health and satisfaction of their occupants. The most prominent strategies that come to mind include:

1. Use Less to Achieve More

The most effective green design solutions are able to address a number of needs with only a few elements. For example, a concrete floor may be simply finished with a colored sealant that reflects daylight for better illumination and eliminates air pollutant emissions from floor coverings. The floor can also be used to store daytime heat and nighttime cold to enhance occupant comfort. Thus, a carefully designed element serves as a structure, a finished surface, distributes daylight, and stores heat and cold, thereby saving materials, energy resources, capital, and operating costs.

2. Incorporate Design Flexibility and Durability

Buildings that are designed to adapt to changing functions over long useful lives reduce life-cycle resource consumption. Durable sustainable structural elements that contain generous service space and are able to accommodate movable partitions can last for many decades, instead of being demolished because they lack the ability to adapt to changing building functions. Durable envelope assemblies improve comfort and reduce life-cycle maintenance and energy costs.

3. Combinations of Design Strategies Must be Carefully Considered to be Truly Effective

Green buildings are incorporating increasingly complex systems of interacting elements. Intelligent green design must consider the impact of the elements on each other, and on the building as a whole. As an example, the need for mechanical and electrical systems is greatly affected by building form and envelope design. Combining strategies like daylighting, solar load control, and natural cooling and ventilation can all work together to reduce lighting, heating and cooling loads. Carefully combining these strategies can save resources and money, both in construction and operation.

4. Take Advantage of Site Conditions

Buildings are usually considered more sustainable when they respond to local microclimate, topography, vegetation, and water resources; they are also usually more comfortable and efficient than conventional designs that rely on technological fixes and ignore their surroundings. Santa Monica, California, for example, has exemplary solar and wind resources for passive solar heating, natural cooling, ventilation, and daylighting, but has meager local water supplies, some of which have recently been polluted. Taking advantage of free natural resources, and conserving scarce high-priced commodities are excellent approaches to reduce costs and connect occupants to their surroundings.

5. Incorporate Preventive Maintenance Instead of Repairing After the Fact

Addressing potential problems from the beginning by applying preventive maintenance makes practical and economic sense. For example, using low-toxicity building materials and installation practices is more effective than diluting indoor air pollution from toxic sources by employing large quantities of ventilation air.

"Smart growth" is another feature of green design that concerns many communities around the country. It relates mainly to the ability to control sprawl, reusing existing infrastructure, and creating walkable neighborhoods. Locating suitable places to live and work within walking distance or near public transport is an obvious advantage toward reducing energy. Likewise, it is more logical and resource-efficient to maintain or reuse existing roads and utilities than building new ones. The preservation of open spaces, farm lands, and undeveloped land strengthens and reinforces the evolution of existing communities and helps maintain their quality of life.

3.3.2 The Integrated Design Process (IDP)

The IDP approach differs from the conventional design process in several ways. In the IDP approach the owner takes on a more direct and active role in the process and the architect assumes the role of team leader rather than sole decision maker. Additional key consultants, including the structural, electrical, mechanical, lighting, and other players, become an integral part of the team from the outset and participate in the project's decision-making process (Figure 3.5). Therefore, from a design perspective, the key process difference between green-building design and conventional design is the concept of integration.

To achieve a successful sustainable building project indeed requires the employment of an integrated design process with clear and precise design objectives, which should be identified early on and held in proper balance during the design process. An integrated approach to design and construction is required to achieve a successful high-performance building. For example, by working collaboratively as a team, the main players

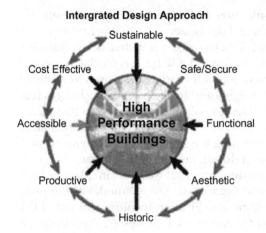

FIGURE 3.5 A diagram showing the various elements that impact the design of high-performance buildings using the integrated design approach.

(architect, engineers, landscape architect, etc.), can maneuver and direct the ground plane, building shape, section, and planting scheme to provide increased thermal protection, and reduce heat loss and heat gain. By reducing heating and cooling loads, the mechanical engineer is able to reduce the size of mechanical equipment necessary to achieve comfort. Moreover, the architect, lighting, and mechanical engineers can work in unison to design, for example, a more effective interior/exterior element such as a light-shelf, which serves not only as an architectural feature, but can also provide needed sun-screening to reduce summer cooling loads while at the same time allowing daylight to penetrate deep into the interior. This results in a more efficient environmental performance in addition to on-going operational savings. If first costs exceed initial estimates, it should be quickly offset by decreased life cycle dollar and environmental costs.

In the IDP process, the decision to identify a leader for the project normally lies on the shoulders of the owner/client. The project owner will identify the need for a person to undertake the role that is proficient and capable of leading a team to design and build the project on the basis of specific requirements in the form of a project brief for space and budgetary capacity. The project brief accompanying this planning activity should describe existing space use, include realistic estimates of both spatial and technical requirements, and contain a space program around which design activity can develop. Depending on a project's size, type, and complexity, there may be a need to employ a construction manager or a general contractor and who may come on board at this point.

Once the pre-design activities are completed, the architect, designer of record (DOR), and other key consultants, in collaboration with the other team members or sub-consultants, may produce preliminary graphic proposals for the project or portions of it via a 3D modeling program (e.g., BIM) or manually. The intention of the preliminary proposals are meant more to stimulate thought and discussion then to describe any final outcome, although normally the fewer changes initiated before bidding the project the more cost effective the project will be. It is crucial to involve all relevant consultants and sub-consultants early in the process in order to benefit from their individual insights and to prevent costly changes further along in the process. A final design will gradually emerge that incorporates the interests and requirements of all project team participants including the owner, while also meeting the overall area requirements and project budget that was established during the pre-design phase.

At this stage a schematic design proposal will be in place which should include a site location and organization, a 3D model of the project, space allocation, and an outline specification including an initial list of systems and components that are part of the final design. A preliminary cost estimate can now be made, and depending on the size and complexity of the project, it may be performed by a professional cost estimator or a computer program at this point. For smaller projects this service may be performed as part of a preliminary bidding arrangement by one or more of the possible builders. On larger projects, the cost estimate can be linked to the selection process for a builder, assuming other prerequisites are met such as experience and satisfactory references.

Following the schematic design, the *design development* phase comes into play. This entails going into greater detail for all aspects of the building, including systems and materials, etc. The collaborative process continues with the architect working hand in hand with the owner and the various contributors. The resulting outcome of this phase is a detailed design, with a consensus from all players and who may be asked to sign off on it. When the project design is developed using an integrated team approach, the end product is usually a design that is highly efficient with minimal, if not zero, incremental capital costs, and reduced maintenance and long-term operating costs.

The development and production of *contract documents* involves converting the design development information into formats that can be used for pricing, bidding, permitting, and constructing the project. An efficient set of contract documents can be achieved by careful scrutiny, and accountability to the initial program requirements as outlined by the design team and the client, in addition to careful coordination and collaboration with the technical consultants on the design team. Design, budgetary, and other decisions continue to be made with the appropriate contributions of the various players. Changes in scope during this phase should be avoided as they can significantly impact the project, and once pricing has commenced can invite confusion, errors, and added costs. Cost estimates may be made at this point, prior to or simultaneous with bidding, in order to assure compliance with the budget and to check the bids.

Once the general contractor is selected during the *construction phase*, other members of the project team must remain fully involved. There will remain many issues that need to be addressed such as previous decisions that may require clarification, or supplier samples and information that must be reviewed for compliance with the contract documents, and proposed substitutions that need evaluation. Whenever proposed changes affect the operation of the building, the owner/client must be informed and approve such changes. Any changes in user requirements may necessitate modifications to the building's design which will require consultation with the other consultants and sub-consultants to assess the implications and ramifications that such changes may create. Any changes must be priced and incorporated into the contract documents.

The ultimate responsibility for ensuring that the building upon completion meets the requirements of the contract documents lies with the design team. The level of a building's success of meeting program performance requirements can be assessed by an independent third party in a process known as *commissioning*. Here the full range of systems and functions in the building are evaluated and the design and construction team may be called upon to make any required changes and adjustments. This is discussed in greater detail in Chapter 12. After the building becomes fully operational, a post-occupancy evaluation may be conducted to confirm that the building meets the original and emerging requirements for its use.

3.3.3 Green Building Design and the Delivery Process

The process of green building design and construction differs fundamentally from current standard practice. Successful green buildings result from a design process that displays a strong commitment to the environment and to health issues. Measurable targets challenge the design and construction team, and allow progress to be tracked and managed throughout development and beyond. Employing computer energy simulations offers the ability to assess energy conservation measures early and throughout the design process. By collaborating early in the conceptual design process the expanded design team is able to generate alternative concepts for building form, envelope, and landscaping, and also focus on minimizing peak energy loads, demand, and consumption.

Computer energy simulation is employed to evaluate a project's effectiveness in energy conservation, and its construction costs. Typically, heating and cooling load reductions from better glazing, insulation, efficient lighting, daylighting, and other measures allows smaller and less expensive HVAC equipment and systems, resulting in little or no increase in construction cost compared to conventional designs. The use of simulations to refine designs and ensure that energy conservation and capital cost goals are met is extremely valuable; and to demonstrate regulatory compliance.

The evaluation of design alternatives is based on of capital cost as well as reduced life-cycle cost. Design alternatives are aimed at minimizing the buildings' construction cost and its life-cycle cost. Assessments include costs and environmental impacts of resource extraction; materials and assembly manufacture; construction; operation and maintenance in use; and eventual reuse, recycling, or disposal. Computer energy simulation is but one tool that is used to incorporate operational costs into the analysis. There are other computer tools also available to help perform life-cycle cost analysis.

For traditional, non-green buildings, the various specialties associated with project delivery, from design and construction through building occupancy, are responsive in nature, utilizing restricted approaches to address particular problems. Each of these specialists typically has wide-ranging knowledge and experience in their specific fields, and they provide solutions to problems that arise based solely on their knowledge and experience in their specific fields. For example, an air-conditioning specialist if asked to address a problem of an unduly warm room will suggest increasing the cooling capacity of the HVAC system servicing that room, rather than investigating the source of the problem. The excessive heat gain could for example, be mitigated by incorporating operable windows or external louvers. The end results, therefore, while often being functional, are nevertheless highly inefficient so that the building ends up comprised of different materials and systems with little or no integration between them.

Green building design and construction typically utilizes an integrated design process (IDP), which uses a multi-disciplinary team of building professionals who usually work collectively together from the pre-design phase through post-occupancy to produce a superior, cost effective, environmentally friendly, high performance building project. The essence of integrated design is that buildings typically consist of numerous systems that are interconnected or interdependent, and all of which impact one another to varying degrees.

Properly engineered and functioning systems help to ensure the comfort and safety of building occupants. They also empower designers to create environments that are healthy, efficient, and cost effective. Integrated design is a critical factor and consistent component in the design and construction of green buildings. This summary description highlights the benefits of integrated design and the main characteristics that differentiate conventional and integrated design processes.

3.3.4 The Integrated Multidisciplinary Project Team

Close collaboration by multi-disciplinary teams is important, from the beginning of conceptual design and throughout the design process and construction. For sustainable projects, the design team usually has to broaden itself to include certain specialists and other interested parties, such as energy analysts, materials consultants, cost consultants, and lighting designers; often, contractors, operating staff, and prospective tenants are also included. This enlarged design team provides fresh perspectives reflecting new approaches and feedback on performance and cost. The design process becomes a continuous, sustained team effort from conceptual design through commissioning and occupancy.

Some of the professionals required to join the project team in this type of process may include the following:

The *owner's representative (OR)* is the person that represents the owner and speaks on his/her behalf. The OR may be hired by the owner as a consultant or may be chosen from within the organization commissioning the project. The selected OR will be required to devote the necessary time needed to advocate, clarify, defend, and develop the owner's interests.

The *construction manager (CM)* is the person or entity who provides construction management services, either as an advisor or as a contractor, and is normally hired on a fee basis to represent the logistics and costs of the construction process, and can be an architect, a general contractor, or specifically a consulting construction manager. It is important for this person to be involved from the beginning of the project.

In the majority of building projects, the architect is required to lead the design team and coordinates with sub-consultants and other experts. The architect is also required to ensure compliance with the project brief and budget. In some cases, the architect has the authority to hire some or all of the sub-consultants; in larger projects the owner may decide to contract directly with some or all of them. The architect usually administers and manages the production of the contract documents and oversees the construction phase of the project, ensuring compliance with the contract documents by conducting appropriate inspections, and managing submissions approvals and evaluations by the sub-consultants. The architect also oversees the evaluation of requests for payment by the builder and other professionals and chairs monthly or bi-weekly site meetings.

Early involvement of the *civil engineer* is an integral part of the project team and essential for understanding the land, soil, and regulatory aspects of a construction project; the civil engineer may be hired directly by the owner or the architect. The role of the civil engineer usually includes preparation of the civil engineering section of the contract documents in addition to ensuring compliance of the work with the contract documents.

Consulting structural, mechanical, and electrical engineers may be commissioned by the architect as part of his work, or sometimes on larger or more complex projects may be employed directly by the owner. These consultants are responsible for designing the structural, heating, ventilation, and air-conditioning, and power and illumination aspects of the project.

Each consultant produces that portion of the contract documents that is within his/her specialty and all participate in assessing their part of the work for compliance with those documents.

The *landscape architect* is often hired as an independent consultant depending on the type and size of the project. The role of the landscape architect is designing the arrangement of land for human use involving vehicular and pedestrian ways and the planting of groundcover, plants, and trees. He/she should be involved early in the design process to assess existing natural systems, how they will be impacted by the project, and ways to accommodate the project to those systems. The landscape architect should have extensive experience in sustainable landscaping including erosion control, managing storm water runoff, green roofs, and indigenous plant species.

Specialized consultants are embraced by the design team as needed and reflected by special requirements of the project. These may include lighting consultants, specifications writers, materials and component specialists, sustainability consultants, and technical specialists. The size, complexity, and specialization of the project will suggest the kinds of any additional experts that may be required. As with all contributors to the integrated design process, specialist consultants should be involved early in the design process to incorporate their suggestions and requirements in the design so as to ensure that their contributions are taken into account to ensure maximum efficiency.

3.4 DESIGN PROCESS FOR HIGH PERFORMANCE BUILDINGS

The world is rapidly changing, and building construction practices and advances in architectural modeling technologies have reached a unique crossroad in history with changing needs and expectations. And with many successful new building projects taking shape throughout the country today, it calls into question the performance level of our more typical construction endeavors, and makes us wonder just how far our conventional buildings are falling short of the mark and what needs to be done to meet these new challenges. High performance outcomes necessitate a far more integrated team approach to the design process and mark a departure from traditional practices, where emerging designs are handed sequentially from architect to engineer to sub-consultant (Figure 3.6). As mentioned above, an integrated approach results in a typically more unified, more team-driven design and construction process that encompasses different experts early in the design setting process. This process increases the likelihood of creating high performance buildings that

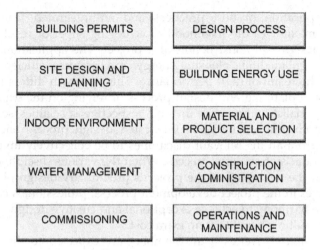

BUILDING PERMITS	DESIGN PROCESS
SITE DESIGN AND PLANNING	BUILDING ENERGY USE
INDOOR ENVIRONMENT	MATERIAL AND PRODUCT SELECTION
WATER MANAGEMENT	CONSTRUCTION ADMINISTRATION
COMMISSIONING	OPERATIONS AND MAINTENANCE

FIGURE 3.6 The main elements of high performance building design.

achieve significantly higher targets for energy efficiency and environmental performance.

In the absence of an interactive approach to the design process it would be extremely difficult to achieve a successful high-performance building, largely because the successful design of a building requires the integration of many kinds of information into a synthetic whole. The process draws its strength from the knowledge and expertise of all the stakeholders (including the owner) across the life cycle of the project, in addition to their early collaborative involvement in recognizing the needs for the building through planning, design, construction, operation, and maintenance of the facility, and building occupancy is part of this process. The best buildings result from active, consistent, organized collaboration among all players, which is why the stakeholders need to fully understand the issues and concerns of all the parties and be able to interact closely throughout the various phases of the project to achieve a successful outcome.

By implementing a team-driven approach, high performance buildings are basically utilizing a "front-loading" of expertise. The process typically begins with the consultant and owner leading a green design charrette with all the stakeholders (design professionals, operators, and contractors) in a brainstorming session reflecting a "partnering" approach that encourages collaboration in achieving high performance green goals for the new building, while breaking down traditional adversarial roles. The design development phase identifies opportunities for collaborating, invites continuous input from users and operators in order to advance progress, and initiates commitment to decision making.

The application of best practices and an integrated team-driven approach maximizes the likelihood of achieving superior results in the building design and construction of a project. The application of integrated design methods elevates energy and resource efficiency practices into the realm of high performance. This approach differs from the conventional planning and design process of relying on the expertise of various specialists who work in their respective specialties somewhat isolated from one another. The integrated design process encourages designers from all the relevant disciplines to be collectively involved in the design decision-making process. When the various disciplines come together at the beginning of the project's preliminary design phase and at key stages in the project development process, professionals can work together in harmony to achieve exceptional and creative design solutions that yield multiple benefits at no extra cost.

With design charrettes team members are expected to discuss and address problems beyond their field of expertise. Charrettes can be instrumental in complex situations where the interests of the client often conflict and when they are represented by different factions. Charrette participants understand the issues and resolution encourages them to commit to the schematic solutions. Although final solutions may not necessarily be produced, often important interdependent issues are studied and clarified. Conducting a facility performance evaluation to confirm that all the designated high-performance goals have been met and will continue to be met over the life cycle of the project is also an important consideration. Retrocommissioning is another factor that should be considered to ensure that the building will continue to optimally perform through continual adjustments and modifications.

As discussed earlier in this chapter, when computer energy simulations are conducted, it should be as early as possible in the design process and continue until the design is complete to offer a reliable assessment of energy conservation measures and to allow the design team to generate several alternative concepts early in the process for the building's form, envelope, and landscaping. Computer energy simulation has proven to be an excellent tool to assess the project's effectiveness in energy conservation, as well as its construction costs. Employing sustainable approaches that reduce heating and cooling loads allows the mechanical consultant to design a more appropriate, more efficient and less expensive HVAC system that in turn results in a minimal increase, if any, in construction cost compared to conventional designs.

There are many positive attributes to simulations, but the most important is to see how a design can be improved to ensure that energy-conservation and capital cost goals are met, and to check that a design complies with all regulatory requirements. Alternative design proposals are readily evaluated either on capital cost or on the basis of reduced life-cycle cost.

The primary aim of exploring alternative designs is to simultaneously minimize both a building's construction cost and its life-cycle cost. But in order to more accurately assess these costs requires a comprehensive approach that includes accurate information on costs and environmental impacts on all aspects of construction, including resource extraction and materials and assembly manufacture. It also requires costs relating to operation and maintenance to final reuse, recycling, and disposal. But computer energy simulation is but one of several tools employed to incorporate operational costs into the analysis; other computer tools are also available to facilitate performing life-cycle cost analysis.

High performance sustainable buildings continue to gain in popularity and are emerging as an important market sector both in the United States and globally. At the same time, this increased demand for high performance buildings has encouraged facility owners, investors, and design professionals to reevaluate their position in the design process. This reassessment of emerging patterns and primary processes on successful high-performance building projects is having a consequential impact on both the private and government sectors. The integrated design process reflects a more profound analysis of the project than might normally be expected of a conventional design, and requires a more concerted effort from design consultants. While this additional work is often reflected in the design fees requested, such fees are really insignificant when compared to the environmental and cost impacts over the life of a typical building.

A Federal Leadership in High Performance and Sustainable Buildings Memorandum of Understanding (MOU) was signed in January 2006, for which the signatory agencies commit to federal leadership in the design, construction, and operation of high-performance and sustainable buildings. An important component of this strategy is the implementation of prevalent approaches to meet certain requirements relating to various activities such as planning, siting, designing, and building, operating, and maintaining high performance buildings. The MOU contains a number of guiding principles to be adopted by federal leadership in high performance and sustainable buildings. These incorporate greater detailed guidance on the principles for optimizing energy performance, conserving water, improving IEQ, integrated design, reducing the impact of materials, and other issues. Since the signing of the MOU, many federal facilities have already succeeded in creating high performance buildings that save energy and reduce the negative impact on the environment and our lives.

The Interagency Sustainability Working Group (ISWG), as a subcommittee of the Steering Committee established by Executive Order (E.O.) 13423, initiated development of the guidance to assist agencies in meeting the high performance and sustainable buildings goals of E.O. 13423, section 2(f). On December 5, 2008, a new guidance on high performance federal buildings was issued. It includes:

- Revised guiding principles for new construction
- New guiding principles for existing buildings
- Clarification of reporting guidelines for entering information on the sustainability data element (#25) in the Federal Real Property Profile
- Clarification and explanation of how to calculate the percentage of buildings and square footage that are compliant with the guiding principles for agencies' scorecard input.

3.5 GREEN PROJECT DELIVERY SYSTEMS

One of the first decisions an owner has to make is selecting the right project delivery system. Each delivery system has its advantages and disadvantages, depending on the type and size of the project under consideration. It is also necessary to have a good understanding of the various systems. Indeed, selection of the right project delivery system is one of the most significant factors to impact a construction project's ability to succeed. Project delivery is simply a process by which all of the processes, procedures, and components of designing and building a facility are organized and incorporated into an agreement that results in a completed project. The process begins by fully stating the needs and requirements of the owner in the architectural program from concept design to final contract documents. There are a wide range of construction project delivery systems. These include the traditional design-bid-build system, the design-build system, the construction management, and other methods. Each of these systems has its advantages and disadvantages, which the owner must consider before making a final system selection.

Barbara Jackson, author of *Construction Management Jump Start*, says "There are basically three project delivery methods: design-bid-build, construction management, and design-build." Jackson goes on to say, "These three project delivery methods differ in five fundamental ways:

- The number of contracts the owner executes
- The relationships and roles of each party to the contract
- The point at which the contractor gets involved in the project
- The ability to overlap design and construction
- Who warrants the sufficiency of the plans and specifications.

Regardless of the project delivery method chosen, the three primary players—the owner, the designer (architect and/or engineer), and the contractor—are always involved."

It is not always easy to determine what project delivery approach is right for a given project. In fact, the single most pressing question in many owners' minds is which is the most appropriate delivery system that will produce the highest quality and most efficient project at the lowest cost and earliest time. To attempt to answer this question, the owner must first define and prioritize how to measure the project's success and choose a project delivery approach that will take the project in that direction. But whichever system is chosen, realistic expectations must be maintained since no project delivery approach is perfect and none can guarantee a perfect project. The project delivery approach that is finally chosen will determine the expected trade-off between the owner's control of the project delivery process and the anticipated risks that come with this decision. This choice also governs the amount of involvement, both in time and expertise, required of the owner to make the project delivery successful from then on.

This has prompted a growing number of owners to engage design and construction professionals as independent advisors to help them make informed decisions and meet these demands. While this owner strategy is known by several different names in the industry, the concept remains the same, i.e., these professionals advise, serve, and represent the owner and have no other interest in the project other than the protection of the owner. It is important here to avoid conflict of interest.

3.6 TRADITIONAL GREEN DESIGN-BID-BUILD PROJECT DELIVERY

The most traditional construction project delivery system today is the design-bid-build (DBB). It has remained the predominately accepted means by which construction projects are developed and delivered. However, this method is somewhat modified and perhaps more complex with the inclusion of green features into the equation.

This traditional delivery method has been the approach of choice for centuries in both public and private construction projects. It remains the project delivery system that is most widely used today and which is still required by some states. The design-bid-build method is well understood by the majority of owners and contractors. With this delivery system, risk is minimized through the owner's control and oversight of both the design and construction phases of the project. The design-bid-build process usually provides the lowest first costs based on submitted tenders, but takes the longest time to execute.

With this traditional project delivery system, the owner contracts separately for the design and construction of the project to a proposed budget. The owner generally contracts directly with a design professional for complete design of the project including contract documents and professional assistance during the bidding stage. The design professional often provides project oversight and continues to administer the construction phase of the project for the owner. This entails reviewing shop drawing submittals, monitoring construction progress, and checking payment requests, as well as answering contractor queries about the construction documents and addressing change order requests. When the plans and specifications (bidding documents) are complete, they are released for bidding and solicitation of tenders to prequalified contractors. Prequalification requires certain information that facilitates the selection of potential constructors. This information includes proof of past experience in similar work, financial capability, a record of exemplary performance by responsible references, and current work load.

Allowing all responsible and qualified contractors to tender on an equal low-bid basis helps eliminate allegations of owner favoritism (whether real or perceived) in the selection process. The design of the project must be complete prior to contractor selection. Once the general contractor is selected (generally through a competitive bid process, which in most cases is the lowest acceptable bidder), the owner enters into a separate contract with the general contractor to build the project. Under the design-bid-build project delivery system, the owner retains responsibility for overall project management and all contracts are generally executed directly with the owner.

Among the advantages of the design-bid-build process is that it provides much needed checks and balances between the design and construction phase of the project. This process is generally perceived to be a fair process for contractor selection. Another advantage from the owner's point of view is the ability to provide significant input into the process throughout the project's design phase. When a lump sum price is committed to between the owner and contractor, the owner can usually rely upon the accuracy of the price and the owner, with the assistance of the consultant designer, is able to compare bids to ensure the best contract price has been obtained. It should be noted that there is no legal agreement between the contractor and the designer of record.

The main disadvantages of the traditional design-bid-build process are that it is a lengthy time-consuming process and the owner often has to address disputes that arise between the contractor and design professionals due to mistakes or other special circumstances. With this process the ultimate estimated cost of construction is unknown until bids are finalized, and the system encourages potential change orders. Moreover, there is generally no contractor buy-in to green processes and concepts. An additional challenge with this system is the possibility that construction bids may exceed the owner's stated budget (because plans and specifications are completed prior to tendering the project), the consequence of which is either abandoning the project or redesigning it. Moreover, the owner is required to make a significant financial commitment up front in order to have a complete design in hand as part of the contract documents before solicitation of tenders.

Petina Killiany, associate vice president of PinnacleOne, a leading construction consulting firm, says that the design-bid-build approach is generally best suited for projects where:

- The owner desires the protection of a well-understood design and construction process
- The owner desires the lowest price on a competitive bid basis for known quantity and quality of the project
- The owner has the time to invest in a linear, sequential, design-bid-build process
- The owner needs total design control.

According to Killiany, the project success factors that owners sacrifice when using the design-bid-build approach are, "First, because there is no input from the contractor during the design phase, their input is lost on what may provide the best value in the trade-off between scope and quality. The construction contract is usually performed on a lump sum basis, any savings are not returned to the owner. Design-bid-build projects normally do not allow for fast track design and construction, and as a result, can take more time than those delivered by other approaches."

It should be noted that should gaps be discovered between the plans and specifications and the owner's requirements, or errors and omissions are found in the design, it is the owner's responsibility to pay for correcting these errors.

3.7 GREEN CONSTRUCTION MANAGEMENT

According to the ASHRAE Green Guide, "The construction manager method is the process undertaken by public and private owners in which a firm with extensive experience in construction management and general contracting is hired during the design phase of the project to assess project capital costs and constructability issues." This project delivery system is a process by which a "construction manager" is added to the construction team to oversee some or the whole project independent of the construction work itself. The construction manager's role can be to oversee scheduling, cost control, the construction process, safety, the CxA, bidding, or the complete project.

Joseph Hardesty of Stites & Harbison, PLLC says, "In many ways, the construction management process is not, by itself, a separate construction delivery system, but is a resource the owner can use to assist in the construction project. The added cost of a construction manager must be weighed against the benefits this consultant brings to the project. Often, the architect can fulfill the role provided by a construction manager. However, depending upon the degree of sophistication of the owner's in-house construction staff, and depending upon the complexity of the project, a construction manager can provide an essential element to the construction project."

Hardesty goes on to say, "A construction manager is most useful on a large, complex project which requires a good deal of oversight and coordination. A construction manager is also helpful to an owner who does not have a sophisticated in-house construction team. A construction manager can help the owner control costs and avoid delays on complex projects."

It should be noted here that there are two basic types of construction management to consider under this method, including: 1. The agency CM, and 2. The at-risk CM (sometimes called CM/GC).

1. The *agency* CM is where the construction manager acts as advisor to the owner on a fee-based service and where the owner separately commissions the general contractor and design of record. With this method the construction manger acts as an extension of the owner's staff and assumes little risk except for that involved in fulfilling its advisory responsibilities. With this method the general contractor remains responsible for the construction work and still carries out construction management functions relative to his/her internal requirements for managing the project to completion. However, the agency CM is not at risk for the budget, the schedule, or the project's performance.
2. The *at-risk* CM delivery approach is not very different from the traditional design-bid-build method in that the construction manager (CM) replaces the general contractor in this scenario during the construction phase and commits to delivering the project on time and within a guaranteed maximum price (GMP). The CM holds the risk of subletting the construction work to trade subcontractors and guaranteeing completion of the project for a fixed price negotiated at some point either during or upon completion of the design process. However, unlike design-bid-build, during the development and design phases the at-risk CM acts mainly as a consultant to the owner.

The owner should weigh the relative advantages and disadvantages of each construction delivery system prior to beginning the project. Petina Killiany lists some of the advantages to an at-risk CM over design-bid-build:

- Because construction can often begin before the design is complete, the overall project duration can be shorter.
- The owner generally gets better estimates of the ultimate cost of the project during all phases of the project.

- The owner benefits from a contractor perspective in making decisions on the trade-offs during the design phase between cost, quality, and construction duration.
- Constructability and design reviews by the contractor prior to bidding often result in better designs and lower trade contractor contingencies and bids.
- The expertise of the construction manager in pre-qualifying trade contractors helps to achieve better performance and workmanship by the trades.
- The architect and contractor working together during the design portion can result in a better team effort after the GMP is established.

Killiany adds that a major disadvantage of the at-risk CM approach relative to design-bid-build is that it may not be permitted by statute to a public owner. Additionally, some owners may not fully understand how to successfully implement this method because, not being the traditional method, they may rely on the advice of the CM when they should in fact be questioning it.

In any case, the owner should consider the size and complexity of the project, the relative importance of cost or schedule, and the in-house expertise the owner has to manage the project before deciding whether this delivery method is suitable or not.

When the project management/CM is engaged in an advisory capacity the service is totally different. For while project owners can't get around risks, it is possible to mitigate them to an acceptable level. Richard Sitnik, a senior project manager with Pinnacle One, says, "When given appropriate responsibility and the ability to provide effective leadership, project managers/CMs as advisor promote project success through informed, experience-based decision making, and well-disciplined and regimented project controls." Sitnik also says that the project manager/CM as advisor can provide a wide range of services to the owner throughout the design, bidding, negotiation, and construction phase of the project. Following are some of the more pertinent services outlined by Sitnik:

- Perform needs assessments
- Provide direction on alternate project delivery systems
- Assist in the selection of appropriately qualified consultants
- Manage governmental agency approvals
- Identify and manage risks
- Anticipate potential problems before they become costly
- Produce master budgets and schedules
- Establish project controls
- Control costs
- Perform quality controls.

A project manager/CM as advisor's maximum value occurs when he/she is engaged very early in the design process to begin establishing controls, including budgets and master schedules. This contributes to greater design efficiency, with fewer change orders in the field and less likelihood of surprises to the owner on bid day. The principal role of the project management/CM as advisor is to prevent delays, cost overruns, and a failure to meet project objectives. This can be achieved by basically providing the owner with total support and impartial advice and counsel, and guiding the owner to make informed decisions, without comprising the ability to coordinate the multiple agendas and sometimes conflicting interests of the design professionals, contractors, and owners.

Sometimes the term "program management" is used which is essentially the same service as project management/CM as advisor with the distinction that program management is the name applied when used on large, complex, and multi-project programs. The benefit to the owner in having a program manager is the expertise and experience these firms have obtained. The program manager can assist the project owner in developing an overall strategy to manage projects within the program, as each building project may differ in its requirements and how it is constructed. When the program manager is overseeing a process consisting of more than one building, the various roles may be allocated in a manner so that, e.g., one building/project may include an architect, general contractor, and a project manager/CM as advisor, while another project in the program may be constructed by a design-builder or at-risk CM.

But whatever the case, construction projects are always likely to contain some risk. Employing a program manager and/or a project manager/CM as advisor should be seriously considered for large projects, particularly in cases where owners are faced with the risks and responsibilities of choosing and implementing project delivery approaches, but lack appropriate in-house technical capability or who need an increase in staff when timeframes restrict the use of internal staff.

3.8 GREEN DESIGN-BUILD PROJECT DELIVERY

The Design-Build Institute of America (DBIA) describes this method as "an integrated delivery process that has been embraced by the world's great civilizations. In ancient Mesopotamia, the Code of Hammurabi (1800 BC) fixed absolute accountability upon master builders for both design and construction. In the succeeding millennia, projects ranging from cathedrals to cable-stayed bridges, from cloisters to corporate headquarters, have been conceived and constructed using the paradigm of design-build."

A distinguishing feature of the design-build approach is that there is only one contract, i.e., the owner contracts with one entity (the designer/builder) that will assume responsibility for the entire project, i.e., its design, supervision, construction, and final delivery. The selection process usually consists of soliciting qualifications and price proposals from various design/builders, usually teams of contractors and designers, before or during the conceptual design phase of the project. The design-build team is generally led by a contractor (often with a background in engineering or architecture), resulting in the owner issuing a single contract agreement to the contractor, who in turn contracts with a designer for the design. According to Killiany, design-build when allowed is generally suited for projects where:

- The owner is willing to forego control of design and does not seek a highly complex design program/solution
- The owner can provide a complete definitive set of performance specifications and program for design for the design/builder to serve as the basis for the design/builder's proposal and the owner's contract with the design/builder
- The owner has realistic expectations for the end-product and a thorough understanding of risk giving up the control of the design
- The owner desires a fast delivery method and is willing to compensate the design-build team for its assumption of risk for design and construction.

3.8.1 The Basics of the Design-Build Process

In the construction industry, design-build is gaining popularity as the project delivery system of choice (Figure 3.7). Many owners prefer the design-build project delivery system to the design-bid-build system because it provides a single point of responsibility for design and construction rather than contracting separately for the design phase and then for the construction with two separate entities. Although it has the advantage of removing the owner from contractor and design disputes, it has the disadvantage of eliminating some of the checks and balances that often occur when the design and construction phase are contracted separately. Additionally, the owner loses much of the project control that exists under a design-bid-build process. The owner also loses the owner/architect advisory relationship that exists in the design-bid-build process.

Having to interact with a single entity has obvious advantages for the owner, such as easier co-ordination and more efficient time management. The design-build contractor or firm will endeavor to streamline the entire design process, construction planning, obtaining permits, etc. Furthermore, with the design-build process activities can easily

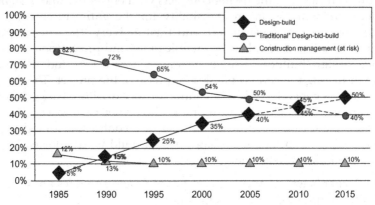

FIGURE 3.7 A graph showing the rising popularity of design-build in non-residential construction in the United States over the years. (Source: Design Build Institute of America.)

overlap—construction on parts of the project can begin even while the design is not finalized. At times, the main contractor may involve other organizations on the project with him, but in such cases too, the contractor will be the one dealing with them and assume responsibility. This overlapping offers flexibility to make changes to the design while construction is in process. With the traditional design-bid-build system, this isn't possible, since construction can begin only once the blueprints are finalized.

3.8.2 The Advantages and Disadvantages of Design-Build

There are many potential advantages for the numerous parties involved in a design-build contract, especially if all the parties understand the mechanics of the process as it applies to their project. Kenneth Strong and Charles Juliana of Gordon and Rees, LLP list some of the advantages and disadvantages below.

A. Advantages

1. Time savings: By combining the selection of a designer and a contractor into one step, the design-build method eliminates the time lost in the DBB process. Further, the design-build contractor is able to start construction before the entire design is completed. For instance, the design-build contractor can start excavation as soon as the foundation and utility relocation design has been prepared.

Meanwhile, the design professional can continue design work for the rest of the project during excavation.

2. Cost savings: Potential costs savings can be realized with the design-build system because it has high value engineering capabilities due to the close coordination between the A/E and construction contractor. Construction contractors have direct and real experience with the cost of purchasing and installing materials and, in the design-build system, can share that experience directly with the design professional during the design phase of the project. This process has the potential to translate into lower costs, which savings can then be passed on to the owner.

3. One point of contact: The one point of contact feature for both design and construction is integral to the design-build system. The advantages of this feature are relative—having only one entity to deal with in many instances will outweigh the oversight benefits an owner would otherwise get from contracting separately with a design professional for the project design.

4. Fewer change orders: A definite advantage of the design-build system is that an owner can expect far fewer change orders on a design-build project. However, if an owner decides it wants a design change during the design-build project, and, that change is not covered by the defined scope of the project, that would be considered an extra. Still, in the design-build system, the owner is not liable for any errors the design professional makes because the design professional is part of the design-build team.

5. Reduced risk to the owner: The shifting of liability for design quality from the owner to the design-build contractor is one of the most significant features of the design-build project delivery system. The advantage to the owner is that the cost of that risk is known from the outset. As the design-build contractor is in a better position than the owner to manage and minimize that risk, this is a significant advantage of design-build contracting.

B. *Possible Disadvantages to Using the Design-Build Method*

1. Loss of control of project design: In the design-build system, the shift in responsibility for the design from the owner to the contractor implicitly includes some shift in control. The owner should evaluate the degree to which this loss of control will affect the success of the project. If the owner has specific needs or requirements, they must be clearly articulated in defining the scope of work, or the owner must accept the risk of paying extra to get what is wanted via the change order process. Change orders issued to revise scope are not inherently less likely or less expensive in the design-build project delivery method.

2. Less project oversight/control of quality: As has been discussed, one of the advantages of the design-build concept is the cooperation between the design professional and the construction contractor because they both are part of the same team: the design-build contractor. However, this feature can also be a disadvantage, as the architect is no longer the owner's independent consultant and is now working with and for the contractor. For owners who do not have their own design-proficient staff, the loss of the architect's input and judgment may expose them to quality control problems. If the owner is one that is used to having the design professional act as its agent, plans should include another entity to take that responsibility.

3. Suitability of design-build teams: In the DBB methodology, while public agencies are bound by state law to hire the lowest responsive, responsible bidder for construction work, they have more flexibility in selecting designers for their projects. In other words, DBB public owners are allowed to take into account in the selection of a designer more than simply which candidate offered the lowest price. In design-build, the public owner loses the latitude it had in DBB in selecting a design firm. True, the risk for adequacy of the design has been shifted to the design-build contractor, but that is little solace to an owner if the finished project is structurally sound but operationally deficient.

In addition to the above, there are other possible disadvantages, such as difficulty in pricing the work. It is difficult to establish a firm price for the project if the design is incomplete, which often reflects the situation when the design-build organization is selected. Costly tendering is another issue. Owners are expected to pay for the efforts by design-build organizations to formulate their tenders which normally may include preliminary design work to be able to estimate prices.

3.8.3 When to Consider Design-Build for a Project

When evaluating whether the design-build methodology would be appropriate for a given project, the following factors should be considered:

- Design-build is an appropriate project delivery system for projects that need to be completed within a short time frame.
- The design-build system offers several cost saving benefits in terms of the budget. Cost savings can be achieved by shifting more cost control responsibility to the contractor. For example, a construction contractor may wish to use certain materials and methods that meet the owner's requirements but were not considered by the designer. Any potential cost savings accrued from the contractor's proposed modifications should be passed on to the owner.

- Perhaps the most significant factor to impact the delivery system selection is the type of project to be constructed. An appropriate candidate for design-build is a project where the performance and form of the finished project is properly described in a scope document. However, in a project where the owner's needs are very specific and specialized, design-build may not be the best method to adopt.

3.8.4 AIA Design-Build Documents

Over the last couple of decades, the design-build project delivery system has steadily increased dramatically in popularity on vertical construction projects. Project owners and contractors, however, have displayed rising concern that the standard AIA forms of agreement for design-build projects do not adequately address their needs. In direct response to these concerns, the American Institute of Architects (AIA) made a commitment to completely overhaul the design-build forms of agreement. The overhaul resulted in the introduction of five completely new forms of agreement and the retirement of the 1996 series (the A191, A491 and the B901).

The new agreements include the A141-2004 Agreement Between Owner and Design-Builder (replaces A191-1996); the A142-2004 Agreement Between Design-Builder and Contractor (replaces A491-1996); the B142-2004 Agreement Between Owner and Consultant where the owner contemplates using the design-build method of project delivery (no 1996 counterpart); the B143-2004 Agreement Between Design-Builder and Architect (replaces B901-1996); the G704-2004 Acknowledgment of Substantial Completion of a Design-Build Project (no 1996 counterpart). The 2004 design-build documents use one agreement for both design and construction, with three possible methods of payment available to the parties, including:

- Stipulated sum
- Cost plus a fee with a guaranteed maximum price
- Cost plus a fee without a guaranteed maximum price.

While the design-build form of delivery system has a number of advantages, owners have nevertheless come to realize that with this system they have less overall control in guaranteeing that the owner's "intent" is clearly articulated and this has been a cause for great concern. Out of this concern emerged a concept known as "bridging," which is defined as the owner's means of conveying its intent to the design-build team. Bridging can take on various forms so that the owner can have a more expansive role and have much more say in the design, or alternatively the owner can assume a more limited role, and simply set forth its intent in a more conceptual manner.

Green Construction Cost Monitoring

4.1 GENERAL

Lenders, owners, contractors, architects/engineers, and material or equipment suppliers are just a few of the players that are involved and participate in the construction of facilities and infrastructure. Lenders are usually presented with conceptual designs and specifications, proformas, construction cost estimates, etc. by the borrower, who is typically the project owner for the primary purpose of providing enough information to be able to make a loan determination. The need for a professional objective review of construction loan commitments and payment requests is obvious, if the lender is to be fully protected.

The construction consultant is hired by the lending institution to examine the conceptual design and specifications or contract documents for engineering soundness and compliance with governmental regulations, and make an assessment of cost comparables for similar projects, in addition to a trade-by-trade breakdown. This estimate is then compared with the borrower's estimate for general agreement.

Consultants should take note that the lender's interest in each property is subject to rights and restrictions stated and articulated in the loan documents. In the case of new construction, the consultant is usually hired by the lender prior to commencement of the project and basically has the responsibility of administering the project to completion. The consultant therefore assumes that satisfactory access to the property, staff, vendors, and documents will be provided by the borrower. In the event the borrower fails to cooperate, the lender will apply its leverage to assist the administrator in securing access and all information necessary to monitor the project and to protect the lender's rights. In no event should the lender's representative seek access to any property, staff, vendor, or documents if the borrower refuses such access or attempts to restrict the lender's representative from performing its contractual duties.

Project Evaluation and Analysis

Upon establishing that the borrower's estimated costs are in line with typical local costs, the consultant proceeds to prepare a comprehensive review of the project plans and specifications to assure the lender that the design is in compliance with good engineering practice. A detailed written report is submitted to the lender describing important aspects of the project and commenting on the following items:

- Completeness of plans, specifications, and related information and their conformance with all applicable building codes and zoning ordinances
- Design of architectural, structural, HVAC, electrical, plumbing, fire-protection systems; elevators, site improvements, and other relevant information
- Borrower's itemized trades cost breakdown
- Soil borings contents, load tests, engineering reports, and environmental impact studies
- Areas of potential complications, which would become a problem to the lender
- Architectural and engineering agreements, material and construction contracts for completeness, function, responsibility, and costs
- Conformity of materials specified with the project's overall quality objectives
- Conformity of project scope and design as outlined in the plans and specifications and the project description as set forth in the loan agreement
- Achieving borrower's projected construction commencement and completion dates.

4.2 FRONT-END ANALYSIS

The lender often requests the consultant to perform a one-time front-end analysis consisting of the following services, which are summarized in the project analysis report (PAR) prepared for the lender.

- A construction documents review
- A construction costs review
- A pre-closing construction progress inspection
- A pre-closing site inspection
- Attendance at the lender's pre-construction meeting.

All front-end analysis work is documented together in a separate report, dated and signed by the consultant(s) performing the front-end analysis. The original front-end analysis report is typically delivered to the lender within roughly three weeks following receipt of the required documents and notice to proceed from the lender.

Construction Documents Review

Two complete, half size, sealed and signed sets of the plans and specs itemized below are forwarded to the consultant. Plans are to be an exact duplicate set of those submitted to the department of buildings and the lender. A list of the drawings from the architect's office should accompany the drawings forwarded to the lender's representative, which references each drawing as to the date of preparation and last revision date. If one of the sets is not stamped "approved" by the building department, a letter should then accompany the plans from the architect's office confirming that the documents are an exact duplicate set of those approved by the building department. Copies of all revised drawings (with revisions indicated) and specification addenda should be forwarded as issued. One complete set of existing building plans and specs (as-builts if available) should also be forwarded. Documents to include:

- Site plans and off-site plans, if any
- Landscape plans
- Zoning sheets
- Architectural and interior design
- Structural plans and calculations
- HVAC
- Electrical
- Plumbing
- Fire protection
- Specifications/project manual
- Parking structure plans and specifications.

The exchange of information required by the construction consultant to begin the loan monitoring includes the following:

Letter of Agreement A letter of agreement between the construction consultant and the lender outlining the extent of services to be performed, fee rate, and method of payment.

Plans and Specifications The construction consultant should review the architect or designer's plans and specifications to confirm that the lender's intended understanding of value will be translated to the contractor. At the site, the construction consultant will confirm that materials specified for the project are in fact used and any modifications are appropriately documented for the protection of the parties.

The Owner's Agreements with Contractors   It is important that the construction consultant is familiar with all the parties that are involved in the project. It should also be noted that there may be other work proceeding in the project that differs from the owner/contractor's agreement listed with the lender; these should be taken into account so that the lender may understand potential liabilities on the project. The AIA® Document A101™ standard form of agreement between the owner and contractor is widely used, where the basis of payment is a stipulated sum.

Survey   The instrument survey should be completed at the appropriate time and submitted to the construction consultant for verification of compliance with zoning requirements.

Title Report/Deed   The construction consultant should review the title report and take note of any special restrictions or conditions that may be placed on the property and confirm that the project conforms to these restrictions.

Contractor's Schedule of Costs   Development of an accurately broken down contractor's schedule of costs into significantly small items will avoid overpayment to the contractor and is one of the most important tasks of the construction consultant. This will also be the data used to determine the value of draws against completed work in place to date. This schedule is intended to work hand-in-hand with the construction schedule.

Confirmation of Utilities   The construction consultant should confirm prior to the release of the financial commitment letter that specified utilities are available to the property or that the contractor has made alternative arrangements.

Building Permit   Before building construction is permitted to commence, confirmation that a full building permit was issued should be confirmed. Sometimes partial building permits are issued and can present a great deal of difficulty for the lender, and therefore, it is important to fully investigate the reasons for partial permit. Changes or conditions requested by the municipality should also be noted.

Release from Special Entities   Releases from special entities should be confirmed before construction begins. This might include special approvals from design review boards, curb cut permits, etc.

4.3 REQUISITION FORMAT

The majority of lenders have developed their own requisition form or format that they would prefer for the borrower to use. The lender should be consulted to see whether such a form exists.

The requisition should be put together with the line items organized in the AIA G702 format and with as many line items as reasonably possible. Where line items contain more than one trade or work scope they should be broken down into the individual subcontracts that will be awarded for same. All of the sub-contractual costs are to be subtotaled prior to adding general conditions, a builder's or developer's fee, and the contingency line items.

With respect to AIA document G702 Application and Certificate for Payment, the AIA says, "The forms require the contractor to show the status of the contract sum to date, including the total dollar amount of the work completed and stored to date, the amount of retainage (if any), the total of previous payments, a summary of change orders, and the amount of current payment requested. G703–1992, Continuation Sheet, breaks the contract sum into portions of the work in accordance with a schedule of values prepared by the contractor as required by the general conditions. (NOTE: The AIA does not publish a standard schedule of values form.) G702–1992 serves as both the contractor's application and the architect's certification. Its use can expedite payment and reduce the possibility of error. If the application is properly completed and acceptable to the architect, the architect's signature certifies to the owner that a payment in the amount indicated is due to the contractor. The form also allows the architect to certify an amount different than the amount applied for, with explanation provided by the architect."

When reviewing the payment requisition the project architect/administrator (the "administrator") is authorized by the lender only to approve funds commensurate with the value of work-in-place at the time of the site visit. The administrator will not approve projected or anticipated values of completion. Figure 4.1 is a sample letter addressed to the lender bank confirming the procedure and monthly document requirements to allow disbursement of funds.

The requisition columns for related work completed during the period covered and the columns relating to the total value of work completed to date should be completely penciled-in prior to the site visit. The requisition should also include the architect-of-record's sign-off.

To support the value of work completed to date for the various line items, the borrower (project owner) submits the subcontractor's schedule of values, prepared by all of the subcontractors for the consultant's review and approval. These schedules should be previously reviewed and approved by the borrower and his CM or GC to guard against front-loading prior to the consultant receiving them. Lump sum amount subcontractor invoices (although necessary) on their own are not deemed adequate to establish value of work completed.

The consultant should be satisfied that the amount requested accurately reflects the value of work-in-place, and also that the line item has a sufficient balance available to cover completion of the remaining work. And unless otherwise directed by the lender, monies and percentages of completion approved will be based solely upon the actual subcontract amount, and not the line item's original budgeted amount. Should there be a buyout savings, this "savings" is to be allocated to the requisition's

Name   Date
Title
Name of Bank
Street Address
City, State Zip
(__) __ - __ (tel)
(__) __ - __ (fax)
Email _____

Re: ABC Project No._____
        Loan Disbursement Requirements
        Name of Borrower
        Name of Project

Dear   :

Attached is a copy of ABC's "Monthly Documentation Requirements".

Please review, revise, and fill in the blanks so as to comply with Name of Bank's disbursement procedure requirements/policies and this project's construction loan agreement (CLA). Your attention is specifically directed to Item Nos. 4, 8, 9, 10, 11, 12, 13, and 14.

Please fax an edited copy of this schedule to this office and ABC will then revise the attached schedule and send same to the borrower as per your requirements.

Please do not hesitate to call me should you have any questions at ext. _____.

Sincerely,

ABC GREEN INTERNATIONAL, INC.

Name
Project Manager

___/_____
enclosure

SK:\PMO\CHECKLST\MONTHLY.LTR

FIGURE 4.1   A sample letter to the lender is shown to approve and comment upon the monthly document requirements from the project owner (borrower) to facilitate normal disbursement procedure requirements.

MONTHLY DOCUMENTATION REQUIREMENTS
Construction Loan Monitoring Services

In accordance with the Construction Loan Agreement (CLA), the following items are to be provided by the borrower to ABC Green International prior to each monthly advance:

1. **Subcontracts & Purchase Orders**	All executed subcontracts and purchase orders greater than $_____ or for which the aggregate cost of services, work, or materials will exceed $_____ .
2. **Payment & Performance Bonds**	Copies of executed bonds are to be provided from the GC or CM and all subcontractors as required by the contract and loan agreements prior to releasing payment for same.
3. **Trade Payment Breakdowns ("TPB")**	Fully detailed TPBs are to be submitted for every subcontract and purchase order. All TPBs are to be approved by the borrower or the CM, should a CM be used.
4. **Revised Budgets**	Any revised direct cost construction budgets or cost-to-complete budgets.
5. **Monthly Job Cost Report**	This schedule will typically indicate: the original budget; buyout status to date by showing actual subcontract amounts; scope changes prior to subcontract; approved, pending, and anticipated change orders; total value of work in place; balance to completion, and the total anticipated project cost. This spread sheet report is to be prepared by the borrower if a GC is used or by the CM and then approved by the borrower.
6. **Payment Requisition**	A direct cost payment requisition prepared in the format of AIA Document No. G702 with the following modifications: line item amounts are to reflect actual subcontract amounts; work completed is to be on a cost incurred basis—not on a percentage of completion; and stored materials are not to be included in the value of work completed, but tracked on a separate Stored Material Inventory Control Schedule. Retainage is to be withheld as per the CLA requirements. The requisition is to be signed-off by the designer-of-record.

FIGURE 4.1   (*Continued*)

MONTHLY PROJECT STATUS REQUIREMENTS—continued

7. Partial & Full Waivers of Lien	Partial waivers are to accompany each subcontractor requisition for subcontracts greater than $_____. Conditional final waivers are to be obtained from each subcontractor and the GC or CM prior to final payment being released.
8. Change Orders	Copies of all change orders complete with sign-off by the architect or engineer-of-record should the change order result in a deviation from the lender's set of approved contract drawings and/or specifications.
9. Change Order Schedule	A separate schedule identifying each change order, the subcontractor, the add or deduct amount, the date received, and whether it has been approved or if it is pending.
10. Stored Material Documentation	Not all lenders allow funding of stored materials. If funding for stored materials is permitted, review the CLA requirements for the caps on a single stored item and the aggregate dollar amount of materials to be funded at any given time. Consult the CLA to determine if corresponding proportional amounts of general conditions and developer's fee are permitted to be released for stored materials prior to the materials actually being installed. Also, determine whether retainage is to be withheld for stored materials. Typical documentation required by the lender in order to receive funding consists of: • Stored Material Inventory Control Schedule • Subcontracts • Bills of sale • UCC statements • Insurance certificates • Inspection and acceptance certificates
11. General Conditions	General conditions are to be funded on a cost-incurred basis, but the amount disbursed is not to exceed the direct cost percentage of completion. Check to determine whether the CLA permits monthly equal payment funding.
12. Retainage	Check to determine whether the CLA requires retainage on such items as stored materials, general conditions; the CM fee, and the developer's fee.

FIGURE 4.1 *(Continued)*

MONTHLY PROJECT STATUS REQUIREMENTS - continued

13. Shop Drawing Documentation	Not all lenders fund for shop drawings; check the CLA. If funding is permitted, have the designer-of-record provide a statement that the shop drawings prepared have been approved. An approved shop drawing invoice complete with back-up documentation for the time expended is also usually required.
14. Trade Mobilization Cost	Not all lenders fund for trade mobilization costs; check the CLA. Some lenders will fund half the up-front mobilization costs leaving the half to be paid upon completion of "de-mobilization" costs. Cost breakdown documentation for the basis of this invoice is usually required.
15. Designer-of-Record's Field Observation Reports	On a weekly basis, all Field Observation Reports are to be forwarded to ABC. Please have ABC placed on the routing list.
16. Monthly Compliance Certification	Certification by the appropriate designers-of-record that the work completed during the payment requisition period was performed in compliance with the lender's set of approved drawings and specifications.
17. As-Built Foundation Plan	Upon completion of the foundation, an as-built foundation survey certified to the bank is to beprepared by a licensed surveyor and accompanied by a statement from the designer-of-record that the foundation was constructed within the required setbacks and to the correct elevations. With respect to a pile foundation, acertified as-driven pile location plan andcertified pile log will be required prior to release of funds in addition to an as-built foundation survey.
18. Construction Schedule	All construction schedule revisions.
19. Testing Reports	Copies of all testing reports. Testing results which deviate from the design requirements should be accompanied with a copy of the designer-of-record's response to same. Please have ABC placed on the routing list.
20. Work Log	A trade-by-trade man-day count synopsis for the payment requisition period.
21. Photos	Copies of prints, if monthly construction progress photos are being taken.
22. Certificates of Occupancy	Whether temporary or permanent, as the space or units are accepted by the local governmental authority.

FIGURE 4.1   (*Continued*)

contingency budget. Ideally, the consultant/administrator typically tries to reach agreement on the value of work-in-place prior to leaving the job site.

Monthly Job Cost Reports

On a monthly basis, or as required by the lender, the borrower should provide a job cost report (JCR) that will detail such information as:

- Actual contract or purchase order costs compared to original budget's line items
- Total of change orders (approved and pending)
- Total estimated project cost

The value of the job cost report is that it provides timely information about the status of the project budget and allows the administrator to take whatever action may be necessary to bring the project into compliance with the original budget. The JCR will essentially provide for each item, the quantity and/or percentage completed to date, the cost of the item to date, the estimated cost remaining to complete the item, and the total cost estimate for the item at completion. The sum totals of these costs are calculated to arrive at an estimate for the entire project.

4.4 SITE VISITS AND OBSERVATIONS

4.4.1 Design Kickoff and Lender's Pre-Construction Meeting

One of the first steps in the design phase is a design kickoff or orientation meeting. This is normally scheduled by the project manager at or about the time that the contract with the design professional is executed and approved, depending upon the project. The main attendees at this meeting generally include the main stakeholders and participants with an interest in the project, including the design team, cost estimator, commissioning agent, and owner. The agenda generally includes introduction of personnel involved in the project, discussion of administrative procedures, discussion of project scope, budget and schedule, and a site visit and walk-through. The design professional/project manager should record attendance and prepare and distribute an agenda and minutes of the design kickoff meeting. The design professional should also prepare a project directory of all participants, including name, title, address, phone, fax, and e-mail address.

Once the project is tendered and following acceptance of the successful bid and subsequent award of the construction contract, the owner, contractor, and lender's representative (consultant/administrator) shall be required to schedule and attend a pre-construction meeting, the purpose of which is to discuss the various requirements of the contract documents and how they relate to the daily operation of the construction project, as well as the lines of authority and communication (Figure 4.2).

Attendance at the lender's pre-construction meeting shall be required unless notified otherwise by the lender. The lender shall coordinate and

AGENDA

Germantown Housing Project
Project Kick-off Meeting

ABC Project No. A30115   **Date: February, 7, 2009**

1. **Project Overview**

   a. What is the development plan?

2. **Project Team**

   a. Who comprises the project team? What is the project extent of involvement?

   b. Table of Organization.

3. **Plans & Specifications**

   a. Are the base building plans and specifications complete?

   b. Any part of the project being completed on a design/build basis?

4. **Construction Period**

   a. When is the projected start and completion dates?

   b. Project duration?

   c. Schedule updates and who will prepare them?

5. **Project Budget and Contracts**

   a. Direct cost budget and what are the components? Is there a detailed direct cost estimate and breakdown; when last updated?

   b. Copies of subcontracts, purchase orders, etc.

   c. What is timetable on Owner/GC/CM Agreement?

   d. Any trade payment breakdowns available?

   e. Any work to be performed with GC's own forces? If so, what are the trades? What percentage of the entire contract amount does such work represent?

   f. What percentage of the direct cost has been secured by subcontract pricing to date?

6.   Existing Building Condition Survey Reports

   a. Envelope/Structural Report

   b. MEP Systems

7.   Permit Status

   a. Are permits issued as a whole or in stages?

   b. Status of documentation review by municipality.

   c. Schedule of necessary permits.

8.   Bonds

   a. Will the GC/CM, if any be securing a 100% Performance and Payment Bond for the project?

   b. Will subcontractors be bonded? If so, who?

9.   Architect/Engineer Field Inspection Reports and Punchlists

   Will architect/engineer field inspections be conducted, at what frequency, and will reports and punchlists be prepared?

10.   Quality Assurance Program

   What type of program is in place for quality assurance in addition to the municipality required controlled testing?

11.   Change Orders

   a. That is the procedure for change order approval?

   b. A separate schedule is requested identifying each change order, the subcontractor, the add or deduct amount or scope change, the date received, and whether it has been approved or is pending review.

FIGURE 4.2   A typical agenda for a pre-construction kickoff meeting. The actual agenda will depend on the type of project, project requirements, and circumstances.

establish the date, time, and place the meeting will be held. The purpose of this meeting is to meet collectively with all parties to the construction project (including the owner's representative, lender's representative, architect, primary engineers, contractor's project manager, and supervisory staff, as well as major subcontractors and vendors) and discuss the status of the work, construction documents, and contractual relationships, and the lender's draw procedures and requirements. Sometimes certain outside agencies may be called in to attend the initial meeting (such as fire marshals and public utility personnel).

The general agenda for this meeting usually covers all of the items listed in the general conditions and supplemental conditions of the contract but in greater detail. The preconstruction meeting offers an opportunity for the main participants to be introduced and get to know one another. Some of the more common issues discussed during the pre-construction meeting are:

- Introduction, individual roles, and accountabilities
- Mobilization and site logistics (site access and security, temporary utilities, temporary facilities)
- Construction coordination issues (RFIs, subcontracts, submittals, shop drawings)
- Schedule issues (notice to proceed, work schedule, sequence of work, liquidated damages)
- Payment issues (application for payment, schedules of value)
- Change orders and additional work
- Dispute issues
- Completion procedures (substantial completion, final inspection, final punch-list, final waivers of lien, final payment)

4.4.2 Preconstruction Documents

Following the preconstruction meeting and prior to commencement of construction the contract administrator or owner and/or architect will ensure that certain documents have been executed between the owner and the contractor, including but not limited to:

- "Notice of commencement" from owner
- "Notice to proceed" from the owner
- Property survey from the owner
- All required permits, licenses, and governmental approvals
- Insurance coverage to be carried by the contractor and all subcontractors
- Bonds—contractor's copies of its performance and payment bonds in addition to proof that subcontractors have furnished surety bonds as required by the contract documents

Notice of Commencement

This is a legal document prepared by the owner's attorney or financial lending institution and recorded with the clerk of the county court. The owner is required to have it recorded and a copy must be posted by the owner at the project site. Financing institutions will require the filing of a notice of commencement as a provision of the loan agreement. The owner, administrator, and/or architect should obtain a photocopy of the notice of commencement for his project files.

The notice of commencement is considered to be one of the most important documents on a construction project and is the first document filed for the lien process, yet its importance is frequently overlooked by contractors.

Failure to pay attention to the notice of commencement can have serious consequences and adversely affect a contractor's ability to recover for the work performed on a project. There are three main issues that contractors need to pay attention to regarding their project's notice of commencement, these are:

- When does the notice of commencement expire?
- Was the bond attached to it?
- When was the notice of commencement recorded?

By posting the notice of commencement at the project site and on public record the name of the owner, the contractor, and surety are provided, so that anyone wishing to file notice to the owner may do so. Any work executed at the project site prior to the notice of commencement is not covered by the lien laws.

Although the notice of commencement is the owner's responsibility, the owner, administrator, and/or architect should so advise the owner of both the need and benefits of such a document. The owner/contractor agreement should stipulate that work is not to commence until the notice of commencement has been issued. Figure 4.3 is an example of a notice to proceed issued by the State of Florida.

Notice to Proceed

The notice to proceed is the document that certifies the contractor of the acceptance of his proposal and officially directs the contractor to commence work within a specified time, such as 10 business days. Work to be executed under the owner/contractor agreement generally begins on the date specified in the notice to proceed document, and as articulated in the general conditions of the contract for construction. The notice to proceed also triggers the project commencement date by establishing the reference date from which the project duration is measured; often the contract will stipulate that work is to be completed within a stated number of calendar days after the contractor receives its notice to proceed.

It is considered good practice for the owner to notify unsuccessful tenderers. The notice to proceed implies that the site is free of encumbrances and therefore is available for the contractor's use. However, if there are unresolved issues, then the owner may issue a letter of intent stating that its intention to contract with the contractors upon resolving outstanding issues.

The owner and/or architect shall recognize that the "date of commencement" is the official date for the start of the construction project and is specifically identified in the notice to proceed. It would often be difficult however for the contractor to start work on the very date the notice to proceed is issued unless the contractor has had adequate time to mobilize his resources, project team and equipment to the site.

NOTICE OF COMMENCEMENT

STATE OF FLORIDA   Permit No. _____
CHARLOTTE COUNTY   Tax Folio No. _____

The Undersigned hereby gives notice that improvements will be made to real property and in accordance with Section 713.13 of the Florida Statutes, the following information is provided in this NOTICE OF COMMENCEMENT.

Legal Description of property (*include street address, if available*):
_____
_____
_____
_____

This Space Reserved for Recording

General Description of improvements: _____

Owner: _____
Address: _____
Owners interest in the site of improvement: _____

Fee simple title holder (*if other than owner*): _____
Address: _____

Contractor: _____
Address: _____

Surety: _____
Address: _____ Amount of Bond: _____

Any person making a loan for the construction of the improvements:
Name: _____
Address: _____

Person within the State of Florida designated by owner whom notices or other documents may be served as provided by Section 713.13(1)(a)7., Florida Statutes.
Name: _____
Address: _____

In addition to himself, owner designates _____ of _____
_____ to receive a copy of Lienor's notice as provided in Section 713.13(1)(b), Florida Statutes.
Name: _____
Address: _____

Expiration date of Notice of Commencement (*the expiration date is one year from the date of recording unless a different date is specified*). _____

_____
Signature of Owner

_____
Printed Name of Owner

The foregoing instrument was acknowledged before me this _____ day of _____ 20 ___by_____
_____ who is personally known to me or who has produced _____ as identification. And who did take an oath.

_____
Signature - Notary Public/ Deputy Clerk

This document prepared by:

_____
Printed Name Notary Public/Deputy Clerk

FIGURE 4.3 A typical notice to proceed document issued by the State of Florida. This is one of the most import documents that the contractor should be aware of and failure to do so can have serious adverse consequences and significantly impact a contractor's ability to recover for work performed on a project.

4.4.3 Walk the Site/Project

Once the project is awarded and the contractor commences work on-site, the project consultant or administrator representing the lender shall periodically visit the site and walk the entire project to observe the construction progress in conjunction with and for the purpose of reviewing each monthly construction draw application for payment throughout the entire duration of the construction project, unless notified otherwise by the lender.

The consultant/administrator shall conduct a separate walk-thru of the project at the time of the regular monthly site visit to determine the percentage of completion and subsequent to review of the draw application for payment. The purpose of this walk-thru is to observe and determine, in detail, the quality of workmanship and materials and conformance to the contract documents.

During the construction process, the consultant is required to conduct periodic site observation reviews. These regular site visits enable the consultant to ascertain whether construction is progressing satisfactorily and in substantial compliance with plans, specifications, and applicable building codes. The consultant's role includes commenting on the quality of workmanship, materials, stored materials, scheduling, and possible issues. Construction lenders require verification by the consultant that requests for payment of construction funds are accurate and suitable for disbursement. Any issues that need addressing or questions needing answers are discussed with on-site personnel as applicable and if the issues are significant, they are reported to the lender. They are also promptly submitted to the borrower for explanation or corrections.

The consultant varies the inspection schedules to meet the lender's needs, from once a week to once a month, as often as needed to verify satisfactory performance and progress at the time of each requisition for payment from the borrower. A written report of each on-site inspection is submitted to the lender within an agreed time frame, and typically includes the following:

- Detailed description of the construction progress achieved since the previous inspection
- Observation of quality of work in place and whether construction is proceeding in general accordance with the approved plans and specifications
- A calculated percentage of work in place, overall and by trade
- Comments on whether the work is proceeding according to schedule and an estimated date as to when the project will be completed
- Annotated photographs of the project (20–40 photographs) showing progress of construction, problem areas, and unacceptable work or conditions
- Unfavorable discrepancies, if any, with recommendations of corrective action.

Each report is presented in a form designed to convey accuracy and offers the lender a feeling of actually "walking through" the site with the consultant.

4.4.4 Photo Documentation

Prior to the advent of digital photography, photos were usually printed to $3\frac{1}{2} \times 5$ inches for inserting into standard reports. However, today digital cameras are almost exclusively used throughout the industry. The objective of photographing a project, whether an existing building or one under construction, is to document representative conditions and use reasonable efforts to document typical conditions present including material or physical deficiencies, if any. Consultants most often use one of two templates depending on the lender's needs. These consist either of two photos per page or six photos per page (Figure 4.4). Captions explaining each photo are helpful to more clearly explain and convey relevant information regarding the project. It is also sometimes helpful to add an arrow pointing to the particular item of interest in the photograph.

Photography is an extremely effective way of recording factual observations. It is frequently said that "a picture is worth a thousand words." Photographs can provide information detail that would be difficult to convey using other mediums. Later, notes or captions can be added to the photographs for further clarification. If dealing with an existing building or project under construction, the first step is to take photographs of the project from various angles with particular attention to detail of work, noting any defects, etc. This is done prior to writing the report, as it will refresh your memory as to what was taking place during your site visit and it will alert you to specific items that may require the attention of both the lender and contractor. The various photos can also be referenced within the "field observations" section of the report.

The photographs should be sorted and placed in a logical manner within a photo template with captions to reflect the various aspects of the project that is to be portrayed and then included in the report. For most assignments (depending on size, complexity, condition of facility, and stage of construction), the number of photographs will range between 20 and 40. The project aspects that should typically be photographed of a project depend largely on the type of project, but would normally include some or all of the following:

1. Site
2. Structural
3. Exterior/building envelope
4. Roof
5. Interior
6. Plumbing
7. Mechanicals
8. Electrical
9. Fire protection/life safety

5. View showing the Elevator Lobby on the 5th floor of XYZ Office Park that was previously leased by Doe International. Base Building and Tenant Improvements are substantially complete.

6. Interior view showing vacant space on the 5th floor XYZ Office Park. Base Building is complete. Latter of Intent resportedly in hand from Global Computer Solutions.

ABC Green International, Inc.   XYZ Project No. A19421

FIGURE 4.4 A, B   Typical templates using two and six photographs per page. Photographs are almost always required for reports, etc.

ABC Highrise Apartments
December 31, 2008

Interior – View of work in progress on Unit 302 on the third floor of the ABC Apartments. About 230 units are now complete and 12 others are in progress. Unit is scheduled to be completed by end of January 8, 2009.

Interior – View showing work in progress on the upgrading of the Bathroom plumbing in Unit 1104 on the eleventh floor. Unit is scheduled to be completed by January 10, 2009.

Interior – View of Kitchen of Unit 1120 on the eleventh floor of the ABC Apartments. Apartment unit is complete and ready for hand over.

Interior – View of Living room in Unit No. 1120 on the eleventh floor of the ABC Apartments, showing status of renovation. Work on unit is complete.

Interior – View of new Office Workspace on the first floor which is scheduled to be completed by third week of January, 2009.

Interior – View of newly created Handicap unit (No. 225) on the second floor of the ABC Apartments. View of Kitchen area. Unit is temporarily being used as an office.

ABC Green International, Inc.

XYZ Project No. A19421

FIGURE 4.4   *(Continued)*

10. Garages/carports
11. Elevators
12. Amenities
13. ADA required elements
14. Stored materials

After the photographs have been organized, prepare the photo sheets and number each photo with required captions. On the photo sheets, the various components of the project should be identified. For multi-story buildings, each photograph should identify the floor or elevation shown. The object is to convey to the lender the project's progress and any other relevant information pertaining to the project. Comments placed on the photographs should convey a thorough familiarity with the project.

4.5 LOAN DISBURSEMENTS—DRAW APPLICATION REVIEWS

4.5.1 Value of Work in Place

The borrower's payment requisitions are reviewed and evaluated usually during the scheduled site meetings on the basis of accurate quantities of work in place and approved. Following the on-site inspection the results are compared with the borrower's requisition for funding for work in place up to the time of the inspection. Any discrepancies should be promptly resolved, preferably prior to submission of the requisition to the lending institution. The main purpose of closely monitoring the flow of construction loan dollars is to ensure that, at any given time during the life of the loan, sufficient funds remain in the undisturbed portion of the loan to complete the project.

In order to calculate the total value of work in place, the contractor must break down the schedule of values into items or quantities of work, which can readily be evaluated by the administrator when estimating the work in place. This breakdown is separate from the schedule of values and does not replace it. The main purpose of the breakdown is to prevent potential disagreements between the contractor and the administrator when evaluating the quantities of work completed. But the value of work in place should be developed prior to writing the monthly report. This is because developing a number for work in place will exemplify where the emphasis has to be placed when writing the body of the report and the summary. By knowing the total value of work in place and the amount approved for the period covered, the administrator will be alerted as to whether the pace of the project is slowing down, speeding up, or whether a particular line item is heading for a cost overrun, etc.

The contractor then proceeds to prepare a certified copy of the application for payment in the format outlined in the contract documents. The administrator is given a copy of the application for payment and verifies that it is correct as per the site review meeting. The contractor shall bring to the review meeting all materials required to properly evaluate the application for payment, including stored material invoices, release of liens, etc. The administrator, owner, and the contractor assess the project's current status along with the contractor's application for payment and agree upon the amount due the contractor as outlined in the contract documents.

Should the administrator and the contractor not agree on an appropriate amount to be disbursed as per the application for payment, the contractor may then prepare and submit an application for payment that he/she considers to be appropriate. The owner and/or architect will then in consultation with the administrator, recommend to the owner the amount that he/she feels should be certified for payment. The certified value of work-in-place is based essentially upon the latest site visit and the latest contractor's application for payment. In Figure 4.5 we see an example of how the current value of work-in-place is calculated and which is typically included in the project status report.

4.5.2 Funding for Stored Materials

Whether or not stored materials will be funded or not hinges on the policies of the lender. More and more lenders have indicated that the question is negotiable. However, the lender must exercise great care. It is an absolute necessity that the funding procedures for stored material comply with the building loan agreement (BLA) requirements. Unless specifically authorized by the lender, the consultant/administrator has no contractual authority to approve stored materials. This should be clarified in the general conditions or supplemental conditions. If funding for stored materials is requested, the consultant/administrator will note and report to the Lender how the materials are protected from theft, the elements, and vandalism. The consultant/administrator should inspect the insurance certificates for same to ensure that the Lender is named as a co-insured. Likewise, the consultant should inspect the invoice, bill-of-sale, a UCC (uniform commercial code) statement or a contract verifying the cost of same.

The problem of stored material funding has become a major concern in recent years and indications in the industry are that it is getting worse. The problem exists because many lending institutions absolutely prohibit payment for materials until they are physically installed into the improvement. The lender's main concern is due to increased exposure in the event of a failure by the general contractor or borrower, since recovery of these materials or their cost has traditionally been difficult, if

Adjusted Direct Cost Budget	$	156,286,487.00 (1)

Total Value of Work Completed To-Date	$	43,429,127.53
Stored Materials	+	-
Subtotal Work Completed and Stored		43,429,127.53 (2)
Less Retainage	-	2,594,340.96
Total Completed Less Retainage	$	40,834,786.57
Less Previous ABC Certification	-	15,376,051.75 (3)
Current Certification	**$**	**25,458,734.82**

Cost-to-Complete BLA Adjusted Direct Cost Budget,		
Including Retainage	$	115,451,700.43

Cost-to-Complete Based upon ABC's		
Recommended budget of $165,000,000	$	124,165,213.43

(1)   Based upon Borrower's Direct Cost Budget as follows:

Off-Site Work	$	300,000.00
On-Site Work		23,861,912.00
Shell/Common Area Construction		59,694,900.00
Speciality Tenant Construction		6,732,600.00
Speciality Tenant Allowances		11,290,100.00
Anchor Tenant Construction		14,640,164.00
Anchor Tenant Allowances	+	39,766,811.00
Total Direct Cost Budget	**$**	**156,286,487.00**

(2)   Work Completed and Stored To-Date as Follows:

Off-Site Work	$	-
On-Site Work		25,000,000.00
Shell/Common Area Construction		13,515,987.16
Speciality Tenant Construction		66,096.81
Speciality Tenant Allowances		442,493.33
Anchor Tenant Construction		737,775.23
Anchor Tenant Allowances	+	3,666,775.00
Total Work Completed To-Date	**$**	**43,429,127.53**

(3)   Includes amounts not certified by ABC.

FIGURE 4.5   An example of how the current value of work in place is calculated.

not impossible. Faced with these problems, the lending institutions have often modified their approach to "materials only" payments so that if materials are suitably stored at a bonded warehouse off site, payments may sometimes be made. The bottom line is whether payment is made for stored materials hinges on the policies of the lender and for many lenders this has become a negotiable matter.

Therefore, in light of shortages, slow deliveries, and rising cost factors, a borrower would be well advised to reach agreement on the policy toward such advances early in the discussions with the lender, preferably prior to closing the loan. Stored materials should be tracked on a separate stored material inventory schedule provided by the lender. It is strongly recommended that the consultant/administrator review the building loan agreement (BLA) regarding the lender's policy of funding for stored materials because funding may be reserved as a lender's business decision. If the lender does decide to proceed with funding the stored materials, a proportionate amount of the general condition or fee monies will not be released for same.

4.5.3 Approved and Pending Changing Orders

The owner is required to provide a schedule and copies of all change orders as well as a schedule of pending change orders. Change orders should not be approved unless they have been previously approved by the owner and there are sufficient funds within the contingency budget to absorb same. After the owner completes the review of the proposal request and approval for the change order is obtained, the administrator shall prepare a change order, as outlined in the contract documents, utilizing AIA Document G701. The change order will be produced in three copies and forwarded to the contractor for signature. The contractor's signature on the change order acknowledges that the work will be completed as described in the change order for the stated amount in the change order (Figure 4.6). Any additional time, if requested, for the change prder work will be incorporated into the change order. Failure by the contractor to request additional time for the change order work will prohibit the contractor from doing so at a later date.

Following the contractor's signature of the change order all three copies will be returned to the administrator or the owner and/or architect for signature and certification. The change order will then be submitted to the owner for final signature and distribution. The owner, administrator, contractor, and owner and/or architect will receive one signed and certified copy of the change order for their records. The change order will then be added to the next monthly application for payment. The change order will be recorded in the project log book along with a copy of the change order, and a copy will also be transmitted to the central file system.

CHANGE ORDER

AIA DOCUMENT G701

Distribution to:

OWNER ☐
ARCHITECT ☐
CONTRACTOR ☐
FIELD ☐
OTHER ☐

PROJECT:
(name, address)

TO (Contractor):

CHANGE ORDER NUMBER:

INITIATION DATE:

ARCHITECT'S PROJECT NO:

CONTRACT FOR:

CONTRACT DATE:

You are directed to make the following changes in this Contract:

Not valid until signed by both the Owner and Architect.
Signature of the Contractor indicates his agreement herewith, including any adjustment in the Contract Sum or Contract Time.

The original (Contract Sum) (Guaranteed Maximum Cost) was ....$
Net change by previously authorized Change Orders ....$ ....
The (Contract Sum) (Guaranteed Maximum Cost) prior to this Change Order was...$
The (Contract Sum) (Guaranteed Maximum Cost) will be (increased) (decreased) (unchanged)
   by this Change Order ....$
The new (Contract Sum) (Guaranteed Maximum Cost) including this Change Order will be ....$
The Contract Time will be (increased) (decreased) (unchanged) by ....   ( ) Days
The Date of Substantial Completion as of the date of this Change Order therefore is ....

ARCHITECT _____   CONTRACTOR _____   OWNER _____

Address _____   Address _____   Address _____

By_____   By_____   By_____

Date_____   Date_____   Date_____

sk projects\clm\template\changeorder.doc

FIGURE 4.6 Sample change order templates.

The project administrator should check the building loan agreement (BLA) for the approval requirements of individual and aggregate amount change orders. Once the change order is approved, a notice to proceed is issued to the contractor. However, before this can happen, the administrator/owner's representative will issue a written directive to the contractor asking for a request for proposal for the change order work within 10 days for the subject work. This will then be followed with a notice to proceed or a notice to proceed immediately with the work.

The notice to proceed will specify the manner in which the owner will pay for the work in question. The options available will be stipulated in the contract documents and usually require:

1. An accepted estimate
2. Time and material
3. Unit costs.

The RFP should also address subcontractor markups, labor rates, and various other requirements regarding change order pricing. The contractor should not commence with the change order work unless the contractor receives a written notice to proceed. The exact wording and format of the notice to proceed form will vary depending on the nature of the work and requested pricing method.

Owners sometimes prefer to organize change order payments to be separate from the application for payment. The contractor may request assurance from the owner that adequate funds are available to pay for the change order before executing the work. Financing and lending institutions generally stipulate that all change orders be processed through their office before executing the work. Contractor's performance and payment bonds and builder's risk insurance need to be adjusted to reflect substantial changes to the contract sum. The contractor is essentially obligated to execute any change order authorized by the owner even if a dispute occurs regarding the actual cost of the change order work or its impact on the project schedule. These matters should be resolved by exercising provisions included in the contract documents.

4.5.4 Lender/Owner Retainage

Retainage is the withholding of certain portions of the monies due a contractor for work in place as an incentive to complete the work, and is one of the line items identified on the contractor's application for payment. It has been the subject of considerable discussion over recent years and it is quite apparent that many lending institutions do not have a uniform practice on either the amount of retainage or the manner in which it is collected.

Lenders generally prefer using a conservative approach in which a full 10 percent is withheld of all items of construction and is kept through the entire construction period. On the other hand, a liberal policy is maintained in which no retainage is withheld on any item. Between the two extremes, a wide variety of practices are prevalent, including holding a retainage on certain items only or reducing the total amount withheld after a certain point in construction has been reached (usually 50 percent of project completion). Experienced and knowledgeable borrowers try to negotiate the most liberal agreement possible at the outset of the loan program. It is also common practice for the general contractor to impose retainage on their subcontractors as well, although most general contractors try to work with their subcontractors to facilitate early payment where possible.

Once the contractor achieves successful execution of the certificate for substantial completion, the owner may make payments that reflect adjustments in the retainage amounts as provided in the contract documents. The reduction in retainage shall be made with the exception of amounts which have been determined to reflect the costs for remaining work to be completed and/or work requiring correction. In these cases, the administrator, upon consultation with the owner and/or architect, shall establish a value for remaining work and suggest that the owner retain three times the value of incomplete work.

The minimum amount retained for each unacceptable item should reflect the estimated cost to have an alternative contractor brought in to complete or correct the item in question. This includes any costs for mobilization and/or equipment required to correct or complete any outstanding construction deficiencies.

4.5.5 General Conditions

Over the years, many standard general conditions formats have been developed by numerous trade and professional organizations. Perhaps the most widely used general conditions format is that published by the American Institute of Architects (AIA), AIA Document A201. One of the advantages of using the AIA standard is that most contractors and architects are familiar with it. However, many state organizations and universities, etc. have their own standard general conditions.

The general conditions are considered to be one of the most essential documents associated with the construction contract. It sets forth the rights and responsibilities of each of the parties such as the owner and contractor in addition to the surety bond provider, the authority and responsibilities of the design professional, and the requirements governing the various parties' business and legal relationships. It is imperative that the contractor knows exactly what is contained in the general conditions and its implications. If it proves too difficult to read or there is simply not enough time, the project may be put at risk. Some of the general clauses contained in the general conditions that can have a direct affect on the success of a project if appropriate attention is not paid to them include:

General Provisions   Includes basic definitions for the contract and roles of the various parties, the work, the drawings and specifications, and other issues such as change orders, punch lists, etc. In addition, it clarifies the ownership, use, and overall intent of the contract documents.

Owner Responsibilities   Among other things, it outlines the services and information the owner is required to supply. For example, the owner shall, at the written request of the contractor, prior to commencement of the work and thereafter, furnish to the contractor reasonable evidence that financial arrangements are in place to fulfill the owner's obligations under the contract. Furnishing of such evidence shall be a condition precedent to commencement or continuation of the work. After such evidence has been furnished, the owner shall not materially vary such financial arrangements without prior notice to the contractor. Also, except for permits and fees which are the responsibility of the contractor under the contract documents, the owner will secure and pay for necessary approvals, easements, assessments, and charges required for construction, use or occupancy of permanent structures, or for permanent changes in existing facilities. Also outlined are the extent of the owner's rights, the owner's right to stop the work, and the right to carry out the work.

Contractor Role and Responsibilities   This section lays out the obligations of the contractor under the contract. For example, the contractor warrants all equipment and materials furnished and work performed, under the contract, against defective materials and workmanship for a specified period (usually 12 months) after acceptance as provided in the contract, unless a longer period is specified, regardless of whether the same were furnished or performed by the contractor or any subcontractors of any tier. It also includes things like supervision and construction procedures, materials, labor and workmanship, patents, substitutions, record drawings, shop drawings, product data and samples, taxes, permits, and contractor's construction schedules. Additionally, the contractor shall, without additional expense to the owner, comply with all applicable laws, ordinances, rules, statutes, and regulations.

Administration of the Contract   This section assigns duties to the architect for the administration of the contract. Specific clauses dealing with the architect's responsibility for visiting the site and making periodic inspections are included. The section also addresses how requests for additional time, claims, and disputes are to be addressed. Generally, the owner's representative (or lender's representative) will administer the construction contract. The architect will assist the owner's representative with the administration of the contract as indicated in these contract documents. The architect will not be responsible for the contractor's failure to perform the work in accordance with the requirements of the contract documents.

Subcontracts and Subcontractor Relations   This section deals with the general contractor awarding subcontracts or specialty contracts for certain portions of the work. The contractor is required here to furnish the owner and the architect, in writing, with the name, and trade for each subcontractor and the names of all persons or entities proposed as manufacturers of products, materials, and equipment identified in the contract documents and where applicable, the name of the installing contractor. By appropriate agreement, the contractor shall require each subcontractor, to the extent of the work to be performed by the subcontractor, to be bound to the contractor by terms of the contract documents, and to assume toward the contractor all the obligations and responsibilities, including the responsibility for safety of the subcontractor's work, which the contractor, by these documents, assumes toward the owner and architect.

Construction by Owner or Separate Contractors   This clause basically states that the owner reserves the right to perform constriction or operations related to the project with the owner's own work force and that the owner has the right to award separate contracts as stipulated in the contract documents. In this respect, no contractor shall delay another contractor by neglecting to perform his/her work at the appropriate time. Each contractor shall be required to coordinate his/her work with other contractors to afford others reasonable opportunity for execution of their work.

Changes in the Work   This section highlights how changes (overhead and profit on change orders; time extensions; inclusions) are authorized and processed. The owner may authorize written change orders regarding changes in, or additions to, work to be performed or materials to be furnished pursuant to the provisions of the contract. The amount of adjustment in the contract price for authorized change orders will be agreed upon before such change orders become effective. Likewise, an order for a minor change in the work may be issued by the architect alone where it does not involve changes to the contract sum.

Time and Schedule Requirements   The contractor acknowledges and agrees that time is of the essence of this contract. The contract time therefore may only be changed by a change order. Contract time is the period of time set forth in the contract for construction required for substantial completion and final completion of the entire work or portions of the work as defined in the contract documents. This part of the contract therefore deals largely with issues relating to project startup, progress, and completion relative to the specific project schedule in addition to issues relating to delay (notice and time impact analysis) and extensions of time to the contract. The general conditions should clarify certain issues, such as how many days after a delay does a contractor have to give notice, and how is the notice to be delivered (verbally, by mail, by registered mail)?

Payments and Completion   The importance of this section is that it identifies how the contractor will be paid and specifies how applications for progress payments are to be made. The contract sum is stated in the agreement and with the authorized adjustments reflects the total amount payable by the owner to the contractor for performance of the work under the contract documents. Before the first application for payment, the contractor shall submit to the architect a schedule of values allocated to various portions of the work, prepared in such form and supported by such data to substantiate its accuracy as the architect may require. This schedule, unless objected to by the architect, shall be used as a basis for reviewing the contractor's applications for payment. The owner's representative may decide not to certify payment and may withhold approval in whole or in part, to the extent reasonably necessary to protect the owner.

Protection of Persons and Property   This section of the general conditions addresses safety concerns for both the owner's property and the personnel on the project. According to the AIA form 201, "The contractor shall be responsible for initiating, maintaining, and supervising all safety precautions and programs in connection with the performance of the contract." It basically means that the contractor shall conduct operations under this contract in a manner to avoid the risk of bodily harm to persons or risk of damage to any property. Moreover, the contractor is required to comply with applicable safety laws, standards, codes, and regulations in the jurisdiction where the work is being performed.

Insurance and Bonds   These issues deal with various insurance and bonding requirements of the parties. For example, the contractor is required to purchase such insurance as will protect the contractor from claims which may arise out of or result from the contractor's operations under the contract and for which the contractor may be legally liable, whether such operations be by the contractor or by a subcontractor or by anyone directly or indirectly employed by any of them, or by anyone for whose acts any of them may be liable. Furthermore, the owner shall have the right to require the contractor to furnish bonds covering faithful performance of the contract and payment of obligations arising as stipulated in bidding requirements or specifically required in the contract documents including but not limited to the contractor's obligation to correct defects after final payment has been made as required by the contract documents on the date of execution of the contract.

Uncovering and Correction of Work   This section deals with acceptance of the work in place by the architect or owner's representative and stipulates when the contractor is responsible for uncovering and/or correcting work that is considered unacceptable. The AIA 201 form here states that "If a portion of the work is covered contrary to the architect's request or to requirements specifically expressed in the contract documents, it must, if required in writing by the architect, be uncovered for the architect's examination and be replaced at the contractor's expense without change in the contract time." The owner may, at his/her sole discretion, accept work that is not in accordance with the contract documents instead of requiring its removal and correction. In such cases, the contract sum will be adjusted as appropriate and equitable.

Miscellaneous Provisions   This section addresses various issues such as successors and assigns, wage rates, tests and inspections, rights and remedies, codes and standards, records, general provisions and written notice.

Termination or Suspension of the Contract   This section deals with the terms under which parties may terminate or suspend the contract. The contractor may terminate the contract if the work is stopped for a period of 30 consecutive days through no act or fault of the contractor or a subcontractor, employees, or any other persons or entities performing portions of the work under direct or indirect contract with the contractor. Similarly, in addition to other rights and remedies granted to the owner under the contract documents and by law, the owner may without prejudice terminate the contract with the contractor under specific conditions. The owner may also, at any time, terminate the contract in whole or in part for the owner's convenience and without cause.

Most lenders require general conditions to be disbursed in direct proportion to the percentage of completion of the sub-contractual costs so that the lender is assured that general condition monies will be adequate throughout the project's duration. In some cases, especially in CM and "cost plus a fee" contracts, the borrower's contract requires that he pays general conditions on either an equal monthly payment or on a cost incurred basis. The lender should be consulted beforehand to determine the lender's funding policy.

4.5.6 Supplemental Conditions

Whereas general conditions can apply to any project of the type being designed and built, supplemental conditions or special conditions usually deal with matters that are project specific and beyond the scope of the standard general conditions, although they generally augment them. These sections may either add to or amend provisions in the general conditions. Below are examples of project-specific information that may appear in this section:

- Safety and security precautions
- Cost fluctuation adjustments
- Additional bond security
- Materials or other services furnished by the owner
- Temporary facilities requirements
- Prevailing wages
- Project phasing or special construction schedule requirements

- Bonus payment information
- Insurance coverage certificates
- Permits, fees, and notices.

4.5.7 The Designer of Record/Administrator Sign Off

The final action to be taken in the construction process is project close-out. But before the designer of record/administrator is able to sign off and the owner takes possession of the building, certain requirements need to be met. For final completion and before final payment is made the following need to be addressed:

- The architect must certify final completion and that all "punch-list" items are complete
- The architect must certify that final inspection has been satisfactorily conducted
- Commissioning has been satisfactorily completed
- A certificate of occupancy is in place
- All warranties, guaranties, and operating manuals have been received
- All liens of GC/CM and trade contractors are released or satisfied and outstanding claims resolved
- Final lien waivers and contractor affidavits are obtained for all work performed—from architect, GC/CM, and trade contractors
- Close-out agreement—complete all outstanding issues with each trade contractor and with the general contractor, and confirm commencement date for guaranty period, etc.

4.6 PREPARING THE PROGRESS STATUS REPORT (PSR)

4.6.1 Draw Applications—Documents Required

The consultant conducts an inventory and review of construction documents which include, but not limited to:

- Plat plan/boundary survey/site plan
- Topography plan
- Environmental site assessment
- Soils investigation report
- Construction plans
- Construction specifications
- Addenda/change orders
- Construction contracts(s)/schedule of values
- Architect's sontract(s)
- Construction schedule

- Building permits
- Foil documents
- Utility letters
- Estimated variances.

4.6.2 Waivers of Lien

The borrower (owner) is required to provide copies of all partial/full waivers of lien from sub-contractors, vendors, etc. to the administrator, normally on a monthly basis, usually with the payment application (Figure 4.7). These partial waivers are to accompany each general contractor and subcontractor requisition on a monthly basis. While most borrowers execute them on a monthly basis as the project progresses, some prefer to wait until the project is completed and submit them with the final request for payment. In some cases the borrower requests that upon signing the contract the subs waive their right to lien the job. Waivers should be properly organized, stating what trade they are for and included as exhibits in the report. As a project approaches completion, copies of all lien waivers should be obtained. For any that are not forthcoming, recommend to the lender that they be obtained or deducted from the retainage. Below are the four main types of lien waivers:

1. Conditional lien waiver on progress payment: This waiver generally specifies that if claimants have indeed been paid to date the waiver provides effective evidence against any lien claim on the property, which makes it the safest waiver for claimants.
2. Unconditional lien waiver on progress payment: This waiver unconditionally releases all claimant rights through a specific date.
3. Conditional lien waiver on final payment: This lien waiver releases all claimant rights to file a mechanics lien if there is evidence that they have been paid to date.
4. Unconditional final lien waiver on final payment: This provides the safest type of lien waiver for owners; it generally releases all rights of the claimant to place a mechanics lien on the owner's property unconditionally. However, claimants should issue this type of release only when they are satisfied that their work is complete and that the payment has cleared their bank. Owners should demand this release when claimant is paid in full.

4.6.3 Testing Reports

Testing results should normally be reviewed on a monthly basis. A set of contract documents should be submitted to the construction testing agency for their review prior to the commencement of any construction. The testing agency will provide a copy of their current rate schedule

PARTIAL WAIVER OF LIENS

Project Description: _____
                                    _____

Period Ending: _____ __,200_

Work Performed: _____

Work Performed by: _____

Under Contract to: _____

Contract Date: _____

Original Contract Amount: $_____
Change Order Amounts: +_____
Adjusted Contract Amount: $_____

Work Completed to Date: $_____
Less Retainage Not Yet Due: -_____
Net Amount Due to Date: $_____
Less Payments Received to Date: -_____
Total Payment Due: $_____

THE UNDERSIGNED (1) acknowledges receipt of the amount set forth above as payments received to date, (2) to the extent of such payments, waives and releases any claim which it may now or hereafter have upon the land and improvements described above in the project description, (3) that the amount of payments received to the date of this waiver represents the current amount due in accordance with our contact and work completed, and (4) warrants that it has not and will not assign any claims for payment or right to perfect a lien against such land and improvements and warrants that it has the right to execute this waiver and release.

THE UNDERSIGNED further warrants that (1) all workman employed by it or its subcontractors upon this Project have been fully paid to the date hereof, (2) all materialmen from whom the undersigned or its subcontractors have purchased materials used in the Project have been paid for materials delivered on or prior to the date hereof, (3) none of such workman and materialmen has any claim or demand or right of lien against the land and improvements described above, and (4) stipulates that he is an authorized officer with full power to execute this waiver of liens.

THE UNDERSIGNED agrees that _____ and any lender and any title insurer may rely upon this waiver.

By   _____

Title   _____

Sworn to me this____ day
of_____, 200__.

_____
Notary Public

NOTE: Return four (4) signed releases to _____ at _____, _____, __, ____ to the attention of _____ . Additional payments will not be made until the signed releases are returned.

FIGURE 4.7   Examples of partial waivers of lien.

for types of test work and also provide a budget estimate for the specific project for the owner's review and to make a determination. Upon the owner's acceptance and approval, the owner and/or administrator will authorize the testing agency in writing to proceed with required tests.

After the administrator reviews the billing with the owner and/or architect it is given to the owner for payment. The contractor is responsible for coordinating and scheduling required testing activities for the project.

Type of testing reports received (concrete, mortar, timber, etc.) should be recorded whether or not they're in compliance with the design specifications. Any failed test results should be brought to the attention of the owner requesting an explanation before deciding what action is to be taken. A letter may be sent to the designer of record noting the failed test and requesting comment, depending on the test's significance.

The owner needs to provide the administrator with copies of all controlled testing reports.

The testing agency normally provides test reports within 48 hours to the administrator, designated engineering consultants, contractor, and owner. It is the administrator's duty to immediately respond in writing to any test reports indicating that the work fails to conform to the contract documents and to ensure that remedial action is taken. Since the contractor is ultimately responsible for the construction means and methods, it becomes the contractor's responsibility to propose a solution to correct any construction deficiencies.

4.6.4 Daily Work Log

Logs are generally used to monitor the day-to-day activities that take place on the job site. The owner should provide a copy of a typical page from the project's daily work log for the day prior to the administrator's site visit and meeting. The contractor's daily log will normally be submitted to the administrator on a weekly basis. It will typically include:

1. Daily activities.
2. General weather conditions.
3. Meetings and important decisions.
4. Conversation and telephone records.
5. Problems or potential delays.
6. Unusual events such as stoppages, and emergency actions.
7. Accidents.
8. Change orders received (pending or implemented).
9. Material delivery, equipment on-site, etc.
10. Visitors.

4.6.5 Construction Schedule and Schedule of Values

Project schedules are to be provided to monitor the progress of the contractor and subcontractors relative to established schedules and to serve as status reports to the owner. Here, the administrator will discuss how the project stands with respect to the owner's construction schedule and/or the target date for completion. Any factors contributing to delay or progress should be mentioned (e.g., good weather, a strike, tight management, etc.). If the initial construction schedule is revised, the owner should submit a copy of the revised schedule to the administrator. Figure 4.8 is a construction schedule to monitor how the work is progressing in relation to the contractor's schedule.

The project construction schedule shall be prepared in accordance with the contract documents in either PERT, BAR, GANTT, or C.P.M. format. The project construction schedule shall be updated each period by the contractor and verified by the administrator to ensure that the schedule is always up to date and accurate. The updated project construction schedule shall form part of each monthly application for payment and shall also be included in the administrator's monthly project status report.

A schedule of values shall be prepared in accordance with the contract documents and which equals the total cost of the project. Line items for the schedule of values shall be divided into the appropriate specification divisions and broken down to reflect material and labor costs for each item. Unit measurement for the materials described in the schedule of values shall be included. These numbers will be required to estimate the value of work in place at any specific time. The schedule of values for the project will ultimately become the source for verifying costs for any

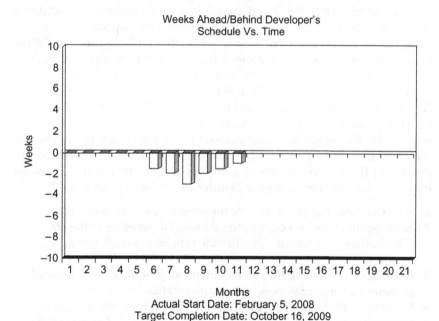

FIGURE 4.8   A typical construction schedule.

additional work and establishing prices for potential modifications to the contract documents (change orders).

4.6.6 Project Progress Meetings

Throughout the construction process various meetings will be held at the job site, most of which are scheduled well in advance. Depending on the type of project and the agreement between the lender and the administrator, these meetings are usually held at regular intervals, usually on either a bi-weekly or monthly basis unless an unscheduled special meeting is called to address special issues. The main purpose of these meetings is to discuss and review the project's progress and to provide a forum in which the main participants (administrator, contractor, subcontractor, architect, engineers, and others) can discuss their concerns, as well as a submitted application for payment. These meetings are usually chaired by the administrator and can be fairly formal in nature with a written agenda. Meetings are normally recorded and the minutes are distributed to all the participants.

4.6.7 Stored Materials Documentation

Materials stored on-site should be in a secured area. Where the contractor is requesting funds for stored materials, a stored materials schedule for such materials should be delivered to the PM and the application for payment and sworn statement of general contractor (AIA Document G702) or equivalent document must list the dollar amounts of all stored materials. The consulting professional must be able to verify all such items.

This section is included only if there are materials for which funding has been requested by the general contractor that are stored either on or off site (Figure 4.9). Where funding is permitted, typical documentation is required by the owner or lender in order to approve funding. And where the contractor seeks payment for materials that have not been incorporated into the improvements and are not stored on-site, the following back-up documents are required in order to process the request:

1. A stored materials statement for the off-site or on-site materials.
2. Evidence of insurance on the stored material (whether at the off-site location or in transit). Identify the type of material, value, and location.
3. Bill of Sale evidencing Borrower's ownership of the stored materials, including a list of materials, value, and location.
4. Letter from the consulting professional or approved third party inspector stating that materials have been sighted and inventoried and that they are suitably stored and marked for the project.

SCHEDULE II

STORED MATERIALS STATEMENT NO.____

BORROWER: _____
PROJECT: _____
ADDRESS: _____
LOAN NO: _____

ITEM NO.	DESCRIPTION OF MATERIALS STORED (ATTACH INVOICES, LISTINGS, AND/OR OTHER PRICE SUPPORTING DOCUMENTATION)	LOCATION WHERE STORED	NAME OF SUBCONTRACTOR/ SUPPLIER	OPENING INVENTORY	ADDITIONS TO INVENTORY	USAGE OF INVENTORY	CLOSING INVENTORY	RETAINED AMOUNT NOT YET DUE

TOTALS (OR SUBTOTALS)

PREPARED BY INSPECTION & VALUATION INTERNATIONAL, INC.

E:\00 GCPM\Chap 04–Green Construction Cost Monitoring\Standardized PMO Documents\Forms, Checklists & Mobilization Docs\StoredMaterials\STORED.xls

FIGURE 4.9   A typical stored material statement (to be modified).

5. Materials stored off-site shall be in an independent bonded warehouse with prior approval of the owner or lender and at no cost to the owner/lender.
6. Any material having architectural finishes will require inspection and acceptance by the architect. In no case will payment be made for bulk marble, granite, etc., that has not been fabricated and inspected.

In addition to the above, the owner or lender may approve funding for stored materials under certain conditions such as in cases where ordering of materials requires a long lead time.

Items that are being fabricated and stored with the manufacturer should also be marked and segregated from the manufacturer's other supplies. Prior to request for payment on materials "in process" of being fabricated, the architect, or other firm or agency acceptable to the lender, should inspect the materials.

4.6.8 Subcontracts and Purchase Orders

As subcontracts and purchase orders are signed during the general course of construction, copies should be provided to the administrator. If this is a general contractor built project where subcontracts are not to be made privy to the owner, then the lender usually requires copies of all major subcontractor trade payment breakdowns prepared on a percentage of completion basis instead of a dollar amount.

4.6.9 Payment and Performance Bonds

A performance bond is usually stipulated in the general conditions to assure the owner that the contractor will complete all obligations set out in the contract, provided that no action by the owner or the owner's agents prevents or inhibits the contractor from the implementation of the contract requirements.

Lending institutions appear to have started to lessen their emphasis on "bonded" contracts. A number of lenders have been disappointed to find they are involved in extended litigation after invoking the very bond they looked to for protection. Moreover, bonding companies often will not bond an owner-builder. Although a number of title companies previously offered a "completion guarantee" designed to assure interim lenders that their project would be completed, these title companies have largely withdrawn such policies due to suffering extensive losses.

The most perilous and defining phase of the delivery of a new project is the actual construction. The construction phase, due to the infinite number of inherent risks associated with the activity, rises far above the other phases of the project delivery when claims, failures, problems, and defaults are taken into account.

The administrator should be aware of both the owner's rights (and Lender's rights) under the contract, as well as the obligations necessary to protect those rights. Should the administrator, while acting as the owner's agent in administering the new project, fail to properly ensure the rights of the owner due to negligence of the requirements, incorrect documentation, or the untimely issuance of notice, the administrator may certainly be held liable by the owner and will be held responsible for any losses, whether moral or financial.

The administrator should also be aware of all requirements necessary to ensure the protection of the payment and performance bonds. As a construction expert and owner's representative, the administrator's role in administering the construction contract requires verification of all notices that are made and that all prerequisites are followed by both the owner and contractor. This process starts by selecting the type of bonds required for the project and as outlined in the contract documents.

No other situation holds greater potential risk than the default of the contractor during construction of the project. For this reason, the successful project administrator must exercise both caution and diligence while reviewing the types of payment and performance bonds required by the contract. Normally, the administrator should specify standard AIA bond forms A312 (performance and performance bonds); these documents are not only respected by the surety industry, but have been thoroughly tested and generally upheld through the judicial process. Separate payment bonds and performance bonds with separate bond numbers and power of attorney offer the owner protection equal to double the face value of the construction project. It may be wise to avoid use of combination payment and performance bonds because of recent legal precedents that have held that the surety was obligated to pay only the face value of the construction project through a combination of claim against both sides of the combined bond.

While the payment and performance bonds offer the owner the appropriate protection against countless scenarios of contractor default, subcontractor non-payment, material or supplier non-payment and liens, the administrator should also recognize the surety and its agents for what they really are—sharp and intelligent businessmen who will carry out all their obligations and responsibilities under the bonds, provided all aspects of the default are in fact the contractor's responsibility. However, the surety may be relieved from their obligations, in whole or in part, by the erroneous actions or negligence of the administrator or the improper actions of the owner, in which case the surety may try to mitigate their losses to the maximum extent possible under the law. Thus, by the time the surety becomes involved in a construction project, the relationships between the owner, contractor, and administrator may very well be uncomfortable to say the least. The surety certainly is not a deep pocketed, sympathetic benefactor that doles out money to those most deserving just because there is a payment and performance bond.

It is imperative that the administrator reads and understands the wording and requirements of each type of bond, for just as every project is unique, so is every situation involving the surety. There is no substitute for studying and fully understanding the requirements of the bond to ensure that each and every action required by the administrator is not only fully implemented but fully implemented within the specified time frames delineated in the bond documents.

It should be noted that the statements and opinions expressed herein reflect a general attitude of surety and its general position when confronted with differing construction situations. However, the reader should be aware that these opinions are generic and not intended to be legal advice.

4.6.10 RFI and Other Logs

Requests for information (RFIs) are issued from the contractor to the administrator or owner for clarification of design information or to present any other questions; they include a summary log containing the status of each request. It is a formal procedure and each inquiry and response should be tracked and documented. It is also critically important to get a quick response. They often originate from a subcontractor, vendor, or craftsman who needs certain information to continue working. Most contractors have a standard format that they use to keep the RFIs consistent.

4.6.11 Permits and Approvals

Before any building construction can commence, the owner is required to apply and receive a building permit for the project from the local authority. This may be the single most critical aspect of the pre-construction process; sometimes it is acquired prior to the bidding process. A delayed permit can cause considerable hardship and project complications. Building permits also contain inspection schedules and zoning ordinances for the project. A copy of the ordinance under which the project was approved as well as copies of all variances, approvals and declaration, and any other zoning information or approvals pertinent to the project should be in the possession of the owner, designer of record, and the administrator. The following permits and approvals need to be provided as well:

- A print of the zoning sheet should be included highlighting all calculations along with evidence that the zoning computations have been approved by the planning board.
- A comprehensive list of all permits necessary to proceed with construction of the suggested improvements prepared by the

designers-of-record. Also, copies of all permits secured to date, with the remaining provided as and when they are available.

- A complete list of all agreements between the borrower (owner) and any governmental agencies and between the borrower and any governmental agencies for the construction of the suggested changes and copies of all agreements.
- Copies of all appropriate approvals/permits from the landmarks commission, historical preservation group, and other related agencies for historical renovation and restoration work.
- Statements of existing certificate of occupancy and/or certificate of occupancy (CO) procedures and any special requirements to obtain a CO upon completion of the work. Also to be noted if any temporary COs have been issued for partial or phased completion.
- Schedule of building code violations if any.
- Copies of utility load calculations prepared by the engineers-of-record and confirmation that existing services are adequate to serve the proposed project in addition to confirmation from the appropriate authority of availability, adequacy, and intent to provide the following utilities for the proposed project:
  - Water
  - Gas
  - Steam
  - Sanitary sewers
  - Storm sewers
  - Electricity.

4.7 FINAL CERTIFICATION AND PROJECT CLOSEOUT

Final acceptance and project closeout is initiated upon receiving notice from the contractor(s) that the work or a specific portion thereof is acceptable to the owner and is sufficiently complete, in accordance with contractor documents, to allow occupancy or utilization for the use for which it is intended. Closeout can take place when all contract requirements, warranty, and closeout documents, along with all punch list items, have been resolved. Normally, a punch list of unfinished and/ or defective work complete with remedial costs is prepared for possible escrow purposes. Upon the owner's request for the final loan advance, the administrator collates and examines all permits, approvals, waiver of liens, and other closeout documents specified in the loan agreement. Upon approval of these documents, the administrator makes another site inspection and verifies and certifies that the work was completed in accordance with the plans, specification, and loan agreement.

Documents typically required for project closeout include:

1. Signed/sealed as-built or record drawings, which show all changes from the original plans.
2. Contractor's certificate of compliance with plans and specifications.
3. Architect's issuance of certificate of substantial completion and/or final acceptance and issuing final certificate(s) for payment after a detailed inspection with the owner's representative is conducted for conformity of the work to the contract documents to verify the list submitted by the contractor(s) of items to be completed or corrected.
4. Architect's certified copy of the final punch list of itemized work stating that each item has been completed or otherwise resolved for acceptance.
5. Determination of the amounts to be withheld until final completion of outstanding punch list items.
6. Notification to owner and contractor(s) of deficiencies found in follow-up inspection(s), if any.
7. Certification from borrower that all closeout requirements, including but not limited to, as-built drawings, warranties, manuals, keys, affidavits, receipts, releases, etc. have been received, reviewed as necessary, and approved for each subcontractor.
8. Certificates of use, occupancy, or operation.
9. Securing and receipt of consent of surety or sureties, if any, for reduction or partial release of retainage or the making of final payment(s) and consents of surety for final payments, if bonds are provided.
10. Final waivers of lien in a form satisfactory to the lender and the title company from all subcontractors, suppliers, and the general contractor and indemnifying the owner against such liens.

4.7.1 As-Built Drawings/Record Drawings

Most contracts stipulate that the contractor maintains a set of as-built (sometimes erroneously called record drawings) as the project progresses. These documents record actual dimensions, locations, and features that may differ from the original contract documents. When the drawings are in an electronic format, the contractor may be required to correct the drawing files and highlight the modifications. An example of this is the location of underground utilities which differs from the original drawings. In such cases, the contractor would be directed (normally in writing) to make the change whereas the design professional fails to make the necessary modification as they are the responsibility of the general contractor. It is usually stipulated that the contractor submits a final complete and accurate set of as-built drawings to the owner prior to receipt of final payment. As-built plans represent the existing field conditions at the completion of a project. Accurate project plans are sometimes needed for possible litigation involving construction claims and tort liability suits.

The resident engineer or architect of a completed project is typically the most qualified individual to note the field changes that occurred (called "redline corrections"). As-built plans are preferably completed within the project using an electronic format such as AutoCad or MicroStation. Having the drawing saved as a CADD file makes it easier to store and update as required.

4.7.2 Contractor's Certificate of Compliance

Certificates of compliance can apply to various issues such as discrimination and affirmative action, subcontractor work, etc. The general contractor must agree and certify compliance with applicable requirements of the contract documents. A temporary certificate of compliance may often be issued for a portion or portions of a building that may safely be utilized prior to final completion of the building. In this respect the contractor also agrees to obtain compliance certifications from proposed subcontractors prior to the award of subcontractors exceeding an agreed sum as per contract documents.

The contractor is required to coordinate the efforts of all subcontractors and obtain any required letters of compliance from the administrator or owner's consultant. The owner is usually required to pay any fee associated with these letters. However, the contractor shall reimburse the owner for any costs resulting from failed tests or inspections conducted to obtain a letter of compliance. This reimbursement should be made as part of a credit change order following the procedures spelled out in the contract documents.

4.7.3 Architect/Administrator's Certificate of Substantial Completion

The date of substantial completion is defined in the contract documents and is considered to be the date that the administrator, owner, and/or architect will certify that the work, or designated portion of the work, may be beneficially occupied or utilized by the owner for its intended use. For this purpose, the AIA G704™ standard form is used for recording the date of substantial completion of the work or a designated portion thereof. This process takes off when the contractor considers the work, or designated portions of the work as previously agreed to by the owner, to be substantially complete. The contractor then prepares and submits to the administrator a punch list of items that remain to be completed or corrected. The AIA G704™ form provides for agreement the time to be allowed for completion or correction of the items and the date when the owner will be able to occupy the project or designated portions thereof. The form will also designate responsibility for maintenance, heat, utilities, and insurance. If the administrator concludes that the work is substantially complete, the AIA form is then prepared for acceptance by the contractor and owner. The failure of the contractor to include any items on the list will in no way alter his responsibility to complete or correct these items per the Contract Documents.

There are variations of the G704™–2000, such as the G704™CMa–1992, Certificate of Substantial Completion, the Construction Manager-Adviser Edition, which serve the same purpose as G704™–2000 except that this document expands responsibility for certification of substantial completion to include both the architect and the construction manager. There is also the G704™DB–2004 which is a variation of G704™–2000 that acknowledges substantial completion of a design-build project. Because of the nature of design-build contracting, in this form the project owner assumes many of the construction contract administration duties performed by the architect in a traditional project. Because there is no architect to certify substantial completion, the AIA Document G704DB–2004 requires the owner to inspect the project to determine whether the work is substantially complete in accordance with the design-build documents and to acknowledge the date when it occurs.

The term "beneficial occupancy" generally means that the project, or portions thereof, are complete to a sufficient degree to allow the owner to utilize the project or portions thereof, for its intended usage. Relevant systems such as the mechanical systems, life safety systems, telecommunications systems, and any other systems which are required to properly utilize the project or portions thereof, shall be complete and in good working order. Items remaining to be completed shall be such that their correction does not inconvenience the owner or disrupt the owner's normal operations.

The owner/lender or its representative (administrator) should be consulted to confirm that there are no other evident construction deficiencies that are not on the contractor's punch list. It is very important to emphasize that responsibility for preparing the original punch list lies with the contractor. If the administrator is requested to make a substantial completion inspection, and it is obvious that the contractor's punch list is incomplete, the inspection shall be discontinued and the contractor advised of this.

Prior to issuing a certificate for substantial completion, the administrator should verify that the following conditions exist:

- Written statement from the contractor that the project, or designated portion thereof, is substantially complete or that construction is sufficiently complete for beneficial occupancy by the owner
- Correctly executed consent of surety for reduction in retainage per the contract documents is in place
- Contractor's punch list with administrator's supplementary comments added. However, prior to any retainage being released the administrator must certify substantial completion and the administrator (or designer of record), general contractor, and owner must agree upon punch list work to be completed
- Temporary certificate of occupancy from appropriate agency and all required permits/approvals are obtained.

Where there is a lender involved in the project, the certificate of substantial completion shall be prepared by the administrator and certified by the owner and/or architect, prior to being submitted to the owner and the contractor for their written acceptance of the responsibilities assigned them in the certificate. The certificate of substantial completion will also establish the dates and responsibilities of any transitional arrangements that will be required between the owner and the contractor.

4.7.4 Architect/Administrator's Certified Copy of Final Punch List

Upon receipt of the contractor's punch list, the administrator will review the list and the completed work to determine whether the list is both accurate and complete. Items requiring correction and/or completion and are not included in the contractor's punch list shall be supplemented by the administrator. The owner should be informed that the items on the punch list shall be rectified and/or completed within the time limit set forth in the certificate of substantial completion. The contractor shall also be advised that any correction and/or completion of punch list items shall be conducted in a manner so as not to adversely affect the owner's occupancy of the facility.

4.7.5 Certificates of Occupancy, Use, or Operation

A certificate of occupancy is a document issued by a local government agency or building department certifying that the building in question complies with all applicable building codes, safety codes, health code requirements, and other laws, and basically stating that the building is in a condition suitable for general occupancy. The procedure and requirements for the certificate of occupancy vary widely from juris-diction to jurisdiction and based on the type of structure. The purpose of this certificate therefore is to document that the use is permitted, and that all applicable safety code and health code requirements have been met.

A use and occupancy certificate is required for the space to be used prior to opening any business. It is also generally necessary both to be able to occupy the structure for everyday use, as well as to be able to sign a contract to sell the space or close on a mortgage for the space. A certifi-cate of occupancy is proof that the building complies substantially with the plans and specifications that have been submitted to, and approved by, the local authority. It basically complements a building permit, the document that is filed by the applicant with the local authority before commencement of construction to signify that the proposed construction will adhere to all relevant ordinances, codes, and laws.

Often a temporary certificate of occupancy (TCO) will be applied for. This grants residents and building owners all of the same rights as a certificate of occupancy, except that it is valid only for a temporary period of time. In New York City for example, TCOs usually expire 90 days from the date of issue, although it is not uncommon for a build-ing owner to re-apply for a TCO, following all the steps and inspections required originally, in order to extend the TCO for another set period of time. Temporary certificates of occupancy are generally sought after and acquired when a building is still under minor construction, but where certain sections or numbers of floors in a building are considered to be habitable (e.g., in a high-rise apartment building), and upon issuance of a TCO, a buildig can legally be occupied or sold.

4.7.6 Final Waivers of Lien

Upon completing the job, the contractor is usually required to com-plete a final lien waiver (Figure 4.10). The general contractor is also required to obtain conditional final waivers from all subcontractors, vendors, and certain individuals prior to final payment being released. A final lien waiver is basically a document from a contractor, subcontrac-tor, material supplier, equipment lessor, or other parties to the construc-tion project stating they have received full payment and waive any future lien rights to the property. It should be noted that in the United States, liens cannot be filed against public property. Moreover, some states only use a conditional waiver on progress payment and an unconditional waiver on final payment.

The mechanics lien process can prove extremely valuable to contrac-tors, subcontractors, material suppliers, and other related parties to a con-struction project in enforcing their claims, if done according to the laws of the various states or the federal government. These parties are entitled to be paid for their material or labor contributions to the improvement of real property. Most lien waiver forms for the process can be obtained online or from local office supply stores.

4.7.7 Miscellaneous Issues

CONTRACTOR/VENDOR FINAL RELEASE AND LIEN WAIVER

The undersigned represents and warrants that it has been paid and has received (or that it will be paid and will receive via proceeds from this pay application) $_____ as full and final settlement under the contract/agreement dated _____ (including any amendments or modifications thereto) (the "**Contract**") between the undersigned and _____ ("**Contractor/Vendor**") for the _____ Project owned by _____ ("**Owner**") (PO Number: _____). In consideration for this final payment, and other good and valuable consideration, receipt of which is acknowledged, the undersigned makes the following representations and warranties:

1. The undersigned and Owner have fully settled all terms and conditions of the Contract (including any amendments or modifications thereto), as well as any other written or oral commitments, agreements, and/or understandings in connection with the Project.

2. The undersigned has been paid in full (or it will be paid in full via proceeds from this pay application) for the labor, services, and materials in connection with the Contract, including all work performed or any materials provided by its subcontractors, vendors, suppliers, materialmen, laborers, or other persons or entities.

3. The undersigned has paid in full (or it will pay in full via proceeds from this pay application) all its subcontractors, vendors, suppliers, materialmen, laborers, and other person or entity providing services, labor, or materials to the Project; there are no outstanding claims, demands, or rights to liens against the undersigned, the Project, or the Owner in connection with the Contract on the part of any person or entity; and no claims, demands, or liens have been filed against the undersigned, the Project, or the Owner relating to the Contract.

4. The undersigned releases and discharges Owner from all claims, demands, or causes of action (including all lien claims and rights) that the undersigned has, or might have, under any present or future law, against Owner in connection with the Contract. The undersigned hereby specifically waives and releases any lien or claim or right to lien in connection with the Contract against Owner, Owner's property, and the Project, and also specifically waives, to the extent allowed by law, all liens, claims, or rights of lien in connection with the Contract by the undersigned's subcontractors, materialmen, laborers, and all other persons or entities furnishing services, labor, or materials in connection with the Contract.

5. The undersigned shall indemnify, defend, and hold harmless Owner from any action, proceeding, arbitration, claim, demand, lien, or right to lien relating to the Contract, and shall pay any costs, expenses, and/or attorneys' fees incurred by Owner in connection therewith.

The undersigned makes the foregoing representations and warranties with full knowledge that Owner shall be entitled to rely upon the truth and accuracy thereof.

DATED:

        (Contractor/Vendor company name)
        By:
        Title:

STATE OF
COUNTY OF

I, a Notary Public for the above County and State, certify that _____ personally came before me this day and acknowledged that he/she is _____ [title] of _____ [company name], and that he/she, as _____ [title], being authorized to do so, executed the foregoing on behalf of _____ [company name]. Witness my hand and official seal this _____ day of_____, 20___.

        Notary Public

My Commission Expires:

NOTICE: THIS DOCUMENT WAIVES RIGHTS UNCONDITIONALLY AND STATES THAT YOU HAVE BEEN PAID FOR GIVING UP THOSE RIGHTS. THIS DOCUMENT IS ENFORCEABLE AGAINST YOU IF YOU SIGN IT, EVEN IF YOU HAVE NOT BEEN PAID.

FIGURE 4.10 Two examples of final waiver of lien formats.

FINAL WAIVER OF LIENS

Project Description:

Contract Date:
Work Performed:
Work Performed by:
Under Contract to:

Listed below is the final information regarding the above contract:

Contract Price	$_____
Net Extras/Deductions	+_____
Adjusted Contract Price	$_____
Amount Previously Paid	-_____
Balance Due-Final Payment	$_____

The undersigned, being duly sworn, deposes, certifies and says that:

(i) He (She) is an officer of, and is duly authorized to make this affidavit, waiver and release on behalf of ("Contractor").

(ii) Contractor has received in full all payments (plus applicable retention) due through the date of this instrument for all labor, services, equipment and materials (sometimes referred to as the "work") furnished to _____ ("Owner") on the job of above project.

(iii) Contractor has paid in full or otherwise satisfied all of its obligations for labor, materials, equipment and services and all other indebtedness associated with the performance of Contractor's work on the Project, including without limitation payment in full to, or other satisfaction of, all persons and entities (the "Subcontractors") which have furnished labor, services, equipment or materials to Contractor.

(iv) In consideration of the payments received, and upon receipt of the applicable retention, Contractor forever waives, releases and relinquishes any and all claims and rights to a mechanic's lien, stop notice, bond right, equitable claim or right to any fund, and right to a labor and material bond or other bond on the Project and all other rights and claims that Contractor has on the Project.

(v) Contractor guarantees to Owner that the work furnished by Contractor (including work furnished by the Subcontractors) on the Project is and, after receipt of the applicable retention, shall be lien free, that the Subcontractors have no right to any mechanic's lien, stop notice, bond right, equitable claim or right to any fund, any right to a labor and material bond or other bond on the Project or other rights and claims with respect to the Project, and Contractor agrees to indemnify _____ and _____ against any claim or lien asserted through or under Contractor with respect to the Project, including without limitation any claim or lien asserted by any person who has furnished labor, materials, equipment or services to Contractor.

(vi) The undersigned further guarantees that all portions of the work furnished and installed by them are in accordance with the contract and that the terms of the contract with respect to these guarantees will hold for the period specified in said contract.

_____

_____

_____

Sworn to me this _____ day of

_____, 200__.

By _____

Title _____

_____
Notary Public

NOTE: Return four (4) signed releases to _____ at _____, _____ to the attention of _____.
Payment will not be made until the signed releases are returned.

FIGURE 4.10   (*Continued*)

1. Commissioning and Warranties

Establish commissioning procedures and dates for the commencement of all warranties. Commissioning and warranty review services consist of:

- Monitoring compliance by GC/CM with commissioning of operating systems, etc.; GC/CM must obtain from trade contractors and provide owner all required warranty documents and operating manuals; as-built (record) drawings; etc.

- Consultation and recommendation to the administrator and owner during the duration of warranties in connection with inadequate performance of materials, systems, and equipment under warranty
- Inspection(s) prior to expiration of the warranty period(s) to evaluate adequacy of performance of materials, systems, and equipment
- Documenting defects and/or deficiencies and assisting the owner in providing instruction to the contractor(s) for rectifying noted defects.

2. Architect's Supplemental Instructions (ASIs)

This form is used by the architect or project administrator to issue additional instructions or interpretations and clarifications or to order minor changes in the work which do not involve added costs and/or time extensions and are often documented by AIA document, Architects Supplemental Instructions G710. It is intended to assist the project architect and administrator in performing their obligations as interpreters of the contract documents in accordance with the owner-architect agreement and the general conditions. The administrator will prepare and issue an architect's supplemental instruction for all additional work that is not included in the contract documents and which will not modify the contract sum or extend the contract time.

All architect supplemental instructions need to be recorded in the project log book. The administrator may prepare the architect's supplemental instruction for the owner and/or architect's signature. The administrator is generally encouraged whenever possible to try and resolve small incidental issues by means of the architect's supplemental instruction. The architect's supplemental instruction shall be forwarded to the contractor for a signature as an acknowledgment that the work described will not modify the contract sum or contract time.

3. Time Extensions

During the course of the construction process the contractor may feel obliged to put forward a request for an extension in the contract time. This can be for one of several legitimate reasons that are totally beyond the control of the contractor, including:

- Inclement weather
- Owner requested changes or additions to the original scope of the work (e.g., change order or construction change directive)
- Delays caused by slow responses to RFIs
- Late material shipments from suppliers
- Slow processing of submittals or shop drawings
- Labor strikes.

The most common reason for time extension requests is inclement weather. The contractor should make allowances for a normal amount of severe weather in the project construction schedule. Time extensions are granted only for abnormally severe weather, defined as weather which was both detrimental to construction activities and more frequent than usually experienced during that time of year. It is important to note that adverse weather during certain phases of construction (e.g., pouring of concrete floors or foundations) can affect the construction schedule more adversely than good weather can benefit the construction schedule during other phases.

The contractor is typically required by the construction documents to notify the administrator, owner, and/or architect of any potential claim for additional time, due to delay, within 20 days of the commencement of the delay. When reviewing claims for time extensions, the contractor's daily log should be reviewed to verify that the bad weather occurred during the specified date and that lost time for that period was actually experienced by the contractor.

When time extensions are requested due to delinquent or late material deliveries, the contractor should be asked to furnish verification of the original date for which the material order was placed. In many cases, the contractor or the subcontractor failed to place the order in sufficient time to ensure that delivery of the material meets the contractor's schedule. Time extensions may consist of simple requests for extensions of time, or the requests may be more complex and include a request for cost reimbursement related to the extra time request. Except for exceptional circumstances, it may be prudent to try and retain all time extension requests until the end of the project.

4. Shop Drawing Submittal and Review Procedure

There are many items associated with construction that cannot be ordered out of a product brochure or off the shelf. In many cases items need to be fabricated in a shop or manufactured specifically for the project. To confirm the owner's intent, "shop drawings" are required which are essentially the supplier's or fabricator's version of information shown on the drawings in the contract documents. Shop drawings are typically submitted by subcontractors or vendors and contain greater detail and configurations of the item in question sufficient to fabricate and erect the item. Once completed, they are submitted to the general contractor who sends them off to the project administrator (or architect) for final approval. Upon approval they are returned to the general contractor and subcontractor for fabrication.

Shop drawings are required for many items such as steel rebar bends, steel beams, trusses, architectural woodwork, and ornamental metalwork. In the case of structural steel for example, shop drawings may include welding details and connections that are not typically part of the structural engineer's drawings. The term submittal often refers to the totality of the shop drawings, product data, and samples; all of these documents are submitted to the owner or owner's representative for approval prior to fabrication and manufacturer of the items they represent. Once approved, submittals may become part of the contract documents and should be incorporated into a submittal log. To be included in the contract documents the shop drawings must have been in existence at the time of the signing of the construction contract and incorporated by reference into the contract, or drawings that were added later as contract modifications and are signed by the owner and the contractor, such as change orders and construction change directives.

Arthur F. O'Leary, author of *A Guide to Successful Construction* and *Learning To Live With This "Necessary Evil,"* says, "To the construction industry, shop drawings seem to be a necessary evil. Contractors find them expensive to produce and architects find them unappealing to review. Both find them time-consuming and costly to administer. We seemingly cannot construct buildings without them; but they have become a perennial source of annoyance and confusion and more importantly, a significant source of professional liability claims against architects.

Undiscovered mistakes in shop drawings will often lead to unexpected or undesired construction results as well as exorbitant economic claims against architects, engineers, and contractors. Some shop drawing anomalies have resulted in costly construction defects, tragic personal injuries, and catastrophic loss of life."

O'Leary goes on to say, "The principal reason architects and engineers need to review the shop drawings is to ascertain that the contractor understands the architectural and engineering design concepts and to correct any misapprehensions before they are carried out in the shop or field. They review shop drawings of any particular trade or component to determine if the contract drawings and specifications have been properly understood and interpreted by the producers and suppliers.

The shop drawings should prove to the architect's satisfaction that the work of the contract would be fulfilled. If the shop drawings indicate that the work depicted will not comply with the intent of the contract drawings and specifications, the architect has an opportunity to notify the contractor before the costs of fabrication, purchase, or installation have been incurred."

The following procedure shall be employed by the administrator for the processing of all shop drawings and related product data for the project:

1. Shop drawings are required to be submitted in the format required by the contract documents. Shop drawings received in any other format may be returned to the contractor with a "not reviewed" note attached to the submittal.

2. Shop drawings shall be initially reviewed by the contractor and stamped accordingly. Shop drawings that do not bear the contractor's approval stamp or contain excessive errors that clearly indicate the contractor's review was inadequate may be returned to the contractor to be resubmitted as required.
3. Submittals are to be date stamped upon receipt and stamped with the standard office review stamp directly below the date stamp.
4. The Standard AIA G-712 Shop Drawing Review form is to be used for submittals. A separate sheet shall be used for each specification division. Log submittals are to be by CSI number and numbered chronologically.
5. Shop drawing submittals shall be as required by the contract documents. All transmittals to the administrator and consultants shall be recorded in the shop drawing log in the same manner as those submittals reviewed by the owner and/or architect. The shop drawing number should be written in the upper corner of the transmittal for ease in tracking.
6. The architect's shop drawing review shall be conducted using a printed copy of the submittal. All correct items and deficient items needing correction are to be marked preferably using different colored markers. All questions during the review process need to be noted.
7. The corrected drawing with appropriate review comments, date stamp and approval stamp with necessary action indicated shall be transmitted back to the contractor. The return submittal shall be recorded in the shop drawing log with a submittal number, date, and status of transmitted item recorded. If the submittal is rejected, this should be noted.
8. Marked-up copies of the submittals should be retained for reference. It is suggested that the copies be marked accordingly (e.g., "mark-up," "final," "rejected," etc.).

When the shop drawings have been approved, they should be distributed to the relevant parties including the administrator, owner, general contractor, and architect of record. Often steps have to be taken in the shop drawing review process to avoid any unnecessary work or assumption of responsibility by the administrator. These include:

- Avoid accepting responsibility for such things as verifying field dimensions, confirming compatibility with other submitted items, etc., as this work is clearly described in the contract documents as the responsibility of the general contractor.
- Normally shop drawings are processed within 10 working days. The contractor should be advised in writing if the review for a particular submittal is anticipated to take longer than this.
- Shop drawings and project samples should be kept in a file cabinet at the administrator's desk and not in the central file system.

- When the shop drawings for specialized equipment are furnished by the fabricator, it would be prudent to take the following precautions:
  - The administrator's responsibilities should be carefully reviewed as they relate to the shop drawing.
  - Statements in the owner/architect agreement that may relieve the architect of the responsibility of design and construction work by others should not be solely relied upon.
  - It is advisable to consider bringing in a specialist (engineer) rather than a fabricator to design and approve the equipment if it presents unusual risks. The specialist's review should be requested prior to approving the equipment.
  - Consider having design and construction details checked by a qualified specialist if they are outside the administrator's expertise.

5. Freedom of Information Letters (FOILs)

FOIL requests are often used as a means with which to check each subject property for building and code compliance, as well as locate its certificate of occupancy (CO). The research log provides space to keep track of the numerous agencies and officials that are invariably contacted to try and locate the municipal departments that record and maintain such information. It is necessary to submit these "foil" requests as soon as possible since most agencies, under the Freedom of Information Act are given anywhere from seven to 10 business days to respond. The information found on foil requests is typically listed by agency, state, municipality, and then building or fire department. Responses to these requests are to be used as exhibits in the project status report. Note that some states such as Virginia stipulate that they are not required to provide such information, even under Freedom of Information statutes, if the person or entity making the request is located outside of the state. In such instances the administrator should be notified immediately (see Figure 5.5).

4.8 QUALITY CONTROL/QUALITY ASSURANCE

Quality control and quality assurance are important concepts, yet most project managers and design professionals have only a vague understanding of their meanings and the differences between them. In fact, both these terms are often used interchangeably to refer to ways of ensuring the quality of a service or product. However, the terms are different in both meaning and in purpose. The ISO 9000 defines quality control as "the operational techniques and activities that are used to fulfill requirements for quality," whereas its definition of quality assurance is "all those planned and systematic activities implemented to provide adequate confidence that an entity will fulfill requirements for quality." This means that quality control refers to quality related activities associated with the creation of project deliverables. Quality control is used to verify that deliverables are of acceptable quality and that they are complete and correct. Quality assurance on the other hand refers to the process used to create the deliverables, and can be performed by a manager, client, or even a third-party reviewer.

Quality assurance is based on a process approach. Quality monitoring and its assurance ensure that processes and systems are developed and adhered to in a manner so that the deliverables are of superior (or at least acceptable) quality. This process is intended to produce defect-free goods or services with basically minimum or no rework required. Quality control, however, is a product-based approach. It checks whether the deliverables satisfy specific quality requirements as well as the specifications of the customers or not. Should the results prove negative, suitable corrective action is taken by quality control personnel to rectify the situation.

Another major difference between quality control and quality assurance is that assurance of quality is generally done before starting a project, whereas the quality control generally begins once the product has been manufactured. During the monitoring process, the requirements of the customers are defined, and based on those requirements the processes and systems are established and documented. After manufacturing the product, the quality control process typically begins. Based on the client requirements and standards developed during the quality guarantee process, the quality control personnel check whether the manufactured product satisfies those requirements or not. Assurance of quality is therefore a proactive or preventive process to avoid defects whereas quality control is a corrective process to identify the defects in order to correct them.

Most activities falling under the scope of quality assurance are conducted by managers, clients, and third party auditors. Such activities may include process documentation, developing checklists, establishing standards, and conducting internal and external audits. Designers, engineers, inspectors, and supervisors on the shop floor or project sight perform quality control activities. Quality control activities are varied and include performing and receiving inspection, final inspection, etc.

Both quality control and quality assurance are to a great extent interdependent. The quality assurance department relies predominantly on the feedback provided by the quality control department. For example, should a recurrent problem occur regarding the quality of the products, then the quality control department provides necessary feedback to the quality monitoring and assurance personnel that there is a problem in the process or system that is causing product quality issues. Upon determining the principal cause of the problem, the quality assurance department then instigates changes to the process to rectify the situation and to ensure that there are no quality issues to worry about in the future. Similarly, the quality control department follows the guidelines and standards established by the quality assurance department to check and ensure that deliverables meet the quality requirements. For this reason, both departments are fundamental to maintaining a high quality of deliverables. And although both quality control and quality assurance are different processes, the strong interdependence between them sometimes leads to confusion.

How the Consultant Functions in the Requisition Process

5.1 GENERAL

In examining the requisition process it will be seen that the smooth flow of funds during the construction process is of the utmost importance to both the borrower and lender. While some banks and consultancy firms have developed their own forms to help process requisitions, an AIA G702 contractor form is commonly used. This is a convenient form with which the contractor can apply for payment and the architect can certify that payment is due (Figure 5.1).

There is an obvious need for a professional objective review of drawings, specifications and construction loan commitments and payment requests to fully protect the lender. In this respect, the construction consultant (consultant) serves the lending institution by examining the conceptual design and specifications or engineering soundness and compliance with governmental regulations, and by making a study of cost comparables for similar projects, in addition to a trade-by-trade breakdown. This resultant estimate is then reconciled with the borrower's estimate for general conformity.

Commonly used Formats

The construction consultant is involved in the direct construction costs (hard costs—total labor and material costs incorporated into the project, including a fee for the general contractor), as well as the indirect costs (soft costs—e.g., land, interest, taxes, design fees, developer's fee, closing costs, etc.) which are part of the total requisition as it relates to the project. The consultant's analysis does not end with reviewing overall costs. The borrower's item-by-item breakdown is also investigated to make sure that it is not "front-loaded." A front-loaded cost estimate, where early trades have been assigned significantly higher costs than are actually estimated, would require large early draws on the loan. Most consultants skilled in cost analysis for lending institutions have little difficulty detecting front-loaded manipulations.

Sustainable design engagements usually need to be conducted with greater breadth and depth than conventional building engagements. Facility managers are being plagued with requirements to employ green strategies and reduce energy and other costs. As such, benchmarking, downsizing, improving productivity and efficiency, streamlining supply, purchasing, and partnering are just a few of the initiatives ongoing in many facility departments today.

Inspect the Property

With existing buildings (whether for renovation or as investments), prospective buyers would be prudent to evaluate the condition and value of each of the properties they are considering to purchase. Professional evaluations or in-depth examinations of a property may well reveal problems or potential problems that are not evident at first glance. For this reason, the buyer is required to be given an opportunity to conduct an independent evaluation. This process, which is known as "due diligence," is an integral part of the buying process.

With respect to the purchase of existing buildings, the knowledge and information possessed by facility managers gives them the power to transform a reactive maintenance program into a proactive one that optimizes future benefits, i.e., maximizes return on investment. This clearly indicates that maintenance protects the capital investment of equipment, systems, and physical plant by prolonging their useful life, and should be considered a valuable investment in the future. Furthermore, to maximize the return on investment of physical plant assets, accurate, sound, and detailed information on the subject elements and the status of the assets managed is required by the facility manager. A comprehensive inventory schedule of the physical plant's equipment, and the components that make up that equipment, is crucial and forms the basis for preventive maintenance, predictive maintenance, space utilization, and capital asset replacement analysis.

Equipment, Systems, and Structures Inspections

Few facilities management activities have as much potential to influence facilities improvements and funding as regularly scheduled inspections. Retro-commissioning is an option that should be seriously considered to see whether all systems are operating as intended. Total building, equipment, and system inspections are an essential part of an

APPLICATION AND CERTIFICATE FOR PAYMENT

AIA DOCUMENT G702

PAGE ONE OF   PAGES

TO (OWNER):   PROJECT:

APPLICATION NO:

Distribution to:
☐ OWNER
☐ ARCHITECT
☐ CONTRACTOR

FROM (CONTRACTOR):   VIA (ARCHITECT):

PERIOD TO:

ARCHITECT'S
PROJECT NO:

CONTRACT FOR:

CONTRACT DATE:

CONTRACTOR'S APPLICATION FOR PAYMENT

Application is made for Payment, as shown below, in connection with the Contract. Continuation Sheet, AIA Document G703, is attached.

CHANGE ORDER SUMMARY	ADDITIONS	DEDUCTIONS
Change Orders approved in previous months by Owner		
TOTAL		
Approved this Month		
Number Date Approved		
TOTALS		
Net change by Change Orders		

1. ORIGINAL CONTRACT SUM .... $
2. Net change by Change Orders .... $
3. CONTRACT SUM TO DATE .... $
4. TOTAL COMPLETED & STORED TO DATE .... $
   (Column G on G703)
5. RETAINAGE
   a. ___ % of Completed Work .... $
      (Column D + E on G703)
   b. ___ % of Stored Material .... $
      (Column F on G703)
   Total Retainage (Line 5A + 5b or
   Total in Column I of G703)
6. TOTAL EARNED LESS RETAINAGE .... $
   (Line 4 less Line 5 Total)
7. LESS PREVIOUS CERTIFICATES FOR
   PAYMENT (Line 6 from prior Certificates) .... $
8. CURRENT PAYMENT DUE .... $
9. BALANCE TO FINISH, PLUS RETAINAGE .... $
   (Line 3 less Line 6)

The undersigned Contractor certifies that to the best of the Contractor's knowledge, information and belief the Work covered by this Application for Payment has been completed in accordance with the Contract Documents, that all amounts have been paid by the Contractor for work which previous Certificates for Payment were issued and payments received from the Owner, and that current payment shown herein is now due.

CONTRACTOR:

By: Date:

State of: County of:

Subscribed and sworn to before me this   day of

Notary Public:

My Commission expires:

ARCHITECT'S CERTIFICATE FOR PAYMENT

AMOUNT CERTIFIED .... $

(Attach explanation if amount certified differs from the amount applied for.)

In accordance with the Contract Documents, based on on-site observations and the data comprising the above application, the Architect certifies to the Owner that to the best of the Architect's knowledge, information and belief the Work has progressed as indicated, the quality of the Work is in accordance with the Contract Documents, and the Contractor is entitled to payment of the AMOUNT CERTIFIED.

ARCHITECT:

By: Date:

This Certificate is not negotiable.   The AMOUNT CERTIFIED is payable only to the Contractor named herein. Issuance, payment and acceptance of payment are without prejudice to any rights of the Owner or Contractor under this Contract.

A	B	C	D	E	F	G		H	I
			WORK COMPLETED						
ITEM NO.	DESCRIPTION OF WORK	SCHEDULED VALUE	FROM PREVIOUS APPLICATION (D + E)	THIS PERIOD	MATERIALS PRESENTLY STORED (NOT IN D OR E)	TOTAL COMPLETED AND STORED TO DATE (D + E + F)	% (G ÷ C)	BALANCE TO FINISH (C − G)	RETAINAGE

FIGURE 5.1   AIA Document G702, Contractor's Application for Payment, which is perhaps the most widely used payment application form in the industry. Some lenders and consultants have developed their own forms to help process contractor requisitions.

overall facilities management plan. Formal inspections or commissioning should not be considered as a replacement for other types of maintenance and operations processes but rather as another process in the network of interdependent processes that make up the facilities management system. Data collection, deficiency reporting, and assessment activity for a piece of equipment is of very little value if the person performing the inspection does not understand how a specific piece of equipment fits into the system of which it is part. The various elements of mechanical and electrical equipment, life safety, and structural components must all be part of interdependent and integrated systems that result in the temperatures, volumes, control, and automation that support the occupants' comfort and function.

5.2 AREAS OF CONCERN TO THE LENDER AND BORROWER

It is necessary for the lender to implement a construction loan monitoring program from the outset to evaluate the construction consultant's experience, established procedures and protocols, and to establish timely communication and service. It should be noted that many of the issues that concern the lender also concern the project owner/potential borrower and often overlap.

The lender's consultant is frequently engaged to advise the lender on handling borrower issues directly related to project construction. Three of the more common concerns expressed by borrowers and contractors involve payment for stored materials, off and on-site, the lender's "retainage" of funds, and cost overruns. The consultant is also required to evaluate the potential risks in funding a specific project.

The general conditions of the contract documents often require the general contractor and major subcontractors to provide performance and payment bonds to add the weight of a surety company to that of the contractor. Lenders will often insist on bonds particularly if the general contractor is considered to be weak.

The best way for a lender to mitigate the risk and protect against unforeseen losses is to require adequate insurance coverage on the improvements. This is discussed in chapter 10. Insurance coverage usually takes the form of an all-risk builder's policy that will provide the funds needed to replace a structure either partially or completely if it is destroyed by fire, floods, or other causes. In the absence of appropriate insurance, such a loss could mean financial disaster. Knowledgeable construction lenders usually insist on coverage in an amount equal to 100 percent of total cost, but it is always advisable to consult an attorney or insurance agent before making a final determination (Figure 5.2).

Worker's Compensation

	Limit of Liability
Worker's Compensation	Statutory Benefits
Employer's Liability	$500,000

Comprehensive General Liability
(including coverage of Contractual Liability assumed by the Contractor under Indemnity Agreement set forth below and completed operations coverage).

	Limit of Liability
Bodily Injury	$1,000,000 per occurrence
Property Damage	$1,000,000 per occurrence
Professional Liability	$1,000,000 per occurrence or claims made form (coverage shall be maintained for three years following expiration of the assignment).

Comprehensive Automobile Liability

	Limit of Liability
Bodily Injury	$1,000,000 per occurrence
Property Damage	$1,000,000 per occurrence

Of note, it is now common practice to limit liability to the Consultant's fee or a sum of say $50,000, whichever is greater.

FIGURE 5.2   Typical insurance coverage to be considered usually includes an all-risk builder's policy that will provide the funds needed to replace a structure either partially or completely if it is destroyed by fire, floods or other causes. Appropriate insurance protects against potential risks that could cause financial disaster.

5.2.1 Inherent Risks and Funding Problems

One of the primary risks inherent to a construction loan is the danger of non-completion of a project. Most construction lenders are fully aware that once the first dollars of the loan are disbursed, there is little possibility of being repaid from the real estate until the project is completed and the value of the real estate can make repayment possible.

Another significant risk that lenders are sometimes confronted with is when the proposed project is not economically feasible and its value upon completion does not provide a means of repayment. This is particularly true when a building that is designed to be high performance does not meet its expectations. When a construction lender fails to consider the real estate and its value as the ultimate source of repayment, it has failed to exercise the prudence and fiduciary responsibility required by the lending institution. Long-term mortgage commitments are comforting, but they exist as a result of the value of the real estate upon completion. It is also important that construction consultants make themselves reasonably available to the lender for telephone conferences, discussions, clarifications, advice, guidance, and recommendations pertaining to any aspect or issue of the project in question without additional costs incurred by the lender or its borrower.

Other inherent risks and problems that may confront the lender when considering project funding include:

- Insufficient funds with which to complete the project may be considered to be the most common problem. Builders can run out of funds for reasons such as: inadequate project capitalization at the start; underestimating the cost of the total project; increased costs during construction; lack of early rentals; and cash drain due to borrower's other projects.
- Business failure of the borrower or principal is another cause of non-completion. Adverse business conditions experienced on other projects by the borrower can seriously jeopardize the well-being of a healthy project.
- A contractor's or subcontractor's failure to perform is another major cause of non-completion. A substantial amount of time and money may be lost if the general contractor or a major subcontractor cannot continue to execute the terms of the contract due to financial problems, increased costs, spreading too thin, lack of material, or non-qualified personnel. Once a contractor starts on a project, it is often difficult for many reasons to be replaced.
- Contractors often encounter problems relating to unavailability of material at critical junctions of a project. For example, such problems may relate to timely delivery of items such as ready-mix concrete during pouring of a concrete floor slab or an inability to find a manufacturer of special-sized components required for constructing a special type of detail. Likewise, if deliveries exceed schedules, this could result in additional project costs and possible claims.
- Strikes, especially during a critical stage of construction, can effectively close down a project and throw a contractor's construction schedule into chaos, causing unacceptable delays or non-completion. The rescheduling of trades involved in a project often causes a loss of momentum that is difficult to regain.
- Destruction of improvements by fire, wind, or other catastrophes is a danger that is frequently overlooked but with which every mortgage lender must live. Damage caused by hazards such as loss of time, material labor, and value created cannot be borne by either the borrower or the lender.
- Major deviations from plans and specs—violation of BLA—and user does not accept or take possession, or permanent lender balks.
- Early buyouts coming in over budget resulting in scope changes.
- With projects seeking green certification there is the risk that the targeted certification (e.g., LEED™ or Green Globes™) is not achieved which may reduce the anticipated value and marketability of the property.

5.2.2 Keeping the Lender Knowledgeably Informed

The purpose of the monthly project status report is to put in writing what has been learned about the project by way of the site visit, review of documents, discussions, etc. during the period covered for the lender representative. This information must be provided promptly, not later than seven days after the site visit. What the bank or lender is most interested in knowing is:

- The value of work in place (excluding stored materials) to allow funding to proceed
- Whether the work performed during the period covered complies with the contract documents
- Whether there will be enough money left in the budget to complete the project
- The status of the construction schedule.

Once the assessment is completed (including the report), the owner is informed of its contents, containing noted deficiencies, recommendations, and cost estimates. The information may surprise the owner, and it may take some time to process prior to being accepted and proceeding to the planning phase. The consultant can answer questions and provide additional explanation if needed.

Upon completion of the final report there are many ways to proceed, including phasing or staging of the work needed based on the information the owner/investor provided regarding the availability and timing of funds. Once the owner/investor is comfortable with the scope and cost of the project, the consultant can develop a plan, help locate good contractors for the owner/investor to consider, and oversee the work once contracts are signed and the project begins. The service the consultant performs depends to a large degree upon who is being represented—the owner or the lender.

If a lending institution finds itself the "owner" of an incomplete project, it is generally placed in an unfamiliar position—that of a "builder." The consultant will take the lender's "builder" position in a faltering real estate construction project and complete the project for the lender. The consultant's services include cost-to-complete reports, renegotiation of relevant contracts, solicitation and analysis of bids, and management of the construction as the lender's agent.

5.2.3 Cost Overruns

Rapidly rising construction costs have adversely impacted many developers of commercial as well as residential projects, particularly if insufficient contingency or profit margin is taken into consideration. For larger projects, it is not uncommon for work being implemented on a project to be based on cost estimates prepared considerably earlier. The specter of substantial cost overruns haunts many projects and is of vital concern to both the borrower and lender. The consultant is often required to verify construction prices in the locale where the project is being constructed.

The consultant of course is not authorized to restructure a loan to provide additional funds. Careful examination of the project's progress as related to the borrower's prepared monthly requisitions can however usually provide an indication that a particular project may be about to experience difficulty. In this respect, the consultant can render an expert opinion to the lender, but the ultimate business decision will rest with the lending institution. And as the lending institution considers the merits of a particular project, it requires from the borrower certain technical data. These are reviewed by the lender as well as the lender's consultant.

It is generally recognized that there are several primary factors that cause project costs overruns, inlcuding:

- Incomplete and inadequate drawings and specifications
- Owner instigated scope change
- Contractor's requested changes orders
- Delays caused by late deliveries, etc.

While the project manager tries to coordinate the general requirements and solve the above problems, there remain several factors that make overruns difficult to avoid. And unfortunately no matter how efficient the cost and control system is, problems may still arise. Following are some of the more common causes of cost problems:

- Faulty comparison of actual and planned costs
- Inadequate estimating techniques and/or standards, resulting in unrealistic budgets
- Unforeseen technical problems
- Out-of-sequence starting and completion of activities and events
- Inadequate or lack of management policy on reporting and control practices
- Inadequate work breakdown structure
- Management induced budgets or bids reduction to be competitive
- Schedule delays that require overtime or idle time costing
- Unrealistic material escalation factors.

5.2.4 Plans, Specifications, and Other Documents

In simplest terms, what is the loan for? Site plans should delineate the extent of the property involved as well as any site improvements such as parking lots, amenities, out buildings, drainage, as well as the major structures. Computations should be provided indicating proposed apartment counts, parking spaces, gross rentable square footage, or other appropriate data. There are numerous sources that an investor or building owner can tap to help find proficient consultants to design a project and provide the required information.

The borrower or consultant should provide a complete set of the latest plans (including site plans, structural, mechanical, electrical, etc.) and specifications and soil reports should be provided. The lender must recognize that fast-track design methods are frequently used by a borrower. However, it should also be recognized that while these methods permit construction to commence prior to the completion of final working drawings for all portions of the work, this depends on the type of contract and delivery system agreed upon. The lender looks to the consultant for comments on the status of the plans and specifications. In the absence of final plans and specifications where fast track methods are being used, the borrower may be requested to submit outline specifications and narrative descriptions for work where plans are not yet finalized. Once the consultant evaluates the materials and documents submitted, an evaluation is prepared for the lender.

Lending institutions usually require that all contracts governing the construction are in place and submitted. These may include the borrower's contract with a general contractor or construction manager as well as typical subcontractor's agreements. Lenders also typically require copies of leases covering rental spaces in shopping centers or commercial buildings. Copies of executed residential leases may also be required.

In order to be assured that adequate capacity exists for the various utilities that serve the proposed project, the lender often requests copies of letters from each utility serving the project generally stating that adequate capacity exists to serve the needs of the new project. The lender also generally requires a copy of the building permit issued by the local municipality.

With older properties in particular, a prospective buyer would want to know the useful life of the existing roof, as replacing it would be a major expense. A prospective buyer would also want to know if the mechanical and electrical systems are working properly and whether a renovation or upgrade is needed. If so, what needs to be upgraded and how much will it cost? Investors are often full of such questions, and for answers a full assessment of the facility and all of its parts is required.

Even well-built, well-designed buildings need regular preventive maintenance and repair. Ongoing assessment involves checklists of needed tasks, keeping up with required maintenance, and being aware of emerging problems. Facility owners can save time and money by knowing their buildings and keeping up on the issues. A small problem can quickly escalate into a major problem if not addressed quickly.

As for historical buildings, the repair or restoration may necessitate investing in a historic structure report. This type of facilities assessment delves into the history of the building, original materials used, and how the building has evolved, changed, or been modified through the years. In this report, the assessor uses history as a basis for making decisions concerning the future.

Another type of facilities assessment is the master plan report where the assessor looks at the possible future space functions and uses of the building. This report incorporates strategic planning aspects into the assessment. Most facilities assessments for existing buildings are led by an architect or engineer, typically supported by a team of professionals with expertise in the areas of mechanical, plumbing, and electrical systems, flooring, roofing systems, windows, structure systems, etc. The building assessor assigns the various components of the building to the person with the appropriate expertise and knowledge.

For property condition assessments (i.e., existing facilities) the more information that is provided to the assessor, the more complete and less expensive the assessment will be. Any drawings or blueprints of the building or additions, maintenance records, historical documentation, records of renovation, or restoration photographs should be forwarded to the assessor for review. If the information is not immediately available, it may be acquired through previous contractors, the local historical society, state or county offices, etc. If the available historical or maintenance data is insufficient, architects may need to conduct "exploratory demolition." This involves breaking through the walls or roof or other areas to investigate what is going on within the structure. For this reason, it may be worth spending extra time trying to retrieve any necessary documentation.

The assessment report should be stored in a safe place for future reference, whether as a historical document for the next assessment or to facilitate work that may be needed to be done at a later point in time. A typical facilities assessment report will include many of the following:

- General introduction
- Executive summary
- Summarized history of the property and maintenance programs, if any
- Description of existing conditions
- Cost estimates
- Value engineering (VE)
- Description of major deficiencies
- Comments and recommendations
- Optional reports from various specialists, such as structural engineer, mechanical or electrical engineer, etc.
- Photographs, drawings (sketches or measured), etc.

5.3 OWNER NEEDS AND EXPECTATIONS

Building today, whether "green" or conventional, is both complex and expensive, and where different people with different needs, interests, and expectations try to work together to design and construct a project for an owner. A building project is normally constructed only once and often provides a challenge for the parties involved to work constructively together during all stages and make the construction process run efficiently without any serious delays, misunderstandings, structural defects, or other complications. It is important to note that many of the risks that an owner faces with a property are similar to those faced by a lending institution. The first step that every building owner should take in this respect is to objectively consider what needs to be built and determine what is feasible to be built on the land and within the budget with which the owner has to work. The importance of conducting a feasibility study is to identify and understand the current and future needs of the building project and translate those needs into a successful building project along with a plan to make it happen. However, it is fairly evident that still more research is needed to gain insights into whether designers and contractors are generally meeting owners' needs and expectations during the design, bidding, and construction of building projects.

For the owner's project to succeed, the owner must both define the project scope, which is a crucial first step toward making the project a success, and set realistic expectations at the design stage. Faced with reduced staff levels, shorter delivery times, and increasingly complex projects, project owners are challenged to reduce the cost, time, and risk associated with the construction-delivery process.

One method that can help projects meet owner expectations is project commissioning. Building commissioning for new buildings is a quality assurance process that is conducted to verify and document that the building's systems function as intended and meet the operational needs of the building owner and building users. Verifying system performance is based on comprehensive functional testing and measurement. Building commissioning provides benefits such as a smoother construction process, reduced operation and maintenance costs, lower energy costs, and satisfied building occupants and tenants. It is also important to ensure that building operations staff members receive appropriate training and resources needed for system operation and maintenance procedures. It should be noted that building commissioning is not standard practice for conventional buildings, although it is often required for green buildings (e.g., LEED certification) and is also becoming more common for large or complicated buildings. Furthermore, the initial costs of commissioning are usually recovered many times over through operational savings, improved staff performance, and avoidance of potentially costly construction problems.

Ideally, building commissioning should begin as early as possible in the design phase of the project, thus allowing the commissioning provider to work closely with the design team and become familiar with the project goals and design intent as decisions are formulated. Plans can be made to effectively incorporate commissioning into the development process. The extent of the commissioning process can vary as well as the roles of those involved in the project. A comprehensive process is generally justified for large, complex projects, and is usually required for projects seeking green certification. The process preferably commences in pre-design and runs through post-occupancy. The commissioning process for smaller buildings is often less complicated, focusing on system balancing, simple functional tests of critical systems, and documentation. This helps ensure success and avoid problems and additional work later in the project. Commissioning can occur in the construction process or after construction is complete, but this makes it more difficult to document the design intent, identify design problems, develop testing plans, and conduct tests.

When implementing new construction and/or a facility change, a cost effective, efficient change process is needed that minimizes disruptions to existing site operations. It is also crucial for project owners to clearly establish and manage project objectives and project expectations from the beginning of project, preferably before meeting the designer, consultant, and contractors. Otherwise it becomes a struggle when modifications are made along the way. Owners often try to minimize the fees paid to architects and engineers, which often impacts their input and the time spent on the contract documents, which can subsequently cause problems during construction.

One risk that project owners face is that of economic feasibility, that is failure of the real estate to achieve the economic success that was envisioned for the project when initially conceived. This may be the result of poor location or design, improper pricing of rental or sale units, underestimating costs, or a combination of all these in addition to other reasons.

The evaluation of proposed real estate projects should be in accordance with their mortgage-ability, and their ability to support a mortgage in an amount equal to the proposed construction loan at the current rate and terms. Furthermore, for a project to be viable it should be capable of carrying any committed long-term mortgage. The risk factor is heightened when, for example, the economic success of the real estate depends on existing firm leases prior to the project's commencement and that reflects the amount of the construction loan sought, but then it is discovered that the credit of the tenants collapse or a tenant's business fails. This could substantially change the economic value of the real estate and cause it to lose its economic feasibility.

Property developers may face risks resulting from external events in addition to the primary risks inherent in the construction process itself. For example, projects can be delayed, stopped, or substantially altered due to community action. Likewise, legislation affecting construction already underway is sometimes enacted, which adds to the project's cost to a degree that it may adversely impact the economic feasibility of the property.

In recent years, environmental issues have become major sources of uncertainty for owners and lenders, particularly for buildings seeking green certification. In addition to the "greenwashing" of materials and performance of green features, many developers are only just beginning to learn how to deal with them. While these risks are real and while they may impact any potential construction loan, they cannot be completely eliminated.

During the tendering phase the project owner should conduct face-to-face interviews with the bidders to discuss directly any changes or modifications that have been made and to negotiate revised prices and costs. A lack of understanding on the owner's expectations can hinder the process.

Project Change Control

When the owner requires changes to be made that will alter the plan, the consultant/project manager must control and track them.

In establishing a cost control system, the best method is to isolate and control in detail the elements that possess the greatest potential impact on final cost, with only summary level control on the remaining elements. The variable with the greatest impact on the final cost of a construction project is usually labor costs. Most of the other cost elements in the project, such as materials, equipment, and overhead are relatively less significant. Good project management planning techniques during the planning processes can prevent potential cost problems later in the project. At a minimum, proper planning will alleviate the impact of these problems should they occur. Figure 5.3 is an example of a project report matrix to monitor and keep track of progress.

Cost control is a process that should be implemented for all types of construction, phases, and activities. Comparison with a cost standard method is more accurate than other methods, although it may be more complicated and expensive. Cost problems are normally caused by a variety of reasons such as inaccurate estimating techniques, predetermined or fixed budgets with no flexibility, schedule overruns, inadequate work breakdown structure (WBS) development, and so on.

Like many other controlling processes, cost control is concerned with monitoring project performance for variances. The project owner should monitor the factors that cause variance to keep their impacts to a minimum. An important thing to remember is that cost control ensures that all appropriate parties will agree to any changes to the cost baseline. This is an ongoing and continuous process that continues to manage cost

FINAL ACCEPTANCE - CONSTRUCTION PROJECT SUMMARY

PROPERTY NAME: **DATE OF VISIT:**

LIST OF CONTACTS:

DESCRIPTION	STATUS	COMMENT
TOTAL UNITS IN PROJECT		
TOTAL BUILDINGS IN PROJECT		
NUMBER BUILDINGS COMPLETE		
TOTAL UNITS COMPLETE & ACCEPTED BY MANAGEMENT		
% UNITS COMPLETE & ACCEPTED BY MANAGEMENT		
PERCENTAGE SITE WORK COMPLETE		
PERCENTAGE LEASING/OFFICE COMPLETE		(describe outstandings)
TOTAL PERCENTAGE COMPLETE		
UNITS IN PROGRESS (if applicable)		
PUNCH LIST GENERATED BY?		(list outstandings)
PROJECT ON SCHEDULE - see attached for delivery dates		
ORIGINAL CONSTRUCTION BUDGET (reference engineering due diligence report)		
CURRENT TOTAL CONTRACT(S) AMOUNT		
CONSTRUCTION $ APPLIED FOR AND APPROVED TO DATE		
% CONSTRUCTION $ APPLIED FOR AND APPROVED TO DATE		
CURRENT CONSTRUCTION DRAW REQUEST		
CURRENT CONSTRUCTION DRAW REQUEST		
TOTAL $ APPROVED AND REQUESTED		
BALANCE CONTRACT(S) $		
ORIGINAL CONSTRUCTION CONTINGENCY (reference engineering due diligence report)		
CURRENT CONSTRUCTION CONTINGENCY AVAILABLE		
TOTAL CHANGE ORDERS APPROVED & REQUESTED		
TOTAL CHANGE ORDERS APPROVED & REQUESTED		(list each and amount)
DEVIATIONS FROM DUE DILIGENCE SCOPE AND DOCUMENTS(reference engineering due diligence report)		(list & describe each item of deviation)
STATUS OF OUTSTANDING DUE DILIGENCE REPORTING RECOMMENDATIONS (reference engineering due diligence report)		(attach separate list)
ADEQUACY OF BALANCE OF FUNDS TO COMPLETE THE PROJECT (if applicable)		
WORK IN-PLACE RECONCILES TO CURRENT APPROVED & REQUESTED FUNDS		
ALL UTILITY METERS INSTALLED?		
CO's ISSUED BY BUILDING OR BY UNIT?		
CO's (required/obtained)		
ADA ISSUES/STATUS		
ADEQUACY OF PRIOR OVERSIGHT – see attached matrix		
FINAL SUBMITTALS IN PLACE – see attached matrix		
ASSESSMENT OF WORK QUALITY		
ENVIRONMENTAL STATUS		

Attachments: Scope of Project, Recommendations, Tabulations with Updates, Photographs

FIGURE 5.3   An example of a project report matrix to help monitor and keep track of a project's progress and status.

changes throughout the project. It is particularly important for contractors to have proper cost control systems. It helps project owners understand and control project costs easier and faster and helps achieve a successful project that meets the owner's expectations.

FINAL SUBMITTALS SUMMARY

Project: _____

Owner: _____

Contractor: _____

COMPLETION SUBMITTAL ITEM	STATUS	COMMENTS
ARCHITECTS STATEMENT OF SUBSTANTIAL COMPLETION		
ARCHITECT'S STATEMENT OF FINAL COMPLETION		
RELEASE OF RETAINAGE BY DEVELOPER and/or LENDER		
FINAL LIEN RELEASE FROM GC		
FINAL LIEN RELEASES FROM SUBCONTRACTORS		
FINAL LIEN RELEASES FROM MATERIAL SUPPLIERS		
CONSENT OF RELEASE FROM SURETY		
OUTSTANDING REGULATORY PERFORMANCE BONDS?		
FINAL CERTIFICATES OF OCCUPANCY		
WARRANTIES HAVE BEEN DELIVERED		

OVERSIGHT TASK/ROLE	IN-PLACE	COMMENTS
Progress Meetings		(how often, who attended, meeting minutes prepared by ___)
Architect		(onsite how often, field reports, change orders, product substitutions approvals, signs off on draws)
Geotechnical Engineer		(client, role, field reports)
Structural Engineer		(onsite, role)
Shop Drawing Review		(who managed, log kept, MEP engineers, structural engineer)
Change Order Process		(who managed, log kept, documents issued)
Concrete and Soils Testing		(name of lab, client, scope, how often, lab reports)
Lender's Consultant		(name, familiar, reports available)
City Inspection		(inspections required)

FIGURE 5.3   *(Continued)*

The use of computers has become imperative in today's construction projects. In fact, it is hard to imagine a project, especially a larger project, moving forward without the use of computers. The use of computers is discussed in other sections of this book, but generally speaking, project managers can rely on project management software and spreadsheet programs to assist them in preparing specifications, 3D model simulation, cost calculation, and earned value, among other things.

5.4 CONFIDENTIALITY OF REPORTS AND DOCUMENTS

5.4.1 Confidentiality and Responsibility

It is generally agreed that all documents submitted to the consultant and all reports both verbal and written produced by the consultant are strictly confidential and for the sole benefit of the lender or its agents and employees (Figure 5.4). Copies of the consultant's reports may be made available

**CONFIDENTIALITY AGREEMENT
BETWEEN
CONSULTANT AND ABC INTERNATIONAL, INC.**

This Confidentiality Agreement is executed this _____ day of _____, 200___ by _____(the "Consultant") and ABC International, Inc. ("ABC"), acting in any capacity.

In consideration of the mutual promises set forth herein and other valuable consideration, the Consultant and ABC agree as follows:

1. **Confidential Information**

   ABC, acting in any capacity, may provide the Consultant with, or allow Consultant access to, certain information not generally known to the public; other independent consultants; other environmental engineering; or engineering, testing or consulting firms conducting similar or the same services as ABC.

2. **No Disclosure**

   Consultant shall not at any time disclose, permit the disclosure of, release, disseminate, or transfer, whether orally or by any other means, ABC's client base or any part thereof, procedures, fees, fee structure, costs of doing business, compensation to ABC's employees or independent contractors/consultants, checklists, or report formats to any other person or entity, whether corporate, governmental, or individual, without the expressed prior written consent of an officer of ABC. Consultant shall return any written information, whether written or paper or computer disk, prepared by the Consultant or consulted by the Consultant, on any assignments conducted by the Consultant on ABC's behalf. Such information includes, but is not limited to: reports, proposals, fees, compensation data or agreements, correspondence, studies, testing results, checklists, client lists, etc. All such information is and shall remain the property of ABC. Consultant is not to retain any copies, whether paper or computer disk, of work conducted on ABC's behalf whether such work is considered draft, preliminary, or final.

3. **Surrender of Documents**

   Upon completion of employment by ABC, Consultant shall not knowingly have in his or her possession, custody or control, any computer disks or hard copies of any ABC reports or forms of ABC's reports, checklists, data, etc. If at any time in the future, Consultant discovers that he is in possession, custody or control of any such material, he will promptly deliver to ABC.

FIGURE 5.4   Example of a typical confidentiality agreement between the consultant and a third party.

4.   **Applicable Law**

This Agreement shall be governed by the laws of the Commonwealth of Virginia.

5.   **Attorney's Fee**

If any legal action or other proceedings of any kind is brought for the enforcement of this Agreement or because of any alleged breach, default or any other dispute in connection with any provision of this Agreement, the successful or prevailing parts shall be entitled to recover all reasonable attorney's fees and other costs incurred in such action or proceedings, in addition to any relief to which it may be entitled.

6.   **Re-Confirmation**

Upon completion of employment or any particular assignment, Consultant may be asked to re-confirm this agreement as a condition to receiving final payment by ABC.

7.   **Entire Agreement**

This Agreement embodies the entire agreement between the parties in relation to the subject matter herein and supersedes all
prior understandings or agreements, oral or written, between the parties hereto.

_____   _____
Consultant   Date

FIGURE 5.4   (*Continued*)

through, and at the discretion of, the lender to permanent lenders, loan participants, and the borrower. The consultant shall not have any legal obligation to a permanent lender or to the borrower, or any agents thereof.

The consultant does not assume or accept any of the design, planning, construction administration, or supervisory responsibility of the architect of record, engineers of record, or the contractor or construction manager, if one is or is not so retained. Moreover, the consultant is not responsible for any of the acts or omissions of the borrower, owner, architect, engineer(s), or contractor(s) in the construction of the stipulated project including, but not limited to, the methods, means, and sequence of construction, safety, compliance to governing codes and ordinances, changes in the work, or substitutions of materials. Services performed herewith reflect the role and responsibilities limited to those as the lender's consultant.

5.4.2 Construction Documents

All construction plans, specifications, work schedules, construction cost breakdowns, construction contracts, construction drawings, and other construction documents will be furnished to the construction consultant (administrator) by the lender. If the consultant receives documents from any other sources, the consultant must identify the document and source to the lender for approval before use in performing any of the services described in the agreement. The consultant shall regard all documents received concerning the project as confidential and prevent disclosure by the consultant or the consultant's employees concerning any specific information seen or heard regarding this project without the lender's prior approval.

5.5 TYPICAL SEQUENCE OF EVENTS UPON BEING AWARDED A PROJECT ASSIGNMENT

5.5.1 Project Preliminaries

The following sequence is actually based on guidelines employed by a very well-known due-diligence consulting firm, Inspection and Valuation International, Inc. (IVI), which I often used during my years with the firm. This is especially appropriate for property condition assessment assignments, but the same basic procedure applies to new construction with minor editing. The process essentially starts with an initial client contact and expression of interest in the award of a project. This if followed by the creation of a project file and assignment of a project number. This file will be maintained throughout the development and awarding of the contract. The sequence of procedure activities is generally as follows:

1. The client typically calls or emails for a quote on an assignment.
2. A standard proposal is prepared with answers for the proposal being saved in the appropriate file (here, IVI uses its own proprietary computer filing system) for the individual client.
3. The information regarding the proposal is entered into the Access database.
4. Upon the proposal being awarded, mobilization of the assignment begins.
5. The assignment is given a project number (anything related to this job is now filed via this project number).
6. The Access database is updated to now show an "awarded" job, the job number, and any other current information with regard to same. If the fees were negotiated, or any other special instructions given, they should be noted here as well. Likewise, if the assignment was not awarded, that information should also be entered into the database so that it is updated to read "not awarded" and any debriefing as to why the assignment was not awarded is noted.

Property records as furnished by the lender and/or borrower are reviewed. Documentary information will generally consist primarily of borrower files that contain descriptive information on construction materials and systems. Some files may have a prior engineering assessment report. If an ADA assessment has been performed, review its principal findings and note the major deficiencies identified. When available, a copy of the ADA report summary should be incorporated into the report's appendix. Drawings and specifications, if available on-site, should only be referred to for descriptive information on construction of the improvements. Plan review is not generally required.

5.5.2 Mobilization Documents

Upon notification by the client to proceed with the Contract work, a project manager is assigned (with larger firms the administrative staff are informed). Below is a brief description of the various documents required to mobilize a property condition assessment and subsequent property condition report (PCR). It is important to remember that the information contained in these documents is used by different employees and at various stages in the completion of a project. Therefore, each document needs to be accurate and updated as promptly as possible. The documents will be listed in the order in which a typical job is mobilized, the number of copies produced, and to whom they are given. Additional copies are easily provided since all documents mobilized for any given job are saved in that specific job number's folder.

Mobilization includes:

- A project manager is assigned to the project
- Preparation of an *assignment data* sheet (the answers saved from the proposal should be used)
- A project file is prepared and includes the following:
  - Contract with scope of work and terms
  - Site contact information
  - Property manager interview dorm
  - Parameter/measurement schedules (for residential buildings)
  - Tenant survey forms (for retail buildings)
  - Condition assessment checklist (for building type: generic; multi-family, hotel, suburban office, high-rise office, shopping center/mall)
  - ADA checklist
- Project file exterior should contain: Project number, project name, city/state, and client

- Assignment data sheet goes into the project file and is given to the project manager
- Project file is sent to the appropriate project manager
- Project manager calls client for *site contact* information if not provided
- Project manager calls site contact to schedule an *inspection date*
- Once inspection date is scheduled, client is faxed a *client site inspection confirmation letter* (which includes the date and time of the site inspection and the project manager)
- Project manager sends to client contact (possibly superintendent, property manager, etc.) the *pre-survey questionnaire and documentation checklist*
- Project manager makes *travel arrangements*
- Project research is initiated which includes the following requests for municipal documentation to be prepared and sent out (see also Chapter 4, Section 4.7.7):
  - *Flood plain map*
  - *Building department FOIL* (Freedom of Information Letter— Figure 5.5a)
  - *Fire department FOIL* (Figure 5.5b)
  - *Zoning department FOIL* (Figure 5.5c)
  - Other inquiries as necessary.

Useful tips for completing FOIAs:

1. The first step is to find the area code of the property and call information to get the numbers for the fire, planning, health, electric, and water departments.
2. Before commencing, call the planning department and check as to whether the property is located in the city or the county.
3. When calling the planning department, ask for the fax number of the building inspector, city planner, and zoning administrator. In addition, ask for the names of the local water and electric companies that service the area. Call them and ask if whether they have aerial photographs for the area available. If they do, ask for what years and get their address and hours of operation.
4. When calling for the fire department, ask to fax to a fire marshal, fire chief, or fire inspector.
5. When calling for the health department, check whether they have an environmental or "green" division; if so speak to someone there.
6. When calling the water department, ensure that this company services the particular piece of property by giving them the address. Then find out where and whom you would need to send the fax to.
7. Same as above for electric, make sure who you send it to is familiar with PCBs and transformers. Also enquire about alternative energy and tax incentives.

ABC Green Design, Inc.

XYZ Los Angeles
2134 Olympic Blvd.
Los Angeles, California 90064
(213) 234-1234

BUILDING DEPARTMENT INFORMATION REQUEST

Date: To:

Subject: Dept.:

                               Tel. No.:

                               Fax No.:

Sender: ABC Proj.No.:

☐ New York	☐ Washington, D.C.	☐ Los Angeles	☐ Chicago
(914) 694-1234 (tel)	(202) 907-1234 (tel)	(213) 896-1234 (tel)	(847).296-1234 (tel)
(914) 694-1235 (fax)	(202) 907-1235 (fax)	(213) 896-1235 (fax)	(847).296-1235 (fax)

ABC Green Design Inc., has been commissioned to conduct a Property Condition Assessment Survey on the above referenced Subject. Please respond to the following documentation/information requests:

1. Does the Subject have any material outstanding building code violations within its file? **Yes ☐ No ☐**
   If "Yes", please fax copies of same.

2. Is there any existing or pending material building or fire/life safety code requirements **Yes ☐ No ☐**
   that the Subject would not be grandfathered and therefore compliance would then be
   mandatory? If "Yes", please briefly explain.

3. Do you have any general or specific knowledge of any physical conditions (site or **Yes ☐ No ☐**
   building) that negatively impact the Subject such as localized flooding, sanitary sewer
   back-up problems, etc.? If "Yes", please briefly explain.

4. Is the Subject within a 100 year frequency flood plain? If "Yes", please identify the **Yes ☐ No ☐**
   Flood Hazard Zone as per FEMA's Flood Insurance Rate Maps.

5. What Building Code is enforced, and what is the local Zoning Ordinance classification of the property?

6. Are there any municipal required procedures or mandated improvements that are triggered by
   a change of ownership/title such as: a re-inspection by the Building Department, the installation of
   sprinklers, installing water conservation devices, etc.? If so, what are they?

7. Please fax us a copy of the Subject's Certificate of Occupancy.

Thank you for your assistance. Should you have any questions or should there be any fees associated with
providing the requested information, please call me.

FIGURES 5.5 Project research includes requests for municipal documentation:
a. Building department FOIL, b. Fire department FOIL, and c. Zoning department FOIL.

8. Find the number for the soil conservation service listed by the county
   name. Call them and find out whether they have aerial photographs
   for the area available. If they do, ask for what years and get their
   address and hours of operation.

9. Also find the number for the central library in the area. The library
   may have city directories available for the area. Often you will need
   to talk to the reference section.

10. Lastly, using one of the mapping programs, locate and map the planning
    department, soil conservation service, public library, and site location.

ABC Green Design, Inc.

XYZ Los Angeles
2134 Olympic Blvd.
Los Angeles, California 90064
(213) 234-1234

FIRE DEPARTMENT INFORMATION REQUEST

Date: To:

Subject: Dept.:

                  Tel. No.:

                  Fax No.:

Sender: ABC Proj.No.:

☐ New York	☐ Washington, D.C.	☐ Los Angeles	☐ Chicago
(914) 694-1234 (tel)	(202) 907-1234 (tel)	(213) 896-1234 (tel)	(847).296-1234 (tel)
(914) 694-1235 (fax)	(202) 907-1235 (fax)	(213) 896-1235 (fax)	(847).296-1235 (fax)

ABC Green Design Inc. has been commissioned to conduct a Property Condition Assessment Survey on the above referenced Subject on behalf of the mortgagee. Please respond to the following documentation/information requests:

1.   Does the Subject have any significant outstanding fire code violations within its file?   Yes ☐ No ☐
   If "Yes", please fax copies of same.

2.   Is there any existing or pending significant fire/life safety code requirements   Yes ☐ No ☐
   that the Subject would not be grandfathered and therefore compliance would then be
   mandatory? If "Yes", please briefly explain.

3.   Do you have any general or specific knowledge of any physical conditions (site or   Yes ☐ No ☐
   building) that negatively impact the Subject such as lack of sprinklers that are required
   by code, inadequate alarm systems, back-up problems, etc.? If "Yes", please briefly explain.

4.   Any general comments or suggested life/safety improvements?

Submitted By: _____ Date: _____

Thank you for your assistance. Should you have any questions or should there be any fees associated with providing the requested information, please call me.

FIGURE 5.5   *(Continued)*

Upon completion of the mobilization phase, the administrative staff shall create an electronic file and copy the following into the file:

- Report outline (some firms generate this from a proprietary computer program and save it in the answer files)
- Import into subdirectory the reserve schedule template, if applicable
- Import the cost schedule template into the subdirectory.

5.5.3 Site Visit/Inspection

Prior to visiting the project site, the project manager shall review the project-working file to understand the scope of work, including reporting requirements and the appropriate protocol. An inspection pack should be prepared, which for a property condition assessment survey should include basic needs such as: moisture meter, combustible gas meter, digital

ZONING DEPARTMENT INFORMATION REQUEST

ABC Green Design, Inc.
XYZ Los Angeles
2134 Olympic Blvd.
Los Angeles, California 90064
(213) 234-1234

Date:	To:
Subject:	Dept.:
	Tel. No.:
	Fax No.:
Sender:	Proj.No.:

ABC Green Design, Inc. has been commissioned to conduct a Property Condition Assessment Survey on the above referenced Subject. Please respond to the following documentation/information requests:

1. Does the Subject have any material outstanding zoning code violations within its file?   Yes ☐ No ☐
   If "Yes", please fax copies of same.

2. Does the placement, quantity or area of signage comply with current zoning requirements?   Yes ☐ No ☐

3. Is there any existing or pending material zoning code requirements/regulations that the   Yes ☐ No ☐
   Subject would be considered an existing non-conforming use? If "Yes", please briefly explain.

4. Was the Subject built as of right?   Yes ☐ No ☐

5. If yes to the above, was a variance necessary?   Yes ☐ No ☐

6. In the event of a catastrophic loss, could the Subject be rebuilt to its current density?   Yes ☐ No ☐

7. Is the Subject within a Zoning District? If "Yes", please identify the Zone/District   Yes ☐ No ☐
   and the requirements.

8. What is the local Zoning Ordinance classification of the property?

9. When was the current Zoning ordinance adopted?

10. Are there any municipal required procedures or mandated improvements that are triggered by a change
    of ownership/title such as: a re-issuance of Zoning Approval by the Zoning Department or Zoning Board of
    Appeals? If so, what are they?

11. Please fax us a copy of the Subject's Zoning Compliance Certificate, if any.

Thank you for your assistance. Should you have any questions or should there be any fees associated with providing the requested information, please call or email me.

FIGURE 5.5   (*Continued*)

camera, tape measure, electric current tester, flashlight, binoculars, screwdriver, and pocket knife.

Observe property components, systems, and elements for evidence of significant physical deficiencies. Physical deficiencies qualifying as significant should be deemed to exist if either of the following are observed: Physical deficiencies which represent a cited or apparent code violation, an immediate life safety or health hazard to the occupants or users of the property, or a fire safety hazard to the property itself. Other physical deficiencies of a lesser nature should also be observed and individually recorded for inclusion in an aggregated cost-to-cure estimate which will otherwise be furnished by the consultant.

Field observations will consist of one or more of the following activities:

1. **Walk-around exterior visual inspections:**
   - Building exteriors—all as visible from grade and accessed roofs and setbacks
   - Principal hard surface areas
   - Representative roofs and roof area
   - Landscaped areas.

2. **Walk-around interior visual inspections:**
   Common area spaces
   - All entries and entry lobbies
   - Toilet rooms (take sample)
   - Multi-tenant area corridors and elevator lobbies (take sample)
   - Egress stairs and exit ways (sample at various levels)
   - Elevator interiors (typical of each bank)

   Leased area spaces
   - Vacant tenant areas (typical; if residential, one of each type)
   - Occupied tenant areas (typical; up to 10 percent of space, no occupied residential units)

   Service areas
   - Typical service spaces and corridors
   - Central service facilities (main equipment rooms and pad areas).

3. **Equipment and system observations.** Random operation of equipment, fixtures, and systems on a sample basis to determine system operability which may include:
   Mechanical equipment (HVAC)
   - Central equipment
   - Typical residential, retail, or office unit installations (5 percent of total)
   - Typical floor or zone installations and equipment (one or two floors)
   - Typical rooftop installations (on floors inspected)
   - EMS system console

   Electrical equipment
   - Central equipment including transformer vaults and pads
   - Typical unit installations (residential, office, retail)
   - Typical floor or zone installations
   - Typical pattern of outlets, switches, jacks, lighting, (5 percent of area)

Plumbing
- Fixtures in toilet operations (a roomsample only)
- Piping and insulation incidental to mechanical areas

Elevator equipment, central equipment, and control room life safety system equipment
- Central consoles and annunciator panels
- Fire pump and central valve stations
- Typical standpipes, sprinklers, smoke detectors, alarm stations, etc. (5 percent of area)
- Central security and surveillance consoles.
4. **Non-destructive, non-invasive testing,** sounding. or detailed observation to be conducted to determine representative conditions.
5. **All deficiencies to be recorded** (whether major physical deficiencies or minor ones).

Upon returning to the office, the project manager should send the client a *debriefing letter* such as illustrated in Figure 5.6. The project manager may also call the client as appropriate to discuss the inspection and findings.

5.5.4 Writing the Report

The writing of a PCA or PSR report requires concentration and attention to detail. One method is to spread out the project materials, particularly the photographs, on a desk and organize them to tell a story. The assessor can then attempt to piece together all the information in a logical form.

The value of comprehensive documentation cannot be overstressed. The degree of confidence in a decision to acquire a facility is directly related to the thoroughness of the documentation review. If the document review is cursory, the chances are much greater that significant issues will be missed. A cursory review could mean that certain building systems or components of these systems are inadequately reviewed or reviewed in such a cursory manner that only the most superficial deficiencies will be identified.

Comprehensive documentation means that review of the available documentation, from soils reports to structural drawings, is imperative for the evaluation to be of significant value. This is particularly appropriate for an in-depth inspection as opposed to a standard walk-through survey. The more vantage points a facility is viewed from, the more likely the chance of noticing deficiencies and other issues that need to be addressed. At times the structural system deficiencies will be discovered by the assessor crawling into the ceiling cavity. Other times, the issue will arise upon review of the construction drawings or specifications. The more documentation

ABC Green Design, Inc. (AGD)
Chartered Architects, Interior Designers, Green Design Consultants
1005 Max Ct. SE, Leesburg, Virginia 20175 Tel: 1 703 851 1111 Fax: 1 703 851 2222
Email: info@abc.green.design.com   www.abc.green.design.com

To: Mr. John Doe, Vice President **Date:** November 25, 2009

Firm: Bank USA Virginia

Fax: (703) 261 3333

From: Sam A.A. Kubba, R.A., Ph.D., LEED AP

Re: Property Condition Assessment
9700 Melvin Road Building
446,000 SF One (1) 2-Story Office Building
Leesburg, Virginia
AGD Project No. SK 2111

Dear Mr. Doe:

Today, we completed our site visit, and are prepared to provide you with a preliminary debriefing as to the property's general physical condition. During our site visit, the weather consisted of light drizzle and the temperature was about 45 degrees F.

Salient Points:

• Overall, the Subject was considered to be in very good condition.

• It was evident that preventive maintenance has been exercised. Based upon our observations and the limited information reviewed to date, maintenance can best be characterized as proactive.

• Only minor physical deficiencies were observed. These will reportedly be attended to by the new tenant during renovation.

• Property management was helpful and very cooperative with our requests for information. They have produced some of the information requested on the Pre-survey Questionnaire and Disclosure Form that we faxed to their offices, which we are to review on Credit Lyonnais's behalf.

We still have to complete our research with the service companies, nor have we prepared any opinions of probable costs to remedy the physical deficiencies observed. Furthermore, we are still awaiting information from the owner. However, should you wish to discuss our observations prior to completing our Property Condition Report, please call me at extension 12.

FIGURE 5.6 A sample PCA debriefing letter sent to the client upon completing the site visit outlining the results of the survey, providing the client with a preliminary debriefing as to the property's general physical condition. Property condition assessments (PCAs) are a necessary prerequisite for the real estate owner and investor.

sources that are reviewed, the more valuable and accurate the survey report commentary and recommendations will be.

The other aspect of comprehensive documentation is the verification of evaluation findings and recommendations. The acquisition assessment is the ideal time to begin what could be referred to as a working operations and maintenance manual. This may be the one time that the facility is thoroughly reviewed and the resulting documentation of this effort can become a benchmark of the facility for all of its useful life.

5.5.5 The Report's Form and Content

The report form is determined by the intended use of the report, the client protocol, the type of property being surveyed, and the necessary level of diligence. Most firms have formulated standard report formats and language to ensure the quality and consistency of the final report product. Basically, for each built property, a property condition report (PCR) is provided after the inspection/survey is completed. For new construction, a property status report (PSR) or property analysis report (PAR) is conducted. The PSR is generally prepared on a monthly basis based on the contractor's application for payment. The PAR is typically a one-time report that is basically an evaluation of a project's feasibility and is completed prior to commencement of construction. As mentioned earlier, most large due diligence firms use proprietary software systems to assist in generating the initial draft report and serve to expedite the process of report preparation, while maintaining quality and consistency.

Good safety practices are of paramount importance and should be exercised at all times. Be aware of your surroundings at all times and avoid dangerous conditions. Avoid entering confined or hazardous spaces and also avoid operating equipment that is otherwise unavailable to the general public. Likewise, do not conduct exploratory probing or testing or be exposed to hazardous materials or conditions and any and all accidents should promptly be reported to the division manager. At the conclusion of the site/facility inspection and before leaving the area of the project site, it is important to organize and review all relevant notes and to complete all field documentation. If information is missing, it can often be easily obtained before leaving the site.

5.5.6 Report Review

Upon completion of the report, it should be checked for typing, spelling, grammar, and conformance with the firm's report style for consistency. The style check should focus on both format and content. The writer is expected to submit a finished product. It is always preferable to submit the finished report to a designated third party to review it for accuracy, errors, etc. The reviewer will examine the report to ensure that the scope of work and all other client and firm's criteria have been satisfied. The reviewer will consult with the assessor on substantive report issues, or if quality problems need to be addressed. The report will be returned to the assessor for rework and resubmission if the quality standards are not met. Once the reviewer passes the report, the report is finalized, set for production, and issued.

5.5.7 Production, Distribution, and Client Follow-up and Closeout

Depending on the size of the firm, the project manager or the administration/production staff collaborate in the production of each report. Once the report is generated, it is then copied, collated, and assembled into a final product. The finished report is then given a final quality check before being sent to the client. A project billing report will be sent to accounting upon release of the report, after which time accounting will send a billing invoice to the client.

The project manager or the administrative staff will call the client within three business days of sending the report to confirm their receipt. The project manager can then field any questions or comments from the client, and revise and resubmit the report to the client if needed. Upon direction from the client, the report will be issued as final.

5.6 THE CONSULTANT'S RESPONSIBILITIES
AND QUALIFICATIONS

5.6.1 General

In the final analysis, the capabilities and qualifications of the assessment team determines the ultimate quality of a PCA survey and report. An *assessor* or *field observer* is the person or team that performs the visual assessments and records the results. The PCR *reviewer* (usually only employed in larger firms) is a qualified individual or team that reviews and checks the accuracy of the assessor's results and is designated to exercise responsible control over the assessor/field observer on behalf of the consultant and to review the PCR. The consultant, who is typically hired by the lender or investor, is ultimately responsible for the PCA process including the accuracy of the PCR. Better PCA results are usually achieved when both assessors and reviewers are involved in the process, particularly when both function as separate individuals or teams. They do not need to be organizationally independent of one another, and in fact are usually part of the same firm.

A consultant's competency relies on many factors that include professional education, training, experience, certification, or professional licensing/registration—both the consultant's field observers and the PCR reviewer. However, no standard can be designed to eliminate the role of professional judgment, competence, and the value and need for experience during the walk-through survey and to conduct the PCA. The qualifications of the assessor/field observer and the PCR reviewer are therefore critical to the performance of the PCA and the resulting PCR. The consultant is ultimately responsible for selecting, engaging, or employing the assessor/field observer as well as the PCR reviewer (if required); therefore, each PCR should include as an exhibit a statement of qualifications of both the assessor/field observer and the PCR reviewer if applicable.

Most clients also require that consultants carry adequate insurance coverage prior to being considered for an assignment. Typically, a proposal submitted by a consultant will not be considered unless accompanied by the appropriate certificates of insurance. Typical insurance requirements are illustrated in Figure 5.2 and in Chapter 10.

5.6.2 Consultant Qualification Requirements

Inspection & Valuation International, Inc. (IVI), a due diligence firm, in conjunction with Standard and Poor's (S&P), an international financial services company, developed certain guidelines relating to consultant qualification requirements, some of which are outlined below. Accordingly, before making a determination on the selection of a consultant, a number of questions should be asked regarding the company's experience, including:

- How long has the firm been in business and what is the year of incorporation or year the partnership was formed? Since inception, has there been a corporate name change? Qualified consultants should be able to prove that they have been in operation for at least three years.
- If the firm has branch offices, provide locations, addresses, number of personnel, manager's name, and telephone numbers.
- List the last 12 building condition survey assignments conducted by your firm, complete with scope of assignment(s), location, client, and client telephone number. Also list the number of property condition survey assignments completed by your firm for each 12-month period over the last three years?
- Are there any pending claims or litigation against the firm? If so, please provide a brief overview as to the basis and status of same.

It is obvious that many new business start-ups may find some difficulty in answering all of the above questions. With respect to personnel, the consultant is to provide resumes of each individual within the firm who will be conducting property condition surveys and/or reviewing completed reports. The resume of the senior project manager (PM) should be provided. The PM will be responsible for report review/quality control, final sign-off, and answering rebuttal questions, if applicable. With regard to all other personnel that will conduct property condition surveys, they shall possess all of the following qualifications:

- Professional engineer's license or architectural registration; no exceptions will be permitted
- Four or more years of experience in specifically conducting property condition surveys on behalf of investors, lending institutions, or government agencies
- The individual signing off on the report (usually the PM) must be a licensed professional engineer or a registered architect.

The consultant should also be required to provide four references with their names, position, company, and telephone number and who are able to opine on the firm's property condition survey reports.

For the majority of commercial real estate that is subject to a PCA, the assessor/field observer assigned by the consultant to conduct the walk-through survey will most likely be a single individual possessing a general, well rounded knowledge of pertinent building systems and components; however, single individuals will seldom possess an in-depth knowledge, expertise, or experience with all building codes, building systems, and asset types. Therefore, when a decision is taken to supplement the assessor/field observer with specialty consultants, system mechanics, specialized service personnel, or any other specialized field observers, it should done in total consultation with the client. Indeed, this decision should be made in accordance with the client's requirements, risk tolerance level, and budgetary constraints. Additionally, the purpose the PCA is to serve, the expediency of report delivery, and the complexity of the subject property should be considered. The level of due diligence conducted during a PCA is often adjusted to the risk tolerance declared by the client.

5.6.3 The Consultant/Administrator as an Independent Entity

Where the project is new construction, the consultant will normally be a person or entity retained by the lender or project owner (the client), to either conduct a PCA or prepare a project analysis report prior to administrating the project's construction phase for the client. In the event the consultant or members of the consultant's staff are employees of, or a subsidiary of, the user, such affiliation or relationship should be disclosed in the executive summary of the report.

With new construction, the consultant/administrator is responsible for the comprehensive and timely administration of all project-related documentation and information. The primary role of the administrator is to represent the lender/owner and/or architect in all project meetings and other field matters relating to the construction of the project. The administrator will make periodic site visits to the project, record observations, and report to the lender/owner all observations related to status of completion, quality of workmanship, and compliance with the contract documents.

The authority and responsibility of the consultant/administrator (lender's representative) is described in AIA Document B352. In addition to AIA B352, the administrator will adhere to the following general policies for construction administration:

1. The consultant/administrator does not have the authority to authorize any deviations from the contract documents without the explicit approval of the client or unless such authority is clearly stated in the contract documents.
2. The consultant/administrator does not have the authority to stop any of the work on the project without the client's written approval.
3. It is not the responsibility of the consultant/administrator to conduct tests.
4. The consultant/administrator is not obligated to resolve construction problems on the spot. The administrator may defer any decisions until adequate consultation is possible with the owner/lender and/or architect or appropriate specialist engineer, despite the contractor's alleged urgency.
5. The basic role of the consultant/administrator is to observe the work and advise the owner, lender, and/or architect of the progress and quality of the work and conformance with the contract documents.
6. Neither the consultant, owner, nor the architect are responsible for construction safety on the project. However, unsafe practices will be noted in the report and brought to the attention of the contractor, architect of record, owner, and lender.
7. The lender's representative (administrator) is required to make a monthly progress review of the project on behalf of the owner/lender; this will preferably be at the time the application for payment is reviewed.

For new construction the lender's representative's role is to periodically (usually to correspond with the contractor's application for payment) visit the job site to observe the work and check to see if the value of the work in place matches the contractor's application for payment. The person responsible for supervising the construction is, in fact, the construction manager or general contractor, although the administrator (representing the lender) checks the monthly payment applications and the project's progress between payment applications. That entity has contracted to complete the work and legally assumes primary responsibility for the work. The administrator is not an "inspector" but rather a representative of the lender with duties that do not encompass the time nor the effort required to provide a comprehensive or exhaustive "inspection" of the work.

In fact, the administrator's responsibilities include the transmission of much paperwork and project documentation, particularly AIA standardized documents. The administrator's role should be clearly defined, but generally consists of taking advantage of the time savings involved in producing computerized master forms for individual AIA documents for the project. Proposal requests, change orders, transmittal letters and architect's supplemental instructions (ASIs) should be prepared in advance (to include all relevant project information such as project name, project address, project number, etc.) and ready for specific information to be inserted.

The owner, the architect, and the contractor form part of the traditional construction triangle and often find themselves in adversarial relationships. The administrator's role is to defend the rights of both the contractor and those of the owner/lender. The lender's representative is typically charged with the impartial administration of the contract documents, and much of the administrator's time is spent reviewing, and then evaluating the contractor's work. It is also helpful to remember that the contractor, owner, and architect share a common objective—the successful completion of the project.

5.6.4 How the Consultant Functions

The typical loan officer employed by a bank or other lending institution is faced with many critical decisions. One of the most difficult issues that must be faced is the method of evaluating the technical phases of the project. For example, is the project structurally sound, and are the costs anticipated by the borrower consistent with actual field costs? Also, who is retained to represent the lender and inspect the project during the construction and preconstruction phases and review the contractor's application for payment?

The consultant is generally retained to review much of the technical data furnished to the lenders. The consultant is also frequently required to provide a cost analysis of the project as early as possible for consideration of the loan. If upon comparing the consultant's estimate with the borrower's proposed costs it is found that the two are reasonably close, the lender may grant cost approval for the project.

After being informed that the projected development costs are reasonable for the intended work, the lender now awaits to be assured that the project has been designed in accordance with good engineering practices and meets all codes and regulations. The interim construction lender is mainly concerned with a viable toxic-free site and general adequacy of design and specifications. This requires the consultant to conduct an investigation of the major elements of the project—soil analysis, environmental audit, verifying adequacy of green features and materials, structural and mechanical plans, and site work. If seeking green building certification (e.g., LEED or Green Globes), compliance of plans, specifications, and credit requirements are needed in addition to meeting applicable building codes and targeted certification requirements. Lack of compliance threatens issuance of a building permit, certificate of occupancy, or green certification, thus endangering the entire project or entry into the project by the permanent lender.

The consultant's report may also contain an overview of the entire project including any special architectural and green features, extent of site work, etc. Projects that involve long-term land development, or development in phases, are of particular importance. Once the loan has been approved, committed and underway, the consultant still needs to answer a number of basic questions remaining in the interim lender's mind. These include:

1. Is the builder diligently following up on the completion of the project? Is the administrator/consultant noting and recording the progress of each of the major trades and building systems? Project activity is measured, including the number of workers on the project, material flow, and whether they meet green rating systems requirements (and are not greenwashed), and the overall conduct of the job.
2. Is the project being constructed in accordance with the approved plans and specifications? If green certification is being sought, is the project meeting requirements and progressing according to schedule? Once construction begins, the consultant is required to undertake periodic field inspections (usually monthly) to assess the project's progress. During these periodic reviews, construction methods and field operations are evaluated. Test reports affecting such factors as concrete strengths, soil compaction, and field welding of structural steel are examined for compliance with specification requirements. For sustainable projects, materials and systems are also assessed.
3. Are the contractor's monthly loan requisitions reflective of the work in place at the site? The obvious purpose of closely monitoring the flow of construction loan dollars to the developer is to make sure that at any given time during the life of the loan, sufficient funds are available in the undisbursed portion of the loan to complete the work. The consultant performs a detailed inspection, noting progress trade-by-trade, and the resulting information is transmitted to the owner and lender. With green-related projects, the lender would want to know if the project does not achieve green certification, would its value still be adequate to pay off the loan.

Other general areas of concern that the consultant may encounter at the project site include portions fully executed by the tenant in rental areas such as office buildings or shopping centers. Costs for completing such tenant areas are frequently funded from separate portions of the loan. The consultant continues to be involved with the project as it approaches completion. The significance of completion to the interim lender is that it triggers its pay-off by the permanent lender. Project closeout and a certificate of occupancy issued by the local municipality usually confirm that this landmark has been reached. To complete all obligations to the lender the consultant may be asked to certify completion, compile costs of remaining work, review punch lists, and other incidental items, as well as confirmation that green objectives have been achieved (if applicable).

5.6.5 Applicable Rules of Engagement

The consultant is neither a direct employee of the lender or owner and can therefore be detached from the day-to-day machinations and servicing of the loan. The borrower and banker are then able to communicate directly using the data provided by the architect/engineer, often resulting in improved communications.

The consultant/administrator shall conduct all communications in a professional and courteous manner at all times. Instructions, letters, and other formal communications shall only be issued after proper authority is received. Relevant telephone calls, conversations, and other informal communications shall be conducted with the clear understanding that the consultant is acting on the client's behalf. The content of significant discussions held in an official capacity shall be recorded in writing and filed accordingly.

It goes without saying that all parties to the contract and the project do not always agree with one another. Particular care should be exercised in such situations and it is not in anyone's interest to allow disagreements to degenerate. The professional's opinion should reflect opinions based on personal training, knowledge, and judgment. It should also be based on knowledge of the facts and issues.

Larger professional organizations typically use a team approach to deliver certain consultancy services. The division manager's role is to provide leadership and guidance to the PCA team, establish policy, maintain client relations, and develop new business. Senior project managers also generally provide leadership and technical guidance to members of the PCA team, review and edit draft PCA reports, and enforce quality control measures, as well as keep management and the client informed of progress. For new projects, the administrator will have the responsibility of monitoring the general progress of construction and chair the monthly or bi-weekly progress meetings, in addition to preparing the project analysis report and project status reports. With larger firms the administrative staff provides support to management and the technical staff, as well as coordinate project assignments, travel plans, technical research, and report production and distribution.

According to the American Society for Testing and Materials (ASTM), the user should arrange for the field observer to have timely access that is complete, supervised, and safe to the subject property's improvements (including roofs). Additionally, any relevant and pertinent documents should be provided by the owner, owner's representative, or made available by the user. Under no circumstances should the consultant/field observer seek access to any parts of the property, property management staff, tenants, vendors, or documents, if the owner, user, or occupant objects to such access or tries to prevent the consultant/field observer from conducting any portion of the walk-through survey, research or interviews, or taking photographs. Any conditions that significantly impede or restrict the field observer's walk-through survey or research, or the failure of the owner or occupant to timely provide access, information, or requested documentation shall be promptly communicated by the consultant to the user, owner, and/or lender. If such conditions persist, the consultant is obligated to state within the PCR or PSR all such material impediments that interfered with conducting the required inspection or assessment.

A client/user should promptly disclose all information in its possession that may assist the consultant's efforts. The client/user should not withhold any pertinent information that may assist in identifying a material or physical deficiency that may include:

- Previously prepared property condition reports
- Any study specifically prepared on a system or component of the subject property
- Any knowledge of actual or purported physical deficiencies
- Any information such as costs to remedy known physical deficiencies.

Should the lending institution find itself as owner of an uncompleted project, the consultant will be required to take on a faltering real estate construction project, review contracts, disbursements, prepare a study on the cost-to-complete the project, etc. The lender will then typically engage the consultant as its agent to complete the project.

For compensation the consultant may ultimately look to the lender or owner/borrower for payment of the agreed fees as per invoice within 60 days of receipt of said invoice and satisfactory acceptance of services rendered.

5.7 AVOIDING CONFLICTS OF INTEREST

The law of conflict of interest is extremely complicated, but suffice it to say that a potential conflict clearly exists any time a prior representation may in any way impair loyalty to a current client.

Consultants typically have to warrant that other than the relationship contemplated by the agreement with the client (e.g., the lender), he/she has no financial interest in the project or with any persons or organizations guaranteeing the loan, the architect of record, the engineers, the contractor, or subcontractors. The consultant will also warrant that he/she has no family relationship with any entities guaranteeing the loan, the architect, the engineers, the contractor(s), or the subcontractors. If, during the term of the agreement, there is a change in circumstance that presents a conflict of interest as outlined above, the consultant shall notify the lender immediately upon discovery.

For example, it would be a conflict of interest for the architect of record who is employed by the borrower to certify the monthly requisitions or provide other certifications to the lender. Moreover, it is not uncommon for the architect to sometimes accept a financial interest in the project in lieu of his fee. This "piece of the action" may compromise the architect's position and firmly place him on the borrower's side. But in any case, since the architect of record is typically employed by the borrower and therefore most often represents the borrowers, lenders most often engage independent consultants to certify the periodic requisitions through completion.

5.8 TYPICAL AGREEMENTS

5.8.1 Typical Loan Agreement

The construction consultant must typically follow the procedures outlined below for inspections and disbursement of funds. Funds are not generally released unless there is strict compliance with the bank's requirements. Also, it must be remembered that not all lending institutions have a standard loan construction procedure kit. Some institutions require the consultant to use his/her own procedures tailored to meet the lender's needs.

To ensure the lender is protected, lending institutions often employ the services of a professional Construction Consultant to monitor new projects and certify the release of funds on a construction project. The types of consultant/lender agreements vary depending upon comprehensiveness and complexity of the services required. Fees are generally based on an agreed per diem for site visits, and a fixed fee for the loan package review and front-end analysis.

The legal form of loan agreements can vary widely from state to state, which can make it difficult if the construction consultant is unfamiliar with the laws of the state in which the project is located. It is extremely important therefore for the construction consultant to become familiar with standard practices in the states in which the projects are to be constructed. The agreement between the lender and the borrower usually starts with the services provided by the consultant that includes a front-end analysis as follows.

SECTION 1: CONSULTANT SERVICES

The consultant services are described in the following two sections: A. "Frontend analysis," and B. "Construction Progress Monitoring."

A. Front-end Analysis

The consultant shall perform a onetime front-end analysis consisting of the following services:

1.00   A construction documents review
2.00   A construction costs review
3.00   A pre-closing construction progress inspection
4.00   A pre-closing site inspection
5.00   Attendance at ABC Bank's pre-construction meeting

All front-end analysis work shall be collectively documented in a separate report, dated and signed by the actual consultant(s) performing the front-end analysis. The original front-end analysis report shall be delivered to ABC Bank's real estate representative within 20 working days following receipt of necessary documents and notice to proceed from ABC Bank.

1.00   Construction documents Review: Conduct an inventory and review of construction documents including, but not limited to:

      a. Plat plan/boundary survey/site plan

      b. Building permits

      c. Utility letters

      d. Topography plan

      e. Environmental site assessment

      f. Soils investigation report

      g. Construction plans

      h. Construction specifications

      i. Construction schedule

      j. Construction contracts(s)/schedule of values

      k. Addenda/change orders

      l. Architect's contract(s)

      m. Variances.

1.01   Identify all documents reviewed, who prepared them, whether or not they are issued for construction, stamped and sealed, or fully executed, list the latest revisions, and, if applicable, determine the status of their development.

1.02   Prepare a detailed description of the existing site topography characteristics, any demolition requirements, and the planned improvements, clearly differentiating between on-site and off-site work.

1.03 Determine whether the plans and specifications are adequate, complete, and sufficiently prepared to complete the proposed construction improvements.

1.04 Describe and comment on setbacks, encroachments, easement requirements, discrepancies, or any objectionable features between the plat plan, legal construction plans and/or specifications.

1.05 Based on review of the construction documents, determine whether or not the planned improvements generally comply with governing codes and ordinances and have been prepared reincorporating accepted standards of practice by a licensed architect and/or engineer.

1.06 Evaluate the quality of the equipment and materials specified, their performance value, adequacy, durability, and comparison to similar projects in the geographical area.

1.07 Review the geotechnical soil investigation report and the soils engineer's recommendations, and evaluate their incorporation into the design and specifications.

1.08 Confirm, based on documents reviewed, the availability of utilities required for this project and identify any inadequacies or restrictions.

1.09 Describe any unusual existing or planned features in the project including access, mobility, or user restrictions and any areas of concern in conjunction with the project.

1.10 Describe the access to the planned improvements (i.e., publicly dedicated roads, ingress, egress, and easements).

1.11 Describe and comment on the type(s), status and any objectionable features of the building permit(s) and any special requirements specified by permits.

1.12 Determine if any code, ordinance, or easement variances are anticipated or required on this project and comment on their approval and/or recorded status.

1.13 Evaluate the development and detail the construction schedule, its critical path deadlines, and determine if the overall completion date can be reasonably achieved.

1.14 Describe the construction contract and the architect's contract by the contract's type of services provided for construction (general construction management, design, etc.) and identify any limitations or restrictions in the general conditions or scope of work that may compromise the quality, value, or controls to achieving a successful project.

1.15 Clearly identify and explain in an itemized fashion any recommended changes, needed clarifications, and any errors, omissions, discrepancies, or inadequacies discovered during the construction documents review.

1.16 Provide followup reviews, comments, and recommendations on issues, responses, and resubmittals resulting from the "original" documents(s) designated and submitted for review by ABC Bank until which time all respective issues are addressed and resolved to ABC Bank's satisfaction.

2.00 Construction cost review: This includes review of the contractor's itemized trades and cost breakdown, the schedule of values, draw schedules, construction contracts, and other necessary documents.

2.01 Determine if the contractor's construction budget is sufficient to complete the planned improvements and adequate to monitor construction progress relative to draws in order to make recommendations to ABC Bank on the disbursements of funds.

2.02 Address each trade and each scheduled value by line item in the contractor's cost breakdown to determine if the budgeted costs are adequate to complete the respective work and reasonably reflect the costs on similar projects in similar markets. Identify and describe the cost review method used, its process and procedures, references used, cost database and/or cost models used, and any assumptions made during the course of the cost review. The cost review may be based on a comparable value/project approach. The consultant shall provide a comparison worksheet accounting for and in support of the conclusions reported.

2.03 Evaluate and explain whether a trade or line item appears to be excessive or inadequate and determine, in the opinion of the consultant, if the overall construction budget is "frontend loaded."

2.04 Describe the status and contract(s) fee type for construction (lump sum, cost plus, guaranteed maximum price, labor only, etc.) and specifically the provisions concerning payment, retainage, liquidated damages, bonuses, shared savings, onsite and offsite stored materials, lien waivers, and change orders.

2.05 Identify and explain by line item any inadequacies, deficiencies, inflation, needed clarification, or discrepancies discovered during the Construction Costs Review.

3.00 A construction progress inspection: This is an "optional service" which may be required prior to closing the loan, at ABC Bank's discretion, on projects where construction is believed to have already commenced. When required, the results of this inspection shall be incorporated into the front-end analysis report.

3.01 Visit the site and walk the entire project to observe the work in place and work in progress to determine, to the extent possible, the work has been performed in a satisfactory manner, and work is in general accordance with the construction plans, and prepare an estimated percentage of completion.

3.02 Review the contractor's latest draw application of payment, verify expenditures to date, and make an evaluation of the cost to complete versus balance to complete in the draw schedule.

3.03 Review lab reports, work schedules, change orders, and progress logs and take photographs while on-site to document and formulate an overall evaluation of the project.

3.04 Attempt to provide the scope of services described in the construction progress monitoring in the relevant sections of these general conditions.

4.00 A pre-closing site inspection: This is an "optional service" that may be required, at the discretion of ABC Bank, to verify that no work has started prior to and/or in conjunction with closing the loan.

4.01 Immediate visit to the site upon notice from ABC Bank of the appropriate date and time to perform the inspection. The purpose of the inspection is to determine if any physical work has started, including mobilization that would potentially qualify for a mechanic's lien position ahead of ABC Bank's lien established upon closing the loan.

4.02 If work, materials, or equipment are observed on-site, then a brief description of what the consultant observed, photographs of the same, and delineation of its general location on the plat plan shall be required.

4.03 A verbal report of the consultant's observations shall be called in to the ABC Bank real estate representative immediately following the inspection. The photographs, flat plan, and affidavit form letter shall be received by the ABC Bank real estate representative within three working days following the inspection.

5.00 Attendance at ABC Bank's pre-construction meeting shall be required unless notified otherwise by ABC Bank. The purpose of this meeting is to meet collectively with all parties to the construction project and discuss the status of the work, construction documents, and contractual relationships, and ABC Bank's draw procedures and requirements.

5.01 The ABC Bank real estate representative shall coordinate and establish the date, time, and place the meeting will be held.

B. Construction Progress Monitoring

1.00   The construction consultant shall visit the site and walk the entire project to observe the construction progress in conjunction with and for the purpose of reviewing each monthly construction draw application for payment throughout the entire duration of the construction project, unless notified otherwise by ABC Bank.

2.00   A verbal and/or fax report shall be made to the ABC Bank real estate representative within 24 hours of completion of draw review and respective inspection. The written construction progress and draw review report, each with original photographs, are to be sent directly to the ABC Bank real estate representative and received within five working days of the on-site progress inspection. The results of the inspections shall be collectively documented in a construction progress and draw review report which shall include the following documented information.

2.01   Determination of the overall project's percentage of completion status including a narrative description of work completed to date, work in progress, and materials stored on-site.

2.02   Review and comment on submitted contractor's application for payment as to whether or not the itemized amounts are a valid representation of the value of work in place and materials stored on-site. The consultant shall conclude with a recommendation to ABC Bank on the amount to fund and/or monies to withhold.

2.03   Determine whether or not the itemized percentages of completion indicated by the contractor on the application for payment are accurate and substantially represent the progress of the observed work in place and materials stored on-site.

2.04   Identify any overruns or savings to line items and make a determination as to the sufficiency of the remaining budget to complete the project.

2.05   Review and verify that the backup invoices are genuine, current, and representative of this project and in support of moneys drawn. Any exceptions are to be clearly identified.

2.06   Review and comment on all proposed changes, change orders, and field orders, including a description of their effect upon the project, their estimated cost impacts, their status of approvals, and scheduled construction implementation.

2.07   Identify the purpose and describe the use of any moneys drawn on or credited to the "contingency" line item.

2.08   Review and comment on the retainage amounts withheld relative to adherence to the terms of retainage determined in the construction contracts and ABC Bank's pre-construction meeting, unless instructed otherwise by ABC Bank.

2.09   Review and comment on all submitted lien waivers and their adequacy and completeness to cover preciously drawn funds; specifically identify any exceptions.

2.10   Determine, in the consultant's opinion, whether or not the work completed to date is in general compliance with the scope of work indicated in the plans and specifications and is within accepted standards of good quality workmanship. The consultant shall clearly identify any discrepancies, deficiencies, or problems to ABC Bank. Any objectionable or questionable work observed shall be discussed with the borrower's contractor and/or architect's on-site personnel for clarification prior to and as part of status reporting.

2.11   Comments on whether or not the on-site stored materials appear to be in good condition and are adequately stored and protected.

2.12   Review and comment on the lab test results from the inspections and tests performed, such as, but not limited to, soil compactions, steel or reinforcement placement, concrete strength, welds, or pressurization capability as required or specified in the construction contract documents.

2.13   Separately and specifically comment on any delays, problems, deficiencies, poor workmanship, deviations from accepted industry standards, or any other matter which in the consultant's opinion does or might adversely affect the value and/or collateral security interest of ABC Bank in the project.

2.14   Review and comment on the construction progress relative to the original work schedule and subsequent revised work schedules, including a determination as to the feasibility of the proposed completion date.

2.15   Review and comment on the status of any required inspections by federal, state, county, or local municipal governing/regulatory agencies.

2.16   Include labeled photographs showing the status of the project and isolate any problem areas or items of concern. Labels should identify the date the photograph was taken, the subject matter, and orientation relative to the site and/or building plans.

2.17   Upon determination that "substantial" and/or "final" completion has been achieved, the consultant shall include the following: comments and status on any known punch list items, final lien waivers, affidavits of payment, retainage releases, and any other outstanding items affecting final disbursements of funds.

SECTION 2: GENERAL

A. Confidentiality and Responsibility

1.00 It is mutually agreed that all documents submitted to the consultant and all reports, both verbal and written, produced by the consultant are strictly confidential and for the sole benefit of ABC Bank or its agents and employees.

2.00 It is understood that copies of the consultant's reports may be made available through, and at the discretion of, ABC Bank, to a permanent lender, if any, loan participants, if any, and the borrower. The consultant shall not have any obligation to a permanent lender or to the borrower or any agents thereof.

3.00 Consultant does not assume or accept any of the design, planning, construction administration, or supervisory responsibility of the architect of record, engineers of record, or the contractor or construction manager, if one is or is not so retained. The consultant is not responsible for any of the acts or omissions of the borrower, owner, architect, engineer(s), or contractor(s) in the construction of this project, including, but not limited to, the methods, means, and sequence of construction, safety, compliance to governing codes and ordinances, changes in the work, or substitutions of materials. Services performed herewith are limited to those as ABC Bank's consultant and described in the relevant of these general conditions.

B. Compensation

ABC Bank acknowledges that consultant's invoices shall be submitted to the borrower. However, the consultant may ultimately look to ABC Bank for payment of the agreed compensation within 60 days of receipt of said invoice and satisfactory acceptance of services rendered.

C. Construction Documents

All construction plans, specifications, work schedules, construction cost breakdowns, construction contracts, construction draws, and other construction documents will be furnished to the consultant by ABC Bank. If the consultant receives documents from any other sources, the consultant must identify the document and source to ABC Bank for approval before use in performing any of the services described in the agreement. The consultant shall regard all documents received concerning this project as confidential and prevent disclosure by the consultant or the consultant's employees concerning any specific information seen or heard regarding this project without ABC Bank's prior approval.

D. Termination

Either ABC Bank or the consultant shall have the right to terminate this agreement at any time upon giving written notice of such termination. In such case, the consultant shall be due compensation in an amount not to exceed more than the required fee for services performed prior to termination.

E. Conflict of Interest

The consultant warrants that other than the relationship contemplated by this agreement, he/she has no financial interest in the project or with any individuals guaranteeing the loan, the architect of record, the engineers, the contractor, or subcontractors. Consultant also warrants he/she has no family relationship with any individuals guaranteeing the loan, the architect, the engineers, the contractor(s), or the subcontractors. If, during the term of this agreement, a condition or situation changes that presents a conflict of interest as described above, the consultant shall notify ABC Bank immediately upon discovery.

F. Consultation

The construction consultant agrees to make him/herself reasonably available to ABC Bank for telephone conferences, discussions, clarifications, advice, guidance, and recommendations pertaining to any aspect or issue of this project without additional costs incurred by ABC Bank or its borrower.

5.8.2 Typical ConsensusDOCS Type Construction Agreement

Terry Wooding, executive vice president, Petra Construction Corp., says, "ConsensusDOCS reflect the future of the industry, where you have the owner, contractor or construction manager, and design team working together in a collaborative environment to deliver the best possible project at a cost-effective price. The documents contain language that tends to make a project work that way."

Construction contracts can make the difference between project success and failure. Below are 10 reasons why ConsensusDOCS contracts, which were collaboratively developed by owners, contractors, subcontractors, specialty contractors, sureties, and design professionals, are considered crucial to advancing the project's best interest. ConsensusDOCS contracts fairly allocate risk while aligning the entire project team with the owner's goals. Fairly assigning risk reduces costly contingencies and adversarial negotiations, ensuring owners get the best contractors, the best prices, and the best project results.

Owner organizations played a key role in developing ConsensusDOCS, the construction industry's only standard contracts endorsed by 22 leading construction industry organizations. ConsensusDOCS were drafted by expert construction practitioners and professionals with hundreds of years of combined experience.

1. Owners Sit in the Driver's Seat

ConsensusDOCS contracts are the first widely-accepted, standard construction contracts written and endorsed by owner organizations.

Owners are given full control to be active participants and ultimate decision-makers on the project, rather than being treated just as check signers.

2. Save Transactional Costs and Time

Rather than negotiating the same one-sided terms or drafting legal documents from scratch, begin your project with an owner-endorsed contractual foundation.

By bringing all parties to the drafting table to reach a consensus on contracts, ConsensusDOCS provides a thorough outline and handles the most difficult part of construction contract negotiations.

ConsensusDOCS allows for easy editing of contract terms at the click of a button, reducing the time-consuming process of revising contracts.

3. Attract the Best Contractors and Get the Best Pricing

Better end results happen through proper planning, and that planning begins with the best contractual foundation with the best team possible.

By using fair documents that do not shift risks inappropriately, contractors don't have to increase prices to account for unknown risk.

ConsensusDOCS' straightforward language provides owners with real benefits by assisting them in acquiring the best prices from the best contractors.

4. Better Project Results

Every provision in the ConsensusDOCS family of contract documents was written to advance the project's success.

ConsensusDOCS help owners maintain their rights to contest claims and change orders, while ensuring constant project progress.

Owners have the most to gain or lose in a project, and saving time and money are integral to better project results.

5. A Balanced Contract Gives Owners a Greater Role

The architect/engineer's authority and corresponding responsibilities are aligned properly with the owner's goals.

ConsensusDOCS contracts allow owners to obtain greater permission to use the design documents for which they paid.

A balanced approach allows projects to proceed in using design documents while respecting intellectual property rights.

6. Prevent, Mitigate, and Resolve Disputes and Claims

Before problems become intractable, innovative dispute resolution and mitigation procedures require parties to meet to resolve disputes early.

Parties communicate directly. If unsuccessful, a third-party or review board provides an objective decision.

ConsensusDOCS contracts provide a strong financial incentive for parties to independently resolve arbitration claims.

7. Establish Positive Working Relationships and Direct Communications

Parties are expressly encouraged to communicate directly and affirmatively agree to act ethically, which provides the basis for a positive team environment.

By establishing positive relationships, parties are better focused on project results and better aligned with the owner's program throughout the construction process.

ConsensusDOCS helps owners resolve contentious claims through innovative claims mitigation procedures, while avoiding unnecessary layers of reporting.

8. Utilizing Building Information Modeling (BIM) and Electronic Communications

The BIM addendum is the first and only standard document to comprehensively address the legal ramifications of using the new transformative technology of BIM.

The BIM addendum defines permissible reliance on the model, addresses intellectual property rights, and facilitates management through a detailed BIM execution plan.

The electronic communications protocol allows secure reliance upon electronically transmitted information while providing management and administration flexibility.

9. IPD Agreement Takes Collaboration to a Higher Level

The ConsensusDOCS 300 Tri-party Collaborative Agreement offers the construction industry's first standard integrated project delivery (IPD) standard contract.

The owner, design professional, and constructor all sign the same agreement and form a core management team to further the project's best interest.

Projects using an IPD and Tri-party approach have already yielded greater efficiency and elimination of waste throughout the construction process.

10. A Balanced Approach to Liability Exposures

Achieving better project results, with fewer headaches and without assuming unnecessary risk, is a goal for all parties.

Owners drafted a mature approach to consequential damages and liquidated damages that provides strong financial incentives for performance.

ConsensusDOCS' provisions trigger parties to communicate to articulate liquidated damages and exclusions from consequential damages.

5.8.3 Typical Agreement Two—Environmental Consulting Agreement

AGREEMENT, dated this _____ day of _____, 2009 between ABC Green Consultants, Inc., a Virginia corporation having an address at 155 Alpine Blvd, Leesburg, Virginia 20175 (the "Company") and XYZ Services Ltd. (hereinafter referred to as the "Consultant").

WITNESSETH

WHEREAS, Company is engaged in the business of providing environmental site assessment services, including site Remediation and action plans on a nationwide basis; and

WHEREAS, Consultant is engaged in environmental consulting operations; and

WHEREAS, the parties wish to set forth the terms and conditions under which Consultant would provide environmental consulting services to Company clients on projects as designated by Company ("Projects"), as well as procure new accounts for Company.

NOW, THEREFORE, the parties agree as follows:

1. Company's Designation of Projects

Company shall notify Consultant of specific Projects which Company wishes to assign to Consultant. Such notice shall set forth the name of the client and the specific nature of the construction services to be performed, including time deadlines, specifications and other relevant information. Such notice shall also state the compensation to be paid Consultant for performing such services which, unless otherwise agreed on, will be a percentage of the fee billed client by Company net the Company's Expenses (procurement of environmental database records, testing services, traveling, overnight mail and communication, etc.); such fee to Consultant for such services shall be paid promptly following Company's receipt of client's payment for such bills. For purpose of this agreement, for Phase I Environmental Site Assessment Services, the Consultant shall be paid a fee of 30%. In addition to the 30% of net fee, Consultant shall be reimbursed for traveling expenses within the parameters set by the Company. The assignment of Projects to Consultant shall be within the sole discretion of Company and the parties acknowledge that the entering into of this Agreement shall not be deemed to obligate Company to assign any particular project to Consultant.

2. Consultant's Acceptance of Project

Within two business days after receipt of the written notice of Project Assignment, or such other period as may be specified in such notice, Consultant shall notify Company in writing as to whether it is willing to accept such assignment and perform the services described therein. The acceptance of any such assigned Project shall be in the sole discretion of Consultant and the parties agree that the entering into of this Agreement shall not be deemed to obligate Consultant to accept any Project which may be assigned by Company in connection with such project. If Consultant rejects any Project assignment, it shall promptly return any materials furnished by Company in connection with such Project. If Consultant accepts such Project it agrees that the services performed shall be governed by the provisions of this Agreement.

3. Consultant's Performance of Service

Consultant's performance of services on Project shall be conducted in compliance with the specifications and requirements of the project and in accordance with general reporting, invoicing, and other procedures as may be designated by Company. While such services shall be performed independently by Consultant without control by Company, it is agreed that in performing such services Consultant shall expressly indicate its relationship with Company rather than emphasizing its independent existence. Consequently, in all communications with clients and other third parties relating to the Project, Consultant agrees to sign documents and otherwise make reference to itself as an Associate Environmental Consultant of the Company. Consultant shall further agree that if requested it shall utilize business cards and/or stationery printed at Company expense or a dedicated telephone line installed at Company's expense. In addition, all billings and invoices for services performed by Consultant shall be done by Company.

4. Business Generated by Consultant

Consultant shall be entitled to a finder's fee of 10% on fees billed by Company for consulting services, net Company's Expenses, rendered for a

Project originated by Consultant for a new client, which Project is accepted by Company. Such fee shall be paid promptly following Company's receipt of Client's payment of such bill. In addition to such finder's fee, Consultant shall be entitled to payment for any services provided for any such Project, the performance and compensation of which shall be in accordance with the terms and provisions of this Agreement.

Such 10% finder's fee will be applied against the first Project awarded to Company as a result solely of the Consultant's efforts, whether such assignment includes one or multiple projects. Consultant will also be entitled to an additional finder's fee of 5% of the next Project awarded by the new client to Company under the same aforementioned terms.

5. Nature of Relationship

The parties agree that the relationship of Consultant to Company is that of an independent contractor and that the services contemplated under this Agreement do not constitute an employer-employee relationship. Consultant recognizes and agrees that its office costs and other expenses of doing business shall be at its sole expense and Consultant further assumes full responsibility for its taxes, social security, unemployment benefits, workers compensation, all as an independent business person or entity. Also, it is expressly acknowledged that Consultant shall have no right or authority to incur obligations or execute documents on behalf of Company.

6. Confidentiality and Surrender of Document

Consultant recognizes that Company may provide or allow Consultant access to certain information not generally known to the public or other firms engaged in the same or similar business activities as those conducted by Company. Consequently, Consultant recognizes and agrees that all such information is the sole property of Company and must be returned to Company upon termination of this Agreement. In addition, Consultant shall not disseminate, or transfer, whether orally or by any other means, Company's client base or any part thereof, procedures, fees, fee structure, costs of doing business, compensation to Company's employees or independent contractors/consultants, checklist, or report format to any other person or entity, whether corporate, governmental, or individual, without the expressed prior written consent of an officer of the Company. Consultant shall return any written information, whether written or paper or computer disk, prepared by the Consultant or consulted by the Consultant, on any assignments conducted by the Consultant on Company's behalf. Such information includes, but is not limited to: reports, proposals, fees, compensation data or agreements, correspondence, studies, testing results, checklists, client lists, etc. All such information is and shall remain the property of Company. Consultant is not to retain any copies, whether paper or computer disk, of work conducted on Company's behalf, whether such work is considered draft, preliminary or final. Upon termination of the Agreement, Consultant shall not knowingly have in his or her possession, custody or control, any computer disks or hard copies of any Company reports or forms of Company's reports, checklists, data, etc. If at any time in the future Consultant discovers that he is in possession, custody or control of any such material, he will promptly deliver to Company.

7. Termination

Either party may terminate this Agreement upon 10 days' written notice, following which neither party shall continue to have the right to refer to the relationship established hereunder in any of its business dealings. If on Termination Date Consultant is performing services on an uncompleted continuing Project, Consultant shall be entitled to receive payment of billings for services rendered on such Project prior to Termination Date, but all other rights and obligations hereunder shall end except as provided in the following sentence. Notwithstanding the foregoing, Consultant agrees, if so requested by Company, that following Termination Date it shall continue providing services on Projects existing at such date pending their completion, in which case Consultant shall be entitled to receive payment of billings for services rendered after such date. In addition to returning all records and information as required upon termination, as set forth under paragraph 6 hereof, Consultant agrees that so long as this Agreement is in effect and for a period of one year from the date of its termination that neither Consultant nor its employees shall actively solicit business from any of the Company's clients.

8. Miscellaneous

A.   Governing Law
      This Agreement shall be governed by the laws of the State of Virginia.

B.   Attorney's Fees
      If any legal action, or other proceedings of any kind, is brought for the enforcement of this Agreement or because of any alleged breach, default, or any other dispute in connection with any provision of this Agreement, the successful or prevailing parts shall be entitled to recover all reasonable attorney's fees and other costs incurred in such action or proceedings, in addition to any relief to which it may be entitled.

C.   Entire Agreement
      This Agreement embodies the entire agreement between the parties in relation to the subject matter herein and supersedes all prior understandings or agreements, oral or written, between the parties hereto.

IN WITNESS WHEREOF, the parties have executed this Agreement as of the date first appearing.

ABC Green Consultants, Inc.

By:_____
John A. A. Doe
President

_____
Consultant

Choosing Materials and Products

Doi: 10.1016/B978-1-85617-676-7.00006-3

6.1 GENERAL

It is important to think green before you begin any design or construction project. This will facilitate the choice of the most appropriate building materials and construction techniques. If LEED™ certification is being sought for a project, then the choice of building materials becomes even more critical. Building materials used in sustainable design usually undergo an extensive network of extraction, processing, and transportation steps required in their creation. Furthermore, the activities needed to create building materials often have a negative impact on the environment by adding to air and water pollution, destroying natural habitats, and depleting natural resources. Thus, building green helps minimize pollution, protect the natural environment, and create a healthy, comfortable, non-hazardous place to live and work. Moreover, incorporating green products into a project does not in any way imply sacrifice in performance or aesthetics, nor does it necessarily entail higher cost.

For maximum benefit, a building's design, and the materials used in that design, need to be appropriate to the site and its geographic location. An all glass envelope may seem like a wonderful design choice for a proposed commercial building. But before deciding to have the glass envelope installed, make sure to consider the physical orientation of the building. If the main facade faces south and isn't properly shaded by overhangs or louvers, you may discover that your new building becomes unbearably hot in the summer and extremely costly to air condition.

Nature provides us with a variety of eco-friendly and naturally occurring materials, such as clay, sand, wood, and stone that are employed in building construction. But apart from the wide variety of naturally occurring materials that are available, we also find an increasing number of synthetic and man-made products being used. Since ancient times, the manufacture of building materials has been an established industry in many countries around the world. Today their use is typically segmented into specific specialty trades, such as carpentry, plumbing, roofing, and flooring. Synthetic materials are often needed to meet certain functional needs which are not always easy to attain with natural materials such as waterproofing or insulation.

A very compelling strategy for minimizing environmental impacts that is being encouraged by several states and government agencies is the repairing and reuse of an existing building instead of tearing it down and building a new structure. Some states, like New York and North Carolina, are providing substantial grants to renovate vacant buildings in rural counties or in economically distressed urban areas. Rehabilitation of existing building components saves natural resources, including the raw materials, energy, and water resources required to build new; it also prevents pollution that might take place as a byproduct of extraction, manufacturing, and transportation of virgin materials; and minimizes the creation of solid waste that could end up in landfills. Green organizations like the USGBC recognize the importance of building reuse, which is why reusing a building can contribute to earning points under LEED Materials Resource Credits.

6.1.1 Defining Green Building Materials

The term green building materials refers to a growing list of products and materials currently on the market and used to build, furnish, and power buildings. To make these lists, materials have to meet certain eco-friendly criteria such as being manufactured from recycled materials or containing low-VOC levels. The more criteria a product meets, the greener it is considered to be. However, many new products have appeared on the market in recent years claiming to be "green," yet upon closer scrutiny there is little to back up such claims. The term "greenwashing" has come into vogue to describe products that purport to contain green characteristics, but in reality lack any proof to substantiate these claims, many of which are misleading. It has become a challenge to specifiers and purchasers to determine the validity and relevance of environmental claims. This is why selecting environmentally preferable materials requires research, critical evaluation, and common sense. Issues such as code compliance, warranties, and the performance of green products, particularly newly introduced materials, need to be carefully considered. Verification that a product meets advertised claims should be using evaluations based on recognized testing procedures and testing laboratories.

Although an exact and precise definition of green building materials may be elusive, in general building materials are called green because they are good for the environment. The ideal building material would have no negative environmental impacts, and might even have positive environmental impacts, including air, land, and water purification. There is no perfect green material, but in practice there are a growing number of green materials that reduce or eliminate negative impacts on people and the environment. It is important therefore for a green consultant to understand a product's benefits and drawbacks and be able to suggest alternatives when needed that may be more environmentally friendly. This data is used to profile individual products from specific manufacturers. Green material directories are essentially databases that provide listings of available products (as say, organized by CSI sections) with the environmental attributes claimed by the manufacturers. It is fairly easy today for a designer or builder to go online and find almost any material, such as on the online GreenSpec®—The Environmental Building News Product Directory and Oikos® Green Building Source Directory, which lists product descriptions for over 2000 environmentally preferable products. It should be noted that not all "green" products meet up to manufacturers' claims.

The concept of producing green materials from recycled materials is very effective and puts waste to good use. Typical examples include carpeting made from recycled soda bottles. Building materials are also classified as green because they can be recycled once their useful lifespan is over, for example, aluminum roofing shingles. If that product itself is made from recycled materials, the benefit of recyclability is multiplied. Building materials are also considered green when they are made from renewable resources that are sustainably harvested, such as flooring made from sustainably grown and harvested lumber or bamboo. The term "green" can also apply to building materials that are durable. For example, a durable form of siding can outlast a less-durable product, resulting in significant savings in energy and materials over the lifetime of a property. Moreover, durable products made from environmentally-friendly materials such as recycled waste offer even greater benefits. Another good example of an environmentally friendly building material is siding made from cement and recycled wood fiber (wood waste).

6.1.2 Materials: Natural versus Synthetic

Natural materials are those that are unprocessed or minimally processed by industry, such as timber or glass. Synthetic materials on the other hand are manufactured in industrial settings after undergoing considerable human manipulations, such as plastics and petroleum-based paints. Both have their uses and their advantages and disadvantages. Certain materials and techniques are better suited for specific geographical locations. Thick adobe walls, for example, have heavy thermal mass and a high diurnal range that help mitigate the dramatic changes in temperature (e.g., between night and day) that take place in the arid Southwest. That style of architecture is also more appropriate for desert climates. This type of construction however would be ill advised in a hot humid climate where lightweight construction designs that encourage cross ventilation and cool breezes are more appropriate. The choice of material is very important to achieve success. There are basically two types of building materials used in the construction industry: 1. Natural, and 2. Synthetic.

1. Natural Building Materials

They are all around us—mud, stone, and timber are among the most basic natural occurring building materials. And because many of these materials are available throughout the world, the costs and pollution associated with the transportation of these materials across the country decreases.

Using natural materials also reduces the amount of toxins in buildings. Additionally, many of these techniques are energy efficient, inexpensive, and easy to build with little required construction knowledge. These materials are being used by people all over the world to create shelters and other structures to suit their local weather conditions. In general, stone and/or brush are used as basic structural components in these buildings, while mud is generally used to fill in the space between, acting as a form of concrete and insulation. The more popular naturally occurring building materials are:

- Adobe
- Cob
- Cordwood
- Earth-sheltered
- Earth bag
- Rammed earth
- Stone, granite
- Straw bale
- Bamboo
- Lumber.

However, there are several challenges to using natural materials in the building of a structure. Natural materials are rarely mentioned in building codes, and most building officials may be unfamiliar with these methods of construction. Furthermore, banks are hesitant to finance alternative methods of construction and insurance may be difficult to obtain, particularly in this difficult economic climate. It may also be difficult to find suitable contractors if backup or consultation is needed.

2. Synthetic Materials

For the vast majority of building construction today, natural materials can neither cope with nor meet the required specifications of today's industrial challenges. Plastic is a good example of a typical synthetic material. The term plastic covers a range of synthetic or semi-synthetic organic condensation or polymerization products that can be molded or extruded into objects, films, or fibers. Plastics vary immensely in heat tolerance, hardness, and resiliency. This adaptability and the general uniformity of composition, as well as the lightness of plastics, facilitates their use in a wide variety of industrial and other applications.

There are several criteria that determine a material's greenness, such as whether the material is renewable and resource efficient in its manufacture, installation, use, and disposal. Other criteria include whether the material supports the health and well-being of occupants, construction personnel, the public, and the environment, whether the material is appropriate for the application, and what if any environmental and economic trade-offs exist among alternative materials.

There remains a substantial amount of research that needs to be conducted to satisfactorily evaluate the best material alternatives for a project. Material selection should take into consideration the impacts of a product throughout its life-cycle, i.e., from inception (raw material extraction) to use, and then to reuse, and finally to recycling or disposal, as the case may be.

It has been found that the most successful applications of environmentally-friendly materials are those that extract multiple benefits from the products selected. The majority of available green building materials and products have one or more of the following properties attributed to them that relate to health and/or environmental issues:

- Durability
- Promote good indoor air quality (e.g. through reduced emissions of VOCs and/or formaldehyde)
- Readily recyclable or reusable when no longer needed
- Can be salvaged for reuse, refurbished, remanufactured, or recycled
- Are locally extracted and processed
- Are salvaged from existing or demolished buildings for reuse
- Less energy is used in extraction, processing, and transport to job site
- Are made using natural and/or renewable resources
- Do not contain ozone depleting substances like CFCs or HCFCs
- Do not contain highly toxic compounds, nor does production result in highly toxic by-products
- Manufactured from waste material such as fly ash or straw
- Are obtained from local resources and manufacturers
- Are biodegradable
- Employ sustainable harvesting practices for wood and bio-based products.

6.1.3 Storage and Collection of Recyclables

For projects seeking LEED certification, a "storage and collection of recyclables" area is required. The intent of this prerequisite is basically to reduce waste sent to landfills by encouraging storage and collection of recyclables. USGBC has presented general guidelines (Figure 6.1) and states that the area should be easily accessible and serve the entire building(s). The area should be dedicated to the storage and collection of nonhazardous materials for recycling—including paper, corrugated cardboard, glass, plastics, and metals.

6.2 LOW EMITTING MATERIALS

As more and more people become aware of the dangers of global warming and the benefits of sustainable design and construction, the

Commercial building minimum recycling
Square footage (sf) Area (sf)
0 to 5000 82
5001 to 15,000 125
15,001 to 50,000 175
50,001 to 100,000 225
100,001 to 200,000 275
200,001 or greater 500

FIGURE 6.1 Guidelines recommended by USGBC for recycling storage areas that are based on overall building square footage. (Source: USGBC.)

more people are seeking out "healthy" buildings in which to live and work in. For some, the issue is living free of allergies, asthma, and other breathing problems, partly caused by use of materials that emit hazardous substances. As property owners and employers, they invest in air-cleaning systems, buy formaldehyde-free furnishings, and use nontoxic paints and finishes. They are concerned about preventing air pollution problems that may be caused by construction techniques and building materials.

6.2.1 Adhesives, Sealants, and Finishes

Sealants have the ability to increase the resistance of materials to water or other chemical exposure, and caulks and other adhesives help control vibration and strengthen assemblies by spreading loads beyond the immediate vicinity of fasteners. Furthermore, both properties enhance durability of surfaces and structures. Many construction adhesives formulas are hazardous in manufacture and often contain more than 30 percent volatile petroleum-derived solvents, such as hexane, to maintain liquidity until application. This can be harmful to workers who become exposed to toxic solvents, and as the materials continue to outgas during curing the occupants can also be potentially exposed to emissions for extended periods of time. With regard to water-based adhesives, they are available from a number of different manufacturers. Industry tests indicate that these products work as well as or better than solvent-based adhesives, can pass all relevant ASTM and APA performance tests, and are available at comparable costs to common solvent-based adhesives.

One of the problems surrounding stains and sealants is that they normally emit potentially toxic volatile organic compounds (VOCs) into indoor air. An effective approach to manage this problem is employing materials that do not require additional sealing, such as stone, ceramic and glass tile, and clay plasters. The toxicity, and the air and water pollution generated through the manufacture of chlorinated hydrocarbons such as methylene chloride, emphasizes strongly for using responsible, effective alternatives, such as plant-based, sealant formulations that are nontoxic or have low-toxicity.

For projects seeking LEED certification, the use of adhesives and sealants with volatile organic compound (VOC) content must not exceed the VOC content limits of South Coast Air Quality Management District (SCAQMD) Rule #1168 limits scheduled for 2007. Aerosol adhesives not covered by Rule 1168 must meet Green Seal Standard GS-36 requirements. It is important to avoid all indoor air contaminants that are odorous, potentially irritating, and/or harmful to the comfort and well-being of installers and occupants.

General LEED requirements stipulate that all adhesives and sealants used on the interior of the building (defined as inside of the weatherproofing system and applied on-site) should comply with the reference standards shown in Figure 6.2.

Architectural applications (SCAQMD 1168)	VOC limit (g/l less water)	Specialty applications (SCAQMD 1168)	VOC limit (g/l less water)
Ceramic tile	65	Welding: ABS (avoid)	325
Contact	80	Welding: CPVC (avoid)	490
Drywall & panel	50	Welding: Plastic cement	250
Metal to metal	30	Welding: PVC (avoid)	510
Multipurpose construction	70	Plastic primer (avoid)	650
Rubber floor	60	Special purpose contact	250
Wood: Structural member	140		
Wood: Flooring	100	**Sealants primers (SCAQMD 1168)**	
Wood: All other	30	Architectural porous primers (avoid)	775
All other adhesives	50	Sealants & non-porous primers	250
Carpet pad	50	Other primers (avoid)	750
Structural glazing	100		
		Aerosol adhesives (GS-36)	**VOC limit**
Substrate specific applications		General purpose mist spray	65%
Fiberglass	80	General purpose web spray	55%
Metal to metal	30	Special purpose aerosol adhesives	70%

Note: VOC weight limit is based on grams/liter of VOC minus water. Percentage is by total weight.

FIGURE 6.2 Adhesives, sealants, and sealant primers: South Coast Air Quality Management District (SCAQMD) Rule #1168. VOC limits are listed in the table and correspond to an effective date of July 1, 2005 and rule amendment date of January 7, 2005. (Source: Based on USGBC.)

6.2.2 Paints and Coatings

Paints consist of a substance composed of solid coloring matter suspended in a liquid medium and applied as a usually opaque protective or decorative coating to a surface. Primers are basecoats applied to a surface to increase the adhesion of subsequent coats of paint or varnish. Sealers are also basecoats but are applied to a surface to help reduce the absorption of subsequent coats of paint or varnish, or to prevent bleeding through the finish coat. Figure 6.3 reflects the allowable VOC levels stipulated by SCAQMD.

Paints have significant environmental and health implications in their manufacture, application, and disposal. Most paint, even water-based "latex," is derived from petroleum and creates air pollution and solid/liquid waste. Volatile organic compounds (VOCs) are typically the pollutants of greatest concern in paints. VOCs emitted from paint and other building materials are associated with eye, lung, and skin irritation, headaches, nausea, respiratory problems, and liver and kidney damage. However, there are renewable alternatives, such as milk paint, that address many of these concerns, although some products are only suitable for indoor applications and at an increased cost. Reformulated low- and zero-VOC latex paints with excellent performance in both indoor and outdoor applications are now available at the same or lower price than older high-VOC products. Paints that meet the GS-11 standard are low in VOCs and aromatic solvents and do not contain heavy metals, formaldehyde, or chlorinated solvents, while meeting stringent performance requirements.

Another type of paint that is solvent-free and that may be used on concrete, stone, and stucco is silicate paint. Silicate paint has many advantages being odorless, nontoxic, vapor-permeable, naturally resistant to fungi and algae, noncombustible, colorfast, light-reflective, and resistant to acid rain. Furthermore, these paints possess extraordinary durability and cannot spall or flake off, and will only crack if the substrate cracks.

6.2.3 Flooring Systems

1. Carpet

The manufacture, use, and disposal of carpet have significant environmental and health implications. Most carpet products are synthetic,

Paint type	VOC limit (grams/liter)
Flat	50 g/l
Non-flat	50 g/l
Primers, sealers & undercoats	100 g/l
Quick-dry enamels	50 g/l

FIGURE 6.3   Allowable VOC levels (grams/liter less water and exempt compounds) according to the South Coast Air Quality Management District (SCAQMD).

usually derivatives of non-renewable petroleum products; and its manufacture requires substantial energy and water, and creates harmful air and solid/liquid waste. However, many carpets are now becoming available with recycled content, and a growing number of carpet manufacturers are refurbishing and recycling used carpets into new carpet. Likewise, redesigned carpets, new adhesives, and natural fibers are now available that emit low or zero amounts of VOCs. For improved air quality selected carpets and adhesives should meet a third party standard, such as the Carpet and Rug Institute (CRI) Green Label Plus or the State of California's Indoor Air Emission Standard 1350.

The main reasons carpets comprised of natural fibers are often preferred to synthetic carpets is because they are more environmentally friendly and they are renewable and biodegradable. Options include wool, jute, sisal, silk, and cotton floor coverings. Biodegradable carpets made from plant extracts and plant-derived chemicals are also available. Among the disadvantages of carpets is that they harbor more dust, allergens, and contaminants than many other materials (Figure 6.4). Better indoor air quality can be achieved with the use of other durable flooring materials such as a concrete finish, cork, linoleum, or reclaimed hardwoods to name but a few potential alternatives.

FIGURE 6.4   A photo of FLOR carpet squares laid in a flexible and practical "tile" format. The tiles are considered "green" because they are manufactured from renewable and recyclable materials. The tiles are available in a range of colors, textures, and patterns. (Source: FLOR Inc.)

2. Vinyl/PVC

Polyvinyl chloride's general properties—abrasion resistant, lightweight, good strength relative to its weight, water resistant, and durable—are key technical advantages for its use in building and construction applications. It can also be made scratch resistant, sunlight resistant, and of almost any color. Furthermore, rigid PVC is inherently difficult to ignite and stops burning upon removal of the heat source. Tests show that when compared to its common plastic alternatives, PVC performs better in terms of lower combustibility, flammability, flame propagation, and heat release. It also deserves special attention because it accounts for almost 50 percent of total plastic use in construction today. For example, new products are continuously being developed like Clearstrength®, Durastrength®, Plastistrength®, and Thermolite®. These are used mainly to produce rigid vinyl compounds which are finding their way into building products such as PVC pipe and pipe fittings, window and door lineals, blinds for windows and doors, vinyl siding, fencing, and decking applications.

Vinyl tends to be inexpensive, in part because vinyl production typically requires roughly half the energy needed to produce other plastics. Products made from vinyl can be resistant to biodegradation and weather, and are effective insulators. However, there are serious environmental concerns regarding the use of vinyl. It is said that vinyl's life-cycle begins and ends with hazards, most stemming from chlorine, its primary component. Chlorine makes PVC more fire resistant than other plastics. Vinyl chloride, the building block of PVC, causes cancer. Lead, cadmium, and other heavy metals are sometimes added to vinyl as stabilizers; and phthalate plasticizers, which give PVC its flexibility, pose potential reproductive risks. Manufacturing vinyl or burning it in incinerators produces dioxins, which are among the most toxic chemicals known to man. Research has shown that the health effects of dioxin, even in minute quantities, include cancer and birth defects. There are many substitutes coming on the market but they may be more expensive or require different maintenance. Nevertheless, there are numerous applications where the substitution of eco-friendly alternatives, particularly for indoor applications where occupants can be directly exposed to off-gassing plasticizers, would clearly be prudent for the sake of occupant health and well-being.

3. Tile Products

Tile is manufactured primarily from fired clay (porcelain and other ceramics), glass, or stone, and provides a useful option for flooring, countertops, and wall applications whose principle environmental benefit is durability. One of the principal advantages of tile is its durability as it can last almost indefinitely even in high-traffic areas, eliminating the waste and expense of replacement. Tile production however is energy intensive,

FIGURE 6.5 An interior view of the installation of crystal micro double loading polished tile producing a dramatic interior finish. One of the principal advantages of ceramic tile is its durability. (Source: Foshan YeShengYuan Ceramics Co.)

although tile from recycled glass requires less energy than tile from virgin materials. Tile's attributes include that it is non-flammable, will not retain liquids, does not absorb fumes, odors, or smoke and, when installed with low- or zero-VOC mortar, can contribute to good indoor air quality.

The production process for ceramic glazed tile is the same as for ordinary ceramic tile, except that it includes a step known as glazing. Glazing ceramic tile requires a liquid made from colored dyes and a glass derivative known as flirt that is applied to the tile, either using a high-pressure spray or by direct pouring. This in turn gives a glazed look to the ceramic tile. The most popular types of tiles in use today are glazed and unglazed floor tiles, wall tiles, and ceramic mosaic tiles which constitute nearly 99 percent of the tile industry's output (Figure 6.5). Due to the industry being so focused on decorative tiles, it has become completely dependent on the economic health of the construction and remodeling industries, which are currently in a somewhat discouraging state.

4. Earthen Building Materials

The use of earthen building materials goes back to before the invention of writing, and includes: adobe bricks—made from clay, sand, and straw; rammed earth—compressed with fibers for stabilization; and cob—clay, sand, and straw that is stacked and shaped while wet. Provided they are obtained locally, earthen building materials can reduce or eliminate many of the environmental problems posed by conventional building materials since they are plentiful, non-toxic, reusable, and biodegradable. Well-built earthen buildings are known to be durable and long lasting with little maintenance as evidenced by the temples and ziggurats of Babylon.

Considerations in Earthen Construction

- Earthen walls are thick and may comprise a high percentage of floor area on a small site.
- Construction is generally labor-intensive, although minimal skill is required.
- Multi-story and cob structures require post-and-beam designs.
- It may be more difficult to obtain necessary permits, although the necessary code recognition, structural testing, etc. is available.
- If labor includes primarily building professionals, the square-foot cost of earthen construction is comparable to conventional building methods.

Benefits of Earthen Materials Include

- Environmental impact is minimal, provided materials come from local sources
- Durable and low-maintenance
- Thermal mass helps keep indoor temperatures stable, particularly in the mild to warm climates
- Biodegradable or reusable
- Can be easy to build, requiring few special skills or tools
- If well designed, provides pleasing aesthetics
- Highly resistant to fire and insects
- Requires no toxic treatments, and does not off-gas hazardous fumes; good for chemically sensitive individuals.

6.2.4 Windows

In many cases, the most often considered environmental impact of windows is how they can be best employed to help reduce a building's energy use. Designers as well as tenants have long recognized windows to be a major source of energy inefficiency within buildings. Thus, during the heating season, poor windows are a major source of heat loss. During the cooling season, they allow heat to filter into the building in addition to allowing solar radiation to enter the building, further heating interior spaces. Throughout the year, poorly installed windows will allow air to infiltrate between the frames and the glazing, and between the frames and the building walls. Also, as windows get older, their performance starts to deteriorate.

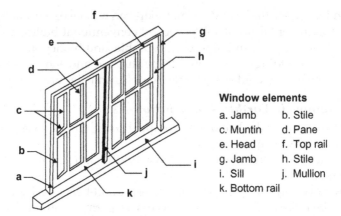

Window elements

a. Jamb   b. Stile
c. Muntin   d. Pane
e. Head   f. Top rail
g. Jamb   h. Stile
i. Sill   j. Mullion
k. Bottom rail

FIGURE 6.6   A drawing showing the various components that make up a typical window.

Nevertheless, windows are an essential element in construction because they provide ventilation, light, views, and a connection to the outside world, even though drafty, inefficient, poorly insulated, or simply poorly chosen windows can compromise the energy efficiency of a building's envelope. The fabrication of window systems, whether made of wood, aluminum, plastic, or steel, as with any other manufactured product, requires energy and will likely generate air pollution. Figure 6.6 shows the important components of a window.

Windows can be designed to offer a variety of glazing options, each with a different thermal performance or resistance (R-value) depending on the specifications. R-values are approximate and vary with temperature, type of coating, type of glass, and distance between glazings. From the lowest resistance to highest:

1. Single glazing and acrylic single glazing have similar "R" values—R = 1.0.
2. Single glazing with a storm window and double glazing have similar "R" values—R = 2.0.
3. Double glazing with a low-e coating and triple glazing have similar "R" values—R = 3.0.
4. Triple glazing with a low-e coating—R = 4.0.

Note that for a conventional, insulated stud wall—R = 14.0.

There are many different types of windows, each having a different cost, insulating ability, and durability as shown below:

- Wood requires continuous maintenance for durability. The wood source should be certified by an accredited organization such as the Forest Stewardship Council (FSC).

- Aluminum and steel are poor insulators, and energy intensive to manufacture. When deciding on using metal-framed windows, recycled content should be sought and also frames with "thermal breaks" to limit the loss of heat to outdoors should be sought.
- Fiberglass is strong, durable, and has excellent insulating value, but is energy intensive to manufacture.
- Vinyl provides good insulation features, but is highly toxic in its manufacture, and also if burned. High-efficiency windows typically utilize dual or triple panes with low-e (low emissivity) coatings and gas fill (typically argon) between panes to help control heat gain and loss. Factory applied low-e coatings on internal glass surfaces are more durable and effective than films. High-quality, efficient windows are widely available from local retailers. To make an informed choice, consider only windows that have the National Fenestration Rating Council (NFRC) ratings. The EPA Energy Star® label for windows can be a useful summary of these factors.

6.2.5 Miscellaneous Building Elements

1. Gypsum Wall Board

Also known as drywall or plasterboard, gypsum wall board is manufactured in the United States and Canada to comply with ASTM Specification C 1396. This standard must be met whether the core is made of natural ore or synthetic gypsum. Gypsum wall board is a plaster-based wall finish that is available in a variety of sizes; 4 feet wide by 8 feet high is the most common. Thicknesses vary in 1/8-inch increments from 1/4 inch to 3/4 inch.

Due to its ease of installation, familiarity, fire resistance, non-toxicity, and sound attenuation, gypsum wall board, known by its proprietary names Dywall® and Sheetrock® is ubiquitous in construction. Gypsum wall board is a benign substance (basically paper-covered calcium sulfate), but it has significant environmental impacts because it is used on a vast scale; domestic construction uses an estimated 30 billion square feet annually.

The main advantages of gypsum board include low cost, ease of installation and finishing, fire-resistance, sound control, and availability. Disadvantages include difficulty in applying it to curved surfaces, and low durability when subject to damage from impact or abrasion. Reclaimed gypsum board can easily be recycled into new gypsum panels that conform to the same quality standards as natural and synthetic gypsum, but doing this may not be practical because gypsum is an inexpensive material that can require significant labor to separate and prepare it for recycling. Gypsum board should be purchased in sizes that minimize the need for trimming (saving time and waste).

2. Siding

Siding is an external protection element that provides protection for wall systems from moisture and the heat and ultraviolet radiation of the sun. Selecting siding that is reclaimed, recyclable, or incorporates recycled material will reduce waste and pollution. However, there are many types of siding and the environmental impacts of siding products vary considerably. Types of siding include fiber-cement siding (which is fire-resistant, pest-resistant, and emits no pollutants), metal siding, wood siding, cement stucco, and composite siding.

6.2.6 Roofing Systems

The principal role of a roof is to keep the weather outside a structure's envelope and protect the structural members and interior materials from deterioration. Weather resistance, dependability, and durability are the most critical characteristics of roofing materials. The extraction, manufacture, transport, and disposal of most roofing materials pollutes air and water, depletes resources, and damages natural habitats. It is estimated that roofing materials comprise an estimated 12 to 15 percent of construction and demolition waste. An environmentally sustainable roof must be durable and long lasting, but may also contain recycled or low-impact materials. Roofs that are environmentally friendly can provide aesthetically pleasing design options, reduced life-cycle costs and environmental benefits such as reduced landfill waste, energy use, and impacts from harvest or mining of virgin materials. Studies show that a well-installed roof with a 50-year warranty can reduce roofing waste by 80 to 90 percent over its lifetime, relative to a roof with a 20-year warranty.

Factors that affect the choice of a roofing system are durability, waste reduction, and potential liability. Other considerations that will impact the type of roof chosen should include:

- The roof's ability to resist the flow of heat from the roof into the interior, whether through insulation, radiant barriers, or both
- The roof's capacity to reflect sunlight and re-emit surface heat (Cool roofs can reduce cooling loads and urban heat-island effects while providing longer roof life)
- The roof's ability to reduce ambient roof air temperatures through evaporation and shading, as in the case of vegetated green roofs
- Roofs that are recyclable and/or have the capability of being reusable, to reduce waste, pollution, and resource use are preferable.
- Roofs comprised of membranes that do not contain bromine or chlorine are preferable.

1. Traditional Roofing

For roofs to comply with local fire codes they may require protective ballast such as concrete tile. Existing PVC and TPO (thermoplastic poly-olefins) roofing membranes, as well as underlying polystyrene insulation, can sometimes be recycled, and this practice is expected to become more prevalent as federal construction specification requirements generate increased demand. For residential and commercial roofs there are several options regarding the type of roof and materials used:

- Composition shingles can get 50-year warranties, which are better than 20- to 40-year products and can be recycled.
- Fiber cement tiles are durable and fire/insect-proof, but are heavy and not renewable or biodegradable. They may be ground up and used as inert fill at demolition.
- Clay or cement tiles are very durable and made from abundant materials, but are heavy and expensive.
- Metal is a durable and fire/insect-proof, recyclable material. It typically contains recycled content, but its manufacture is energy intensive and causes pollution and habitat destruction.
- Recycled plastic, rubber, or wood composite shingles are durable, lightweight, and sometimes recyclable, but not biodegradable .
- Built-up roofing membranes are not generally made from renewable resources, but some may contain recycled content. Built-up roofing durability depends largely on the structure, installation, flashing, and membrane chosen.
- Vegetated green roofs are most commonly installed on roofs with slopes less than 30 degrees.
- Wood shakes are a biodegradable material, but are flammable and not very durable. It is not typically considered to be a "green" option for fire-prone areas.

2. Green Roofing

Vegetated roofing is one of the more significant developments in sus-tainable building design. Depending on load capabilities and other appli-cation-driven requirements, green roofs can be planted with herbs, grasses, flowers, even trees, in an exciting variety of colors, textures, scents, and heights. Emerging new technologies that are helping to promote green building are increasing efforts to make useable space of existing and/or new rooftops to provide additional living space. The key to creating these spaces is to use lightweight and recycled materials and have a plan for storm water management. In this respect traditional drainage systems using pipe and stone are not plausible. Green roof systems are a natural way of providing additional clean air through the transference of CO_2 and oxygen between the plants and vegetation with the atmosphere. Contemporary green roof

FIGURE 6.7   A photograph of the 20,300 square foot vegetated roof on the Chicago City Hall building, Chicago, IL. The type of green roof system employed is the single source provider. (Source: greenroofs.com LLC.)

designs generally contain a mixture of hard and soft landscaping such as the green roof of Chicago's city hall built in 2001 (Figure 6.7).

It is very important that the selected drainage/retention layer is capable of supporting any type of landscape—from roadways and paths, to soil and trees, so as to permit excess water to drain unobstructed underneath (Figure 6.8).

Some important benefits attributed to green roofs include:

- Improved insulation qualities and more moderate rooftop temperatures mean reduced cooling and heating requirements, saving energy and money. Savings on energy heating and cooling costs depend on the size of the building, climate, and type of green roof.
- Reduce storm water run-off, which in turn reduces the stress on urban sewer systems and decreases run-off related pollution of natural waterways.
- Absorption of dust and airborne pollutants.
- Air quality improvement—lower rooftop temperatures help to enhance the microclimate of surrounding areas and mean less smog from the heat island effect.
- Extended life of roof system, which is protected from ultraviolet radiation, extreme temperatures (that cause a roof system to expand and contract), and mechanical damage. (Plant species, soil depth, and root-resistant layers are carefully matched to ensure the roof membrane is not damaged by the roots themselves.)
- Lightweight extensive systems can be designed with dead loads comparable to standard low-slope roofing ballast. Structural

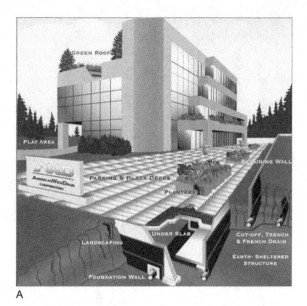

A

Rain or sprinkler

Growing medium
Sand and gravel
Root barrier fabric
Insulation
Roof membrane
Structural support

Sand and gravel system

Rain or sprinkler

Growing medium
Root barrier fabric
Drain core
Separation fabric
Insulation
Roof membrane
Structural support

Amergreen system

B

FIGURE 6.8　**A.** An artist's rendering showing commercial application of green elements (Source: American Wick Drain Corporation), **B.** A drawing showing two types of roofing systems: 1. A sand and gravel system, and 2. The Amergreen Roofing System (Source: American Wick Drain Corporation), and **C.** A drawing of a Quad-lock roofing system detail (Source: Quad-lock Insulating Concrete Forms).

reinforcement may not be necessary, and cost can be comparable to conventional high quality roofing options.
- Noise pollution reduction. It is estimated that noise levels in a building can be reduced by as much as 40 decibels. Sound waves that are produced by machinery, traffic, or airplanes can be absorbed, reflected, or deflected. The substrate tends to block lower sound frequencies and the plants block higher frequencies.

1. Quad-lock 6'' tie blue
2. Quad-lock wire top tie
3. Blocking
4. Quad-lock metal J-track
5. Flashing (Fasten to J-track)
6. Gravel or paver
7. Concrete, wood or steel curb
8. Vegetation
9. Growing medium
   (Sopraflor type X or L)
10. Microfab
11. Sopradrain 10-G
12. Microfab double layer
13. Cap sheet membrane
14. Base sheet membrane
15. Quad-deck panel

16. Welded wire mesh
17. Quad-deck beam reinforcement
18. Rebar chair
19. Quad-deck Z-strips for ceiling finish
    attachment
20. Void holes for utilities
21. Drain
22. Interior wall finish
23. Quad-lock 8'' tie yellow
24. Exterior wall finish
25. Quad-lock panels
26. Reinforced concrete core
27. Primer

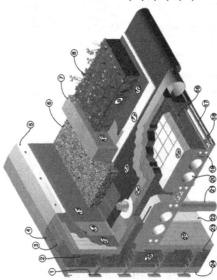

FIGURE 6.8   (Continued)

6.2.7 Wood

There are several types of wood used in building construction including:

1. *Pressure Treated Lumber*

The popularity over the decades for pressure treated lumber has been partially due to its resistance to rot and insects. But with rapid changes taking place in the treating industry, it is more important than ever to ensure that the treated wood meets standard specifications. The U.S. Department of Commerce American Lumber Standard Committee (ALSC) accredits third-party inspection agencies for treated wood products. The ALSC Website (www.alsc.org) can provide updated lists of accredited agencies. There are several arsenic-free preservative formulations on the market, all of which rely heavily on copper as their primary active ingredient. These include acid copper chromate, alkaline copper quat (ACQ), copper azole, and copper HDO.

2. *Certified Forest Products*

Forest certification was launched over a decade ago to help protect forests from destructive logging practices. Where salvaged or reclaimed wood is not available or applicable (i.e., structural applications), specify products that are certified by an approved and accredited organization such as the Forest Stewardship Council (FSC) or the Sustainable Forestry Initiative (SFI). This can also help in achieving LEED or Green Globes certification. Third-party independent certification of sustainable practices has essentially beat out the many unverified company claims, industry-backed compliance schemes, and even government regulatory mechanisms as a way to build trust among customers, buyers, and shareholders.

3. *Engineered Wood Products*

The term engineered lumber, also called *composite wood* and *man-made wood*, consists of a range of derivative wood products that are manufactured by pressing or laminating together the strands, particles, fibers, or veneers of wood with a binding agent to form composite materials. It basically refers to a family of engineered wood panels that includes particleboard, medium density fiberboard (MDF), and hardboard. The superior strength and durability of engineered lumber allows it to displace the use of large (and increasingly unavailable), mature timber. It is also less susceptible to humidity-induced warping than equivalent solid woods, although the majority of particle and fiber-based boards require treatment with a sealant or paint to increase water penetration resistance. These products are engineered to meet precise application-specific design specifications that are tested to meet national or international standards.

Using engineered lumber instead of large dimension rafters, joists, trusses, and posts can save money and reduce total wood use by as much as 35 percent. The wider spacing of members possible with engineered lumber also has the advantage of increasing the insulated portion of walls. Other advantages include their ability to form large panels from fibers taken from small diameter trees, and small pieces of wood and wood that has defects can be used in many engineered wood products, especially particle and fiber-based boards. Engineered wood products also have some disadvantages; for example, they require more primary energy for their manufacture than solid lumber. Furthermore, the adhesives used may be toxic.

4. Structural Sheathing

Sheathing is the structural covering of plywood or oriented strand board (OSB) that is applied to studs and roof/floor joists and serves as a base for finish flooring or a building's weatherproof exterior. Sheathing is considered to be the second most wood-intensive element of wood-frame construction.

One of the environmental problems associated with engineered wood sheathing materials is that the wood fibers are typically bound with formaldehyde-based resins. Interior grade plywood typically contains urea formaldehyde (UF), which is less chemically stable than the phenol formaldehyde (PF) found in water-resistant exterior grade plywood and OSB. This advantage makes exterior grade plywood preferable for indoor applications, as it emits less toxic and suspected carcinogenic compounds. There are other alternatives to these wood-intensive conventional and engineered materials. Fiberboard products rated for structural applications such as Homasote's 100 percent recycled nailable structural board are alternatives to plywood and OSB.

Designs that combine bracing with non-structural sheathing can provide necessary strength while enhancing insulation and reducing wood requirements. Structural insulated panel construction provides interior and exterior sheathing as well as insulation in pre-cut, factory-made panels. Also by taking into consideration disassembly when preparing the design, sheathing materials can be readily reused or recycled.

5. Medium Density Fiberboard (MDF)

MDF is a composite wood product traditionally formed by breaking down softwood into wood fibers, often in a defibrator, combining it with wax and a synthetic resin binder such as urea formaldehyde resins (UF) or other suitable bonding system, and forming panels by applying high temperature and pressure. Additives may be introduced during manufacturing to impart additional characteristics. But all MDF is not the same and will vary in texture, density, color, etc. depending on the material it is made of. Today many MDF boards are made from a variety of materials.

These include other woods, scrap, recycled paper, bamboo, carbon fibers and polymers, steel, glass, forest thinning, and sawmill off-cuts.

Many manufacturers are being pressured to come up with greener products and have now started testing and using non-toxic binders. New raw materials are being introduced such as straw and bamboo which are becoming popular fibers because they are a fast growing renewable resource. Although MDF is highly toxic to manufacture, it does not emit volatile organic compounds (VOCs) in use. Trim waste is significantly reduced when using MDF compared to other substrates. Stability and strength are important assets of MDF, which can be machined into complex patterns that require precise tolerances.

Many of MDF's qualities make it an ideal replacement for plywood or particle board. The product is dense, flat, stiff, has no knots, and is easily machined. It consists of fine particles and provides dimensional stability without a predominant grain as is the case with wood. Unlike most plywoods, MDF contains no voids, and will deliver sharp edges with no tearout. MDF is well damped acoustically, thus making it a suitable material for speaker enclosures. MDF is widely used in the manufacturing of furniture, kitchen cabinets, moldings, millwork, door parts, and laminate flooring. Medium density fiberboard MDF panels are manufactured with a variety of physical properties and dimensions, providing the opportunity to design the end product with the specific MDF characteristics and density needed.

6. Agricultural Residue Panels ("Ag")

One of the more recent developments in North American composites has been the introduction of boards made from agricultural residues. Increasing constraints on residue burning have been a prime motivator for its introduction. Its manufacture entails compressing agricultural residue materials with non-formaldehyde glues; the panels provide an excellent alternative to plywood sheets 3/8 inch and thicker and can be used in much the same way as medium density fiberboard. They are also suitable alternatives for OSB and MDF for interior walls and partitions work. Agricultural residue ("ag-res") boards are made from waste wheat straw, rice straw, or even sunflower seed husks. Ag boards are both aesthetically pleasing and often stronger than MDF, and just as functional.

7. Homasote™

Homasote™ is a panel board consisting of 100 percent paper fibers converted from recycled wastepaper. Added to this are ingredients that give it the superior weather-resistant features. Homasote® board has actually been in production longer than plywood and OSB, and is used in residential and commercial construction. Among its applications are exterior vertical sheathing, paintable interior panels, sound control, roof decking, concrete forming, expansion joint, insulation, and, under the

Pak-Line® brand, industrial packaging. Agricultural straw panels are sometimes difficult to find due to a combination of intense demand, lack of supply, and the time required to bring a manufacturing facility online. Some of Homasote's physical properties include being weather-resistant, structural, insulating, extremely durable board with two to three times the strength of typical light-density wood fiberboards.

8. Advanced Framing Techniques

Also known as optimum value engineering (OVE), advanced framing techniques refers to a variety of new framing techniques and strategies that have evolved and which are designed to optimize the amount of lumber used in the construction of a wood-framed house, so as to create more space for insulation in exterior walls, thereby reducing material cost and use of natural resources. Consequently, cold spots, which are susceptible to condensation and mold growth, are eliminated. The extraction, manufacture, transport, and disposal of lumber deplete resources, damage natural habitats, and pollute air and water. OVE advanced framing techniques include studs spaced at 24 in o.c.; 2-foot modular design that reduces cut-off waste from standard-sized building materials (Figure 6.9); in-line framing that reduces the need for double top plates; building corners with two studs; and using insulated headers over exterior building openings or even using no headers for non-load bearing walls. However, to be successful, advanced framing techniques must be considered early in the design process.

9. Structural Insulated Panels (SIPs)

These are high performance building panels used in floors, walls, and roofs for residential and light commercial buildings. The panels are typically made by sandwiching a core of rigid foam plastic insulation between two structural skins of oriented strand board (OSB) or plywood. Other skin material can be used for specific purposes. SIPs are manufactured under factory controlled conditions and can be custom designed for each home, resulting in a building system that is strong, energy efficient, and cost effective. The resultant of this simple sandwich forms a strong structural panel that is significantly more energy efficient yielding increased R-values as compared to traditional framing. In addition to the excellent insulation properties of SIPs, they offer airtight assembly, noise attenuation, and superior structural strength.

There are several materials used for SIP cores. Expanded polystyrene (EPS) is perhaps the most common core material. It requires less energy to manufacture than other options and is more recyclable than polyurethane or polyisocyanurate. Many products offer a one-hour fire rating when installed with 5/8 inch or thicker gypsum sheathing. EPS foam is expanded with pentane, which does not contribute to ozone depletion or global warming and is often recaptured at the factory for reuse.

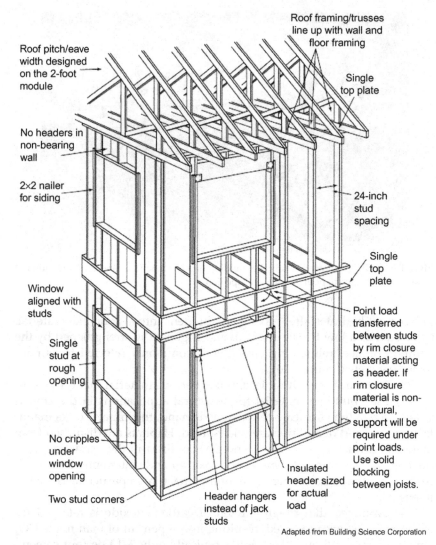

Roof framing/trusses line up with wall and floor framing

Roof pitch/eave width designed on the 2-foot module

Single top plate

No headers in non-bearing wall

2×2 nailer for siding

24-inch stud spacing

Single top plate

Window aligned with studs

Point load transferred between studs by rim closure material acting as header. If rim closure material is non-structural, support will be required under point loads. Use solid blocking between joists.

Single stud at rough opening

No cripples under window opening

Two stud corners

Header hangers instead of jack studs

Insulated header sized for actual load

Adapted from Building Science Corporation

FIGURE 6.9 An isometric drawing illustrating advanced framing techniques used in residential construction. Waste is reduced by spacing studs at 24 in o.c. and using a 2-foot modular design that reduces cut-off waste from standard-sized building materials. To be successful, advanced framing techniques must be considered early in the design process. (Source: Adapted from Building Science Corporation.)

6.2.8 Concrete

Concrete is made up of three basic components: water, aggregate (sand, crushed stone, or gravel) and a binding agent such as cement. The most common form of concrete consists of mineral aggregate such as stones, gravel and sand, cement (usually in powder form), and water.

FIGURE 6.10   A photo showing the building of a concrete floor and the installation of rebar during a concrete pour. (Source: Wikipedia.)

The cement hydrates after mixing and hardens into a stone-like material. Concrete has a low tensile strength and is generally strengthened by the addition of steel reinforcing bars: this is commonly referred to as reinforced concrete (Figure 6.10).

Concrete is a strong, durable, and inexpensive material that is the most widely used building material for structural applications in the United States. Due to the vast scale of concrete demand, the impacts of its manufacture, use, and demolition are widespread. Habitats are disturbed from materials extraction; significant energy is utilized in the extraction, production, and shipping of cement; and toxic air and water emissions result from cement manufacturing. Cement manufacture in particular is energy intensive.

It is estimated that roughly 1 ton of carbon dioxide is released for each ton of cement produced, resulting in 7–8 percent of man-made CO_2 emissions. And although concrete is typically only 9–13 percent cement, it accounts for 92 percent of concrete's embodied energy. Cement dust contains free silicon dioxide crystals, the trace element chromium, and lime, all of which can adversely impact worker health. Mixing concrete requires a great deal of water, and generates alkaline wastewater and run-off that can contaminate waterways and vegetation.

Concrete's impact on the environment can be substantially reduced by substituting alternative pozzolan ash (industrial by-products such as fly ash, silica fume, rice husk ash, furnace slag, and volcanic tuff) for the cement. Fly ash, a residue from coal combustion, is quite popular as a cement substitute that generally decreases porosity, increases durability, and improves workability and compressive strength, although the curing time is increased. Fly ash generally constitutes 10–15 percent of standard mixes, but many applications will allow substituting of up to 35–60 percent of cement, and with certain types of fly ash (class C) cement can be completely replaced for some projects.

The amount of concrete used may be reduced in non-structural applications by trapping air in the finished product or through the use of low-density aggregates. As trapped air displaces concrete it enhances insulation value and reduces weight and material cost, without compromising the durability and fire-resistance characteristics of standard concrete. Low-density aggregates such as pumice, vermiculite, perlite, shale, polystyrene beads, or mineral fiber provide similar insulation and weight-reduction benefits.

Concrete's setting time can be reduced by the addition of accelerators to the concrete to accelerate early strength. The amount of reduction in setting time varies depending on the amount of accelerator used. Calcium chloride is a low cost accelerator, although specifications often specify a non-chloride accelerator to prevent corrosion of reinforcing steel. However, when hot weather conditions require a slower setting time, retarding admixtures are added. Many retarders also function as a water reducer.

1. Pervious/Porous Concrete

Pervious concrete pavement is a unique and effective means to address important environmental issues and support green, sustainable growth. By capturing storm water and allowing it to seep into the ground, porous concrete is instrumental in recharging groundwater, reducing storm water runoff, and meeting U.S. Environmental Protection Agency (EPA) storm water regulations. Up to 75 percent of urban surface area is covered by impermeable pavement, which inhibits groundwater recharge, contributes to erosion and flooding, conveys pollution to local waters, and increases the complexity and expense of storm water treatment. One of the main characteristics of pervious paving is that it contains voids that allow water to percolate through to the base materials below. It also reduces peak storm water flow and water pollution and promotes groundwater recharge. Pervious paving may incorporate recycled aggregate and fly ash, which helps reduce waste and embodied energy. Pervious paving is suitable for use in parking and access areas having a compressive strength of up to 4000 psi. It also mitigates problems with tree roots and the percolation area encourages roots to grow deeper. Enhanced heat exchange with the underlying soil can decrease summer ambient air temperature by 2–4 degrees Fahrenheit.

2. Concrete Formwork

Cast-in-place concrete requires formwork to define its shape as it cures (Figure 6.11). The most common materials used for formwork are plywood and milled lumber, which contribute to construction waste and

FIGURE 6.11   A photo of carpenters setting concrete formwork for the high level waste facility pit walls. (Source: Bechtel Corporation.)

the impacts of timber harvest and processing. Wooden formwork can be made from salvaged wood and typically can be disassembled and reused several times.

Form-releasers or parting agents are materials that help separate the forms from hardened concrete. Form-releasers prevent concrete from bonding to the form, which can deface the surface when forms are disassembled. Traditional form releasers such as diesel fuel, motor oil, and home heating oil are carcinogenic, limiting the potential for reuse of wood formwork because it exposes construction personnel (and potentially occupants) to volatile organic compounds. They are now prohibited by a variety of state and federal regulations, including the Clean Air Act. Water-based form release compounds that incorporate soy or other biologically-derived oils will significantly reduce health risks to construction staff and occupants, and often simplify the application of any required finishes or sealants. Furthermore, soy-based options are often less expensive than their petroleum-based counterparts.

FIGURE 6.12   A photo depicting application of an insulating concrete formwork Quad-lock system. This can be considered a "green" material as it is durable, produces little or no waste during construction, and greatly improves the thermal performance of concrete walls. (Source: Quad-Lock Building Systems Ltd.)

3. Insulated Concrete Forms (ICFs)

At their most basic level, ICFs serve as a forming system for poured concrete walls and consist of stay-in-place formwork for energy-efficient, cast-in-place, reinforced concrete walls. The forms are hollow, lightweight, interlocking modular components that are dry-stacked (without mortar) to create a formwork system into which concrete is poured. The forms lock together and serve to create a form for the structural walls of a building. Concrete is pumped into the cavity to form the structural element of the walls. Insulated concrete forms (ICFs) use an insulating material as permanent formwork that becomes a part of the finished wall. ICFs can be considered "green" materials as they are durable, produce little or no waste during construction, and dramatically improve the thermal performance of concrete walls (Figure 6.12).

Standard concrete is a dense material with a high heat capacity that can be used as thermal mass, reducing the energy required to maintain comfortable interior temperatures. However, concrete lacks good insulation qualities and standard formwork is waste intensive, so that toxic materials are frequently needed to separate formwork from the hardened product. ICFs address these weaknesses by reducing solid waste, air and water pollution, and potentially reducing construction cost. ICF wall systems have superior thermal qualities that enhance their usefulness for passive heating and cooling; comfort is also enhanced and energy costs are reduced. ICFs also offer the structural and fire-resistance benefits of reinforced concrete; structural failure due to fire is rare to nonexistent. Due to the addition of flame-retardant additives, polystyrene ICFs tend to melt rather than burn, and interior ICF walls tend to contain fires much better than wood frame walls, thereby improving overall fire safety.

6.3 BUILDING AND MATERIAL REUSE AND RECYCLABLE MATERIALS

Reusing materials instead of dumping them in landfills can save valuable resources, and actively seeking out and buying products with recycled content can help the environment. Moreover, where possible and appropriate, designers and contractors should always try and employ local materials to avoid excess transportation and environmental costs. Also, products such as doors, cabinets, glass, and metal can be salvaged and reused. Using salvaged materials can significantly reduce costs, and their quality is high. The use of biodegradable materials also goes a long way toward helping the environment. Biodegradable material is generally organic material such as plant and animal matter, and other substances originating from living organisms, which breaks down organically and which may be returned to the earth with none of the damage associated with the generation of typical waste materials. Everyday examples of biodegradable material are: wood, cotton, straw, corn, plants, and animals.

6.3.1 Building Reuse

The rehabilitation of old buildings highlights the many successful commercial redevelopments being executed in many cities around the world. This reuse offers the potential to lower building costs and provide a mix of desirable building characteristics. However, the reuse of existing structural elements is not always appropriate, particularly with large commercial projects, and basically depends on many factors, including fire safety, energy efficiency, and regulatory requirements; all of these should be taken into account for reuse in an existing building. The intent is to reuse building materials and products in order to protect and reduce demand for virgin material resources and to reduce waste, thereby reducing impacts associated with the extraction and processing of virgin resources. For green project certification purposes (e.g., LEED Rating System) building and material reuse can achieve credits.

6.3.2 Material Reuse

Material reuse is essentially the salvage and reinstallation of materials in their original form, whereas recycling is the collection and remanufacture of materials into a new material or product, typically different from the original. Reusing materials slated for the landfill has become one of the most environmentally sound ways to build because the extraction, manufacture/transport, and disposal of virgin building materials pollutes air and water, depletes resources, and damages natural habitats. Construction and demolition are estimated to be responsible for about 30 percent of the U.S. solid waste stream. Real-world case studies by the Alameda County Waste Management Authority, for example, have concluded that more than 85 percent of that material, from flooring to roofing to packaging, is recyclable or reusable.

6.3.3 Recyclable Materials

Recycling is the process of recovering used materials from the waste stream and then incorporating those same materials into the manufacturing process. Recycled content generally refers to that portion of material used in a product that has been diverted from the solid waste stream, and is the most widely cited attribute of green building products. Recycled content is usually determined by calculating the weight of the recycled material divided by the total weight of the material or product and expressed as a percentage by weight. For the purpose of LEED certification, the recycled content "value" of a product is assessed by multiplying the recycled content percentage and the cost of the product.

The intent of recycling materials is to protect virgin resources by increasing demand for building products that use recycled content. If the materials are diverted during the manufacturing process, they are referred to as *pre-consumer* recycled content (sometimes also known as post-industrial). If they are diverted after consumer use, they are termed *post-consumer*. Post-consumer content is generally viewed as offering greater environmental benefit than pre-consumer content.

In recent years numerous federal, state, and local government agencies have implemented "buy recycled" programs aimed at increasing markets for recycled materials. These programs usually have a specific goal of supporting recycling programs to reduce solid-waste disposal, and many communities in the United States now offer curbside collection or drop-off sites for certain recyclable materials. Collecting materials however is only the first step toward making the recycling process work. Successful recycling also depends on manufacturers producing products from recovered materials and, in turn, consumers purchasing products made of recycled materials.

Recyclability is the ability of a product to be recycled and can only describe products that can be collected and recycled through a recognized process. Some manufacturers stretched this definition well beyond credibility by those who base the claim on a laboratory process. Sometimes there is an attempt to justify this lax approach with new products by noting that they aren't expected to enter the waste stream in large quantities for years. However, many national and international companies seek an environmental marketing edge by advertising the recycled content of their products, which is often undocumented or uncertified.

Such claims come under the general jurisdiction of the Federal Trade Commission (FTC), which first published definitions for common environmental terms in its Green Guides in 1992. The LEED Rating System and Green Globes offer credit for recycled-content materials, referencing definitions from ISO 14021. These definitions contain numerous gray areas, which manufacturers often tend to interpret in their own favor. Third-party certification of recycled content is useful in maintaining a high standard and offering the ability to substantiate any claims that manufacturers make.

In some industries, recycling pathways have become very reliable. Products like engineered lumber and fuel pellets have created a reliable demand for waste from the forest products industry, including sawdust and woodchips. Many of these materials may have been part of the waste stream at some point, but the active market for them means that they are nearly always diverted. Internally recycled material comes from scraps leftover in a company's manufacturing process. It may include substandard products that are scraped and remade after being rejected by the company's quality control division.

Waste has a price tag, and we all must bear it. Moreover, the extraction, manufacture/transport, and disposal of building materials clogs our landfills, pollutes our air and water, depletes resources, and damages natural habitats. The California Integrated Waste Management Board (CIWMB) notes that construction and demolition are responsible for about 28 percent of California's solid waste stream. In excess of 85 percent of that material, from flooring to roofing, is either reusable or recyclable. In addition to construction and demolition waste, the material in our recycling bins, our used bottles, paper, cans, and cardboard, are also considered to be suitable raw materials for recycled content products.

Keeping a material or product out of the landfill is only the first of several steps to putting "waste" back into productive use. The material must be processed into a new, high-quality item, and that product must be sold to a builder or homeowner who recognizes its benefits. Benefits of recycled content materials include reduced solid waste, reduced energy and water use, reduced pollution, reduced greenhouse gas emissions, and a healthier economy. The following materials are readily recyclable and generally cost less to recycle than to send to landfills:

- Cardboard
- Metals
- Window glass
- Acoustical ceiling tiles
- Asphalt roofing
- Clean wood (includes engineered products)
- Tile containing recycled glass
- Land clearing debris
- Plastic film such as sheeting, shrink wrap, packaging
- Plastic and wood-plastic composite lumber from plastic and wood chips
- Fluorescent lights and ballasts
- Insulation, such as cotton made from denim, or newspaper processed into cellulose
- Carpet and carpet pad made of plastic bottles or sometimes from used carpet
- Concrete containing ground up concrete as aggregate, or fly ash
- Countertops made with everything from recycled glass to sunflower seed shells
- Drywall made with recycled gypsum
- Homasote® wall board made from recycled paper.

From an environmental and economic point of view avoiding waste generation is considered to be a far better practice than recycling. The main benefits of waste reduction are minimizing energy use, conserving resources, and easing pressure on landfill capacity.

6.3.4 Salvage Material

By salvaging materials from renovation projects and specifying salvaged materials, material costs can be reduced while adding character to projects and maximizing environmental benefits, such as reduced landfill waste, reduced embodied energy, and reduced impacts from harvest/mining of virgin materials.

Certain materials require remediation prior to reuse or they should not be reused at all. For example, materials contaminated by hazardous substances such as asbestos, arsenic, and lead paint must be treated and/or disposed of properly. Avoiding materials that can cause future problems is critical to long-term waste reduction, in addition to improving the general health and well-being of our communities. In addition to being environmentally friendly, reclaimed wood has many applications including for flooring, trim, siding, furniture, and in some cases, as structural members.

Recycling, salvage, and disposal calculations: For green certification, a system should normally be developed for tracking the volume or weight of all materials salvaged, recycled, and disposed of during the project. At the end of the project, or more frequently if stipulated in the specifications, submit a calculation documenting that the project achieved a 50 percent or 75 percent diversion rate. This rate will depend on the targeted goals set for the project.

6.4 CONSTRUCTION WASTE MANAGEMENT

The general application of a construction waste management plan is to minimize the amount of materials going to landfills during construction by diverting the construction waste and demolition and land clearing debris from landfill disposal. It also helps redirect recyclable recovered resources back to the manufacturing process and redirect reusable materials to appropriate sites. From the outset project waste should be recognized as an integral part of overall materials management. Any plan should require regular submittals tracking progress. The plan should also show how the required recycling rate is to be achieved, including materials to be recycled or salvaged, cost estimates comparing recycling to disposal fees, materials-handling requirements, and how the plan will be communicated to the crew and subcontractors. The premise is that waste management is a part of materials management, and the recognition that one project's waste is material available for another project leads to an efficient and effective waste management process.

To be successful waste management requirements should be spelled out early in the design process and be the topic of discussion at both preconstruction and ongoing regular job meetings, to ensure that contractors and subcontractors are fully informed of the implications of these requirements on their work prior to and throughout the construction process.

6.5 REGIONAL MATERIALS

The matter of indigenous materials purchase is an economic issue in addition to an environmental issue. The main intent here is to reduce material transport by increasing demand for building products made within the region. This is achieved by increasing demand for building materials and products that are extracted and manufactured within the designated region. It will help reduce the environmental impacts emanating from transportation (trucking, ship, barge, and rail), in addition to supporting the regional economy.

Regional materials should also be carefully considered for achieving credits toward project certification. For example, LEED requires that a minimum of 10 or 20 percent (based on cost) of total building materials

CHS

FIGURE 6.13   A diagram of a map program that is capable of drawing any required radius for any chosen location. (Source: Free Map Tool.) This program can be used to identify materials within a 500-mile radius sometimes required for certain green credits.

and products be extracted, harvested, recovered, or manufactured regionally within a radius of 500 miles of the project site (Figure 6.13). Either the default 45-percent rule or the actual cost of materials purchased may be used. Excluded from the calculations should be all mechanical, electrical, plumbing, and specialty items such as elevator equipment.

6.6 CERTIFIED FORESTRY PRODUCTS AND RAPIDLY RENEWABLE MATERIALS

Often lumber being used for construction must be certified by a reputable accreditation agency. This helps the environment, as does the use of rapidly renewable materials, which reduces the use and depletion of finite raw materials and long-cycle renewable.

6.6.1 Certified Wood

Today, there are more than 50 forest certification programs worldwide. Forest certification is basically a means for independent organizations to develop standards of good forest management, and independent auditors to issue certificates to forest operations that comply with those standards. This certification verifies that forests are well-managed, as defined by a particular standard, and ensures that certain wood and paper products come from responsibly managed forests. Various rating systems such as LEED and Green Globes encourage such environmentally responsible forest management programs.

The concept of forest certification was launched over a decade ago to help protect our national forests from destructive logging practices. Forest certification was to provide a recognized seal of approval and a means of notifying consumers that a wood or paper product comes from forests managed in accordance with strict environmental and social standards.

One of the first forest conservation programs was developed by the Scientific Certification Systems (SCS) in 1991 and has since developed and grown to emerge as a global leader in certifying forest management operations and wood product manufacturers. The Forest Stewardship Council (FSC) in 1996 accredited SCS as a certification body, enabling it to evaluate forests according to the FSC Principles and Criteria for Forest Stewardship. Through a well-developed network of regional representatives and contractors, SCS offers timely and cost-effective certification services around the world. There are many other certification organizations such as the Sustainable Forestry Initiative (SFI) and the Canadian Standards Association (CSA).

6.6.2 Rapidly Renewable Materials

Rapidly renewable materials consist of products made from plants that are typically harvested within a 10-year cycle or shorter. Typical examples of rapidly renewable material include bamboo, cork, insulation, linoleum, straw bale, and wheat board.

Bamboo has become an important alternative resource to other types of wood commonly used in the United States and abroad. It was previously intuitively used as a basic material for making different household objects and small structures. But ongoing research and engineering efforts are enabling bamboo's true value to be realized as a renewable, versatile, and readily available economic resource.

Bamboo is a giant grass commonly found in the tropical regions of Asia, Africa, and South America and comes in 1500 varieties that produce hard, strong, dimensionally stable wood. It has been used by ancient civilizations in the Middle East and elsewhere as both a building material and for furniture construction. Bamboo is a fast growing woody plant and is one of the most versatile and sustainable building materials available. It grows remarkably fast and can reach maturity in months in a wide range of climates and is exceedingly strong for its weight. It can be used either structurally or as a finish material. Bamboo is characterized as a totally renewable resource; you can clear-cut it, and it grows right back. Bamboo canes are beautiful when exposed and may also be used for paneling, furnishings, and cabinetry as well as house construction (Figures 6.14 and 6.15).

FIGURE 6.14   **A.** Kitchen cabinets made from bamboo. **B.** A photo showing bamboo used as exterior siding. (Source: Bamboo Technologies.)

Bamboo has good strength properties, consisting of extremely strong fibers that have twice the compressive strength of concrete and also roughly the same strength-to-weight ratio of steel in tension. The strongest bamboo fibers also have greater shear strength than structural woods, and they take much longer to come to ultimate failure. However, this ability of bamboo to bend without breaking makes it unsuitable for building

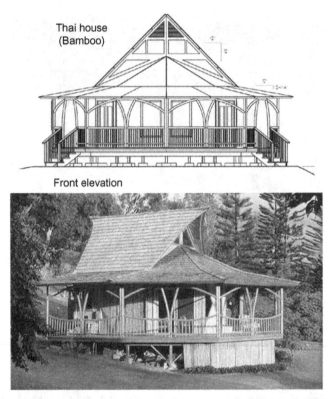

FIGURE 6.15   A bamboo house that can be purchased from Bamboo Technologies in kit form. (Source: Bamboo Technologies via GreenHomeBuilding.com.)

floor structures due to our low tolerance for deflection and unwillingness to accept a floor that feels "alive." Properly maintained, bamboo floors can last for decades.

Cork is a natural, sustainable product harvested from the bark of the cork oak tree. The tree grows predominately in Europe and North Africa and has a life span ranging from 150 to 250 years. Once the tree has reached the correct level of maturity (typically 25 years), the first harvest of cork bark is removed from the tree. Cork is becoming increasingly popular due to its extraordinary combination of beauty, durability, insulation, and renewability. Modern cork floors are durable, fire resistant, provide thermal and acoustic insulation, and are soft on the feet. They are typically covered with an acrylic finish, but may be covered with polyurethane for bathroom or kitchen applications. Cork floors can last for decades and the material is biodegradable at the end of its useful life (Figure 6.16A,B).

Cork has many useful properties such as lightness, impermeability to liquids, resistance to wear, rot, and temperature extremes, and compressibility.

A

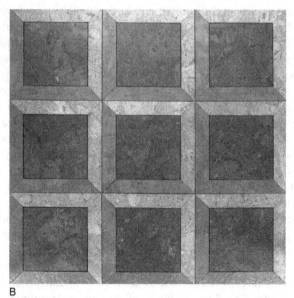

B

FIGURE 6.16   **A.** A residential interior using cork flooring. **B.** Cork pattern detail. (Source: Globus Cork.)

Cork is often used in flooring, making an excellent natural alternative to carpet. It also has excellent acoustical properties with an ability to reduce echo and sound transmission. Furthermore, while carpet can attract and hold indoor pollutants in its fibers, cork is easier to thoroughly clean, and is inherently resistant to mold and mildew, sheds no dust or fibers, and is naturally antistatic. These hypoallergenic properties, combined with thermal and acoustic insulation, allow cork floors to provide the majority of the benefits of carpet, without its liabilities. Furthermore, the benefits of cork extend beyond human health; they include reduced landfill waste, low embodied energy and local availability for many products, excellent aesthetics, and reduced impacts from the harvest or mining of virgin materials. In fact for over 100 years cork has been used throughout the world in homes, libraries, churches, universities, and other projects.

Insulation helps to protect a building's occupants from heat, cold, and noise; in addition, it reduces pollution while conserving the energy needed to heat and cool a building. Well-insulated building envelopes are primary considerations in comfort and sustainability. Environmentally preferable insulation options offer additional benefits, such as reduced or eliminated health risks for installers and occupants, reduced waste and pollution in manufacture and installation, more efficient resource use, recyclability, and enhanced R-values.

Advanced framing techniques increase the wall area covered by insulation, thereby increasing a whole wall's effectiveness. Framing conducts far more heat than insulation, in a similar way that most window frames conduct more heat than double-paned glass. An additional layer of rigid insulation between framing and exterior sheathing can help improve the whole-wall R-value by insulating the entire wall, and not just the clear space. In non-breathable wall designs, closed-cell rigid foam with taped seams can provide an effective vapor barrier. Wall finish is critical in providing both personal physical comfort and maintaining building construction.

Fiberglass is often the insulation of choice mainly for economic reasons. Fiberglass insulation should be formaldehyde-free with a minimum 50 percent total recycled content (minimum 25 percent post-consumer). Some products are manufactured with heavier, intertwined glass fibers to reduce airborne fibers and mitigate the fraction of fibers that can enter the lungs. Like all glass products, fiberglass insulation is made primarily from silica heated to high temperatures, requiring significant energy and releasing formaldehyde. During installation or other contact short term adverse effects may be experienced including irritation to eyes, nose, throat, lungs, and skin. Longer term effects are controversial, but OSHA requires fiberglass insulation to carry a cancer warning label.

Recycled cotton insulation and cellulose (recycled newspaper) insulation are sometimes used as environmentally preferable alternatives. Both cotton and cellulose are treated with borate, which is not toxic to humans, and makes both materials more resistant to fire and insects than fiberglass. Sprayed polyurethane foams expand to fill cracks, and also makes a good alternative, providing insulation, a vapor barrier, and additional shear strength.

Linoleum is often mistaken for vinyl, yet it is very different in nearly every aspect. Linoleum is a highly durable resilient material used mainly for flooring. It is made from all-natural materials, a mixture of linseed oil, wood flour, powdered cork, and pine resin, which is pressed onto a jute fiber backing. All natural, linoleum requires less energy and creates less waste in its production, and can be chipped and composted at the end of its useful life. And because it is made from only natural materials, linoleum is completely biodegradable. Maintenance of linoleum is also less labor intensive and less costly because it does not require sealing, waxing, or polishing as often as vinyl. It is enjoying resurgence in popularity thanks in part to the focus on eco-friendly materials. Although linoleum is most commonly found in sheet form, it is also available in square or rectangular tiles and in a wide range of colors. The material is very resilient and able to stand up well to extended use, and if properly maintained linoleum floors can last up to 40 years.

Building with bales of straw is enjoying a renaissance and is becoming mainstream in several parts of the country, especially in the Southwestern United States. Straw bale construction consists of using compressed blocks (bales) of straw, either as fill for a wall cavity in non-load bearing walls or as a structural component in a load-bearing wall. In this case, the bales themselves can support the weight of the roof. A post and beam framework that supports the basic structure of the building, with the bales of straw used as infill, is the most common non-load bearing approach. This is also the primary method that is permitted in many jurisdictions, although many localities have specific codes for straw bale construction, and some banks are even willing to lend on this technique. Straw is a renewable resource that acts as excellent insulation and is fairly easy to build with.

The bales are typically finished on the interior and exterior sides with stucco, plaster, clay, or other treatment, which adds to its structural properties. Both load- and non-load bearing straw bale designs divert agricultural waste from the landfill for use as a building material with many exceptional qualities and applications. Building codes have recently been developed for both "post and beam" and load-bearing straw bale construction.

Building with bales of straw has become popular in many parts of the country and there are now thousands of straw bale domestic and commercial applications in the U.S. However, with load-bearing straw bale buildings care must be taken to consider the possible settling of the straw bales as the weight of the roof, etc. compresses them. Care should also be taken to assure that the straw is kept dry, or it will eventually rot. Other possible concerns with straw bale walls are infestation of rodents or insects, so the skin on the straw should be made to resist these. In many cases, straw bales provide an excellent "alternative" building material that helps to reduce or eliminate many environmental problems because they use non-toxic, reusable, and biodegradable elements that build durable, comfortable, healthy places to live and work (Figure 6.17).

FIGURE 6.17 A residential interior where straw bale is the primary component. (Source: StrawBale Innovations, LLC.)

Wheat board is a distinctive fiber composite material that is created from a rapidly renewable resource, wheat-straw, and has superior panel properties compared to wood-based composite panels. It has greater moisture resistance, is stronger and lighter in weight, and has a smooth surface that cuts, sands, and routes with standard woodworking tools. Wheat board panels are manufactured in 4 foot by 8 foot sheets and in ½-inch, ¾-inch, and 1-inch thicknesses. The material is environmentally friendly, comes from a variety of resources, and is suitable for furniture manufacturer. This material has no formaldehyde and can be used to create, among other things, quality furniture and cabinets. Produced in sheets, this durable substance can be filled, sealed, painted, stained, or varnished. It can also be shaped in a wide variety of designs.

It is a viable substitute for wood and benefits the environment by reducing deforestation and also lessening both air pollution and landfill use.

6.7 GREEN OFFICE EQUIPMENT

According to the Department of Energy, office equipment accounts for approximately 16 percent of an office's energy use. The use of computers, printers, copiers, and fax machines adds up, but simply turning your computer's sleep mode on when it is not in use can save energy. More and more offices are turning green as an increasing number of people search for ways to reduce their environmental footprint, both at home and at the office. In fact, the impact of energy costs directly affects the bottom-line of both building owners and tenants alike. Energy represents 30 percent of operating expenses in a typical office building, which makes it the single largest and most manageable operating expense in the provision of office space. The cost of energy to power an appliance over time is typically many times the original price of the equipment. By choosing the most energy efficient models available, money can be saved and a positive impact on the environment achieved.

All products and services have some impact on the environment, which may occur at any or all stages of the product's life-cycle—raw material acquisition, manufacture, distribution, use, and disposal. Most retailers typically carry efficient, durable appliances and office equipment. Appliances and equipment with the Energy Star® label are preferred, although these are not necessarily the most efficient of all available models. Energy Star products, however, do usually perform significantly better than federal minimum efficiency standards. According to a 2002 EPA report, ENERGY STAR labeled office buildings generate utility bills 40 percent less than the average office building. Furthermore, many states offer substantial rebates and/or incentives for the purchase of energy and water saving appliances. There are several issues to consider prior to selecting equipment, such as:

- Choose appliances and equipment that use the least amount of energy and/or water.
- Appliances and equipment should preferably have the Energy Star label.
- Appliances and equipment should be durable and meet long-term needs.
- Sealed combustion and direct vent furnaces and water heaters improve indoor air quality.
- Choose natural gas appliances for space and water heating as gas is often more cost-effective and able to reduce overall energy use. However, natural gas, like other fossil fuels, is not a renewable resource.
- Use of occupancy sensors in offices should be considered to minimize unnecessary lighting, as well as "smart" power-strips that combine an occupancy sensor with a surge protector. When space is unoccupied, smart power-strips will shutdown devices such as monitors, task lights, space heaters, and printers that can be safely turned off.

With relation to residential construction ENERGY STAR offers homebuyers many features they seek in a new home, in addition to energy-efficient improvements that deliver better performance, greater comfort, and lower utility bills. To earn the ENERGY STAR label, a home must meet strict guidelines for energy efficiency as set by the U.S. Environmental Protection Agency. Such homes are typically 20–30 percent more efficient than standard homes. There is also the potential to achieve LEED and Green Globes credit points for using "green" office equipment and appliances. When applying for project certification, you should constantly consult the USGBC and Green Globes Websites (http://www.usgbc.org; http://www.thegbi.org).

6.8 LIFE-CYCLE ASSESSMENT OF BUILDING MATERIALS AND PRODUCTS

Life-cycle assessment (LCA) is a methodology for evaluating the environmental impacts of a material through its entire "life-cycle"—from its initial production through to its eventual reuse, recycling, or disposal. LCA attempts to identify and quantify all relevant environmental impacts for materials so that comprehensive comparisons can be made. However, the LCA tools that are currently available are not widely utilized by most stakeholders, including those designing, constructing, purchasing, or occupying buildings. LCA should not be confused with life-cycle cost (LCC), which the National Institute of Standards and Technology (NIST) defines as "the total discounted dollar cost of owning, operating, maintaining, and disposing of a building or a building system" over a period of time. A key aspect of moving toward more performance-based outcomes in sustainable design is the use of life-cycle assessment (LCA) to determine the embodied environmental effects of materials, rather than relying on singular material properties such as recycled content or distances traveled after the point of manufacture.

This method of analysis allows for comprehensive and multidimensional product comparisons. It takes into account the impacts of every stage in the life of a material, from first costs, including the cost of planning, design, extraction/harvest, and installation, as well as future costs, including costs of fuel, operation, maintenance, repair, replacement, recycling, or ultimate disposal. It also takes into account any resale or salvage value recovered during or at the end of the time period examined. However, life-cycle assessment has its limitations. For example, such issues as uncertainty, risk related to toxic releases, and site-specific resource extraction effects are not handled. Also initial failure to address the ease with which the built environment can be safely maintained can lead to unnecessary costs and risks to health and safety at a later date. Moreover, there are several factors that impact the decision making process and not only the initial cost of the project.

Building owners often apply the principles of life-cycle cost analysis in decisions they make regarding construction or improvements to a facility. From the building owner who opts for aluminum windows in lieu of wood windows to the strip mall developer who chooses paving blocks over asphalt, both owners are taking into consideration the future maintenance and replacement costs in their selections. Related to LCA, life-cycle costing (LCC) is the systematic evaluation of financial ramifications of a material, a design decision, or a whole building. LCC tools can help calculate payback, cash flow, present value, internal rate of return, and other financial measures. Such criteria are helpful in understanding how a modest up-front cost for environmentally preferable materials or design features can provide a very sound investment over the life of a building.

Although the lowest life-cycle cost (LCC) is considered the most straightforward and easy-to-interpret system of economic evaluation, there are several other commonly used measures that are used such as net savings (or net benefits), savings-to-investment ratio (or savings benefit-to-cost ratio), internal rate of return, and payback period. All of these systems are consistent with the lowest LCC measure of evaluation if they use the same parameters and length of study period. Building economists, architects, cost engineers, quantity surveyors, and others might use any or several of these techniques to evaluate a project. The approach to making cost-effective choices for building-related projects can be quite similar whether it is called cost estimating, value engineering, or economic analysis. The actual costs of the project can be better seen when an open book accounting system is used and is shared by the entire project team.

The LCCA should be performed early in the design process while there remains an opportunity to refine the design in a manner that results in a reduction in life-cycle costs (LCC). The first and most challenging task of an LCCA, or any economic evaluation method for that matter, is to determine the economic effects of alternative designs of buildings and building systems and to quantify these effects and express them in dollar amounts. According to the Sustainable Building Technical Manual, when viewed over a 30-year period, initial building costs have been shown to generally account for approximately just 2 percent of the total, while operations and maintenance costs equal 6 percent, and personnel costs reflect the lion's share of 92 percent.

There are general costs that are associated with acquiring, operating, maintaining, and disposing of a building or building system. Building-related costs usually fall into one of several categories, including:

- Initial costs include purchase, acquisition, and construction costs
- Operation, maintenance, and repair costs
- Replacement costs
- Finance charges—loan interest payments
- Fuel costs
- Non-monetary benefits or costs
- Residual values including resale or salvage values and disposal costs.

Only costs within each category that are relevant to the decision making and significant in amount are required to make a valid investment decision. To be relevant costs need to differ from one alternative compared with another; costs achieve significance when they are large enough to make a credible difference in the LCC of a project alternative. All costs are entered as base-year amounts in today's dollars; the LCCA method accelerates all amounts to their future year of occurrence and discounts them back to the base date to convert them to values that reflect the present.

Elaborate and detailed estimates of construction costs are not required for preliminary economic analyses of alternative building designs or systems. Such estimates are usually not available until the design is relatively advanced and the opportunities for cost-reducing design modifications have been missed. LCCA can be repeated throughout the design process whenever more detailed cost information becomes available. Initially, construction costs are estimated by reference to historical data from similar facilities. They can also be determined from government or private-sector cost estimating guides and databases. Detailed cost estimates are prepared at the submittal stages of design (usually at 30 percent, 60 percent, and 90 percent) based on quantity take-off calculations. These estimates rely mainly on cost databases such as the *R. S. Means Building Construction Cost Database*. Testing organizations such as ASTM International and trade organizations also have reference data for materials and products they test or represent. Perhaps the best and most logical way for the owner/developer to avoid cost overruns is to maintain the following:

- Reasonable objectives, not subject to modification during the course of the project
- A complete design that meets planning and statutory requirements and will not require later modification
- Clear leadership and appropriate management controls
- An appropriate risk allocation and sufficient contingency that is unambiguous and clear
- Realistic project estimates that are not unduly optimistic
- A project brief that is comprehensive, clear, and consistent
- A payment mechanism that incentivizes the parties to achieve a common and agreed upon goal
- Coordinated design that takes into account maintenance, health, safety, and sustainability.

Project Cost Breakdown

7.1 GENERAL

Successful development projects are grounded in painstaking analysis and rigorous planning, which means that a comprehensive compilation of construction-related cost information is a pivotal tool for assessing the merits of a proposed construction undertaking. There are many software packages on the market like "Projectmates" with capabilities ranging from document management and scheduling to financial budgeting and change order management. Projectmates software contains over 40 different modules for almost every type of construction project, and can save

Doi: 10.1016/B978-1-85617-676-7.00007-5

considerable time, as well as increase accountability among the various project participants. Also, while owners may choose to delegate responsibilities to other professionals on the project team, such decisions are left to their discretion and control.

Project managers and estimators are fortunate that there is a wide range of computer aided cost estimation software systems available, ranging in sophistication from simple spreadsheet calculation software to integrated systems involving design and price negotiation over the Internet. While such software involves costs for purchase, maintenance, training, and computer hardware, the user will experience significant benefits. In particular, cost estimates may be prepared more rapidly and with less effort. Professor Chris Hendrickson, co-director of the Green Design Institute, depicts some of the more common features of computer aided cost estimation software, which include the following:

- Databases for unit cost items such as worker wage rates, equipment rental or material prices. These databases can be used for any cost estimate required. If these rates change, cost estimates can be rapidly re-computed after the databases are updated.
- Databases of expected productivity for different component types, equipment, and construction processes.
- Import utilities from computer aided design software for automatic quantity-take-off of components. Alternatively, special user interfaces may exist to enter geometric descriptions of components to allow automatic quantity-take-off.
- Export utilities to send estimates to cost control and scheduling software. This is very helpful to begin the management of costs during construction.
- Version control to allow simulation of different construction processes or design changes for the purpose of tracking changes in expected costs.
- Provisions for manual review, over-ride, and editing of any cost element resulting from the cost estimation system.
- Flexible reporting formats, including provisions for electronic reporting rather than simply printing cost estimates on paper.
- Archives of past projects to allow rapid cost-estimate updating or modification for similar designs.

The construction industry continues to be riddled with procurement problems arising from an impractical division between the design and construction process, a lack of organization among subcontractors, strained relationships between design professionals and construction team members within the integrated project team, antiquated design and construction methods, preventable delays, and an inferior quality end product. However, in recent years the construction industry has demonstrated some improvements in application of new technology and methods in which clients have been the driving force for change leading to the development of improved and more sophisticated services (e.g., project management, facilities management, better handle on cash flow, and alternate procurement methods). This need for change originated from the desire for a more competitive and efficient industry. Education programs have also been created to merge construction and design to emphasize multi-skilled trades.

When it comes to green building, a Davis Langdon study, "Costing Green: A Comprehensive Cost Database and Budgeting Methodology," compared the square-foot construction costs of 61 buildings pursuing Leadership in Energy and Environmental Design (LEED™) certification to those of similar conventional buildings without green objectives. Taking into consideration climate, location, and other variables, the study came to the conclusion that for many of the sustainable projects, aiming for LEED certification resulted in little or no impact on the general budget.

Another point worth noting is that construction lending is basically real estate lending and the construction lender should therefore be aware that the primary security of the loan is primarily the real estate to be developed and that in order for the loan to be repaid the development has to be completed. Experience shows that few real estate borrowers are able to repay a construction loan from the assets listed in their financial statements. The borrower's professional and financial capabilities are key elements in the loan determination and should be thoroughly looked into before a loan commitment is made.

Professor Hendrickson says, "The costs of a constructed facility to the owner include both the initial capital cost and the subsequent operation and maintenance costs. Each of these major cost categories consists of a number of cost components."

In order to establish the capital cost for a construction project it is necessary to estimate the expenses related to the initial erection of the facility, which according to Hendrickson include:

- Land acquisition, including assembly, holding, and improvement
- Planning and feasibility studies
- Architectural and engineering design
- Construction, including materials, equipment, and labor
- Field supervision of construction
- Construction financing
- Insurance and taxes during construction
- Owner's general office overhead
- Equipment and furnishings not included in construction
- Inspection and testing.

The project owner must also consider the operation and maintenance cost of the project over its life-cycle, for which Hendrickson includes the following expenses:

- Land rent, if applicable
- Operating staff
- Labor and material for maintenance and repairs
- Periodic renovations
- Insurance and taxes
- Financing costs
- Utilities
- Owner's other expenses.

The consequence and significance of each of the above cost elements rely on the project's type, size, and location, as well as the management organization, among many other considerations. As far as the owner is concerned, the ultimate objective is achieving the lowest possible overall project cost that at the same time meets the specified quality and investment objectives of the project.

To estimate the total project cost (TPC) one must include all *hard* and *soft* costs to make the building complete and useable as intended. Costs would include construction costs, construction contingency, architect/engineer fee, project contingency, owner services, and administrative fees. Depending on the general conditions and contract document requirements, the total project cost may also include infrastructure costs, furniture and equipment costs, voice/data costs, instructional technology costs, moving costs, and custodial equipment costs.

Construction cost on the other hand is the fee charged by a general contractor or construction firm for a project. Construction costs per gross square foot will vary from state to state, and based on whether the project is new construction or renovation and the type, size, and complexity of the project (e.g., office, educational, hospital, residential, etc.).

According to Hendrickson, the important thing is for "design professionals and construction managers to realize that while the construction cost may be the single largest component of the capital cost, other cost components are not insignificant. For example, land acquisition costs are a major expenditure for building construction in high-density urban areas, and construction financing costs can reach the same order of magnitude as the construction cost in large projects, such as the construction of nuclear power plants.

From the owner's perspective, it is equally important to estimate the corresponding operation and maintenance cost of each alternative for a proposed facility in order to analyze the life-cycle costs. The large expenditures needed for facility maintenance, especially for publicly owned infrastructure, are reminders of the neglect in the past to consider fully the implications of operation and maintenance cost in the design stage."

Most construction budgets contain a contingency clause for unexpected cost overruns that may occur during construction. While this contingency amount may be included within each cost item, it is preferably included as a single category, namely a construction contingency which is normally a percentage of the project's estimated cost. This contingency amount is based on several factors including the complexity and size of the project and whether it is new construction or renovation, etc. For example, for large new construction projects the contingency is generally about 5 percent of the total cost, whereas for renovations it may be roughly 7 percent. Likewise, for small interior projects, it may be as high as 10 percent. Any remaining contingency amounts can be released upon substantial completion of the project. Figure 7.1 shows a sample project cost breakdown and the main elements to be considered when preparing a budget estimate for a project.

The developer's, builder's, or CM's fee is usually released as a direct percentage of the value of the subcontractual work completed to date.

7.2 BUDGET DEVELOPMENT—AN ANALYSIS

A project budget estimate is a financial plan to design and build a particular project and setting out the estimated costs to complete the project. Regardless of whether the project to be constructed is large or small, a prudent developer will certainly find it necessary to develop a budget for it. The primary purpose of preparing a budget is to understand and control costs and cost overruns. Cost overruns are mitigated by the inclusion of appropriate contingencies in the budget estimate to cover change orders, etc., and these contingency allowances are disbursed as the project proceeds to cover the additional costs. Figure 7.2 is a graph that illustrates the relationship of the contingency as a percentage of the direct cost budget.

Randy White, CEO of White Hutchinson Leisure & Learning Group, says, "Many a project gets into serious trouble when, for whatever reason, the project can't be developed within the budget. Usually, by the time the problem is discovered, it's too late to increase the budget, as financing has already been secured. So to keep the project within budget, critical features end up being compromised, such as the theming, finishes, and the quality of the materials, furniture, and equipment, the things that really matter the most to creating the guest experience. Or certain attractions are eliminated, so the project never performs as originally planned and projections are never achieved. In fact, such last-minute deletions and changes can seriously threaten a project's very long-term survival."

COST BREAKDOWN

ITEMS	DESCRIPTION	BUDGET AMOUNT	COSTS PAID	COSTS TO BE PAID
0100	**GENERAL CONDITIONS**	$	$	$
0101	Architecture and Design			
0102	Permits and Fees			
0104	Testing and Special Inspections			
0105	Soil Engineering			
0106	Structural Engineering			
0107	Survey and Topography			
0108	Tile 24 Compliance			
0109	Insurance and Bonds			
0110	Temporary Fencing and Security			
0111	Supervision			
0115	Temporary Sanitation			
0150	Temporary Utilities			
0199	Contingency (10% of Hard Costs)			
0200	**SITE WORK**			
0210	Clearing and Grubbing			
0211	Demolition			
0221	Site Grading			
0222	Excavation and Backfill			
0224	Erosion Control			
0255	Permanent Utilities			
0260	Paving			
0262	Curb and Gutters			
0263	Walks			
0271	Fences and Gates			
0280	Sewer Connections—Septic Tank			
0300	**CONCRETE**			
0310	Formwork			
0320	Steel Reinforcement			
0330	Concrete Foundation			
0331	Concrete Slabs and Patios			

FIGURE 7.1 An example of a project cost breakdown and budget estimate. To estimate the total project cost (TPC) it is necessary to include all hard and soft costs to fully execute the building as intended. Total project costs include construction costs, contingencies, architect/ engineer fee, owner services, and administrative fees. Depending on the general conditions and contract documents requirements, the total project cost may also include infrastructure costs, furniture and equipment costs, instructional technology costs, moving costs, and other costs.

0335	Concrete Driveways			
0400	**MASONRY**			
0410	Stucco			
0421	Fireplace Masonry			
0422	Concrete Blocks			
0425	Fireplace Facing/Stone Mantel			
0442	Marble			
0445	Exterior Stone Veneer			
0500	**METALS**			
0510	Rough Frame Hardware			
0512	Structural Steel			
0570	Ornamental Stairs and Rails			
0600	**WOODS AND PLASTICS**			
0610	Rough Carpentry Labor			
0611	Framing/Sheathing Materials			
0619	Wood Trusses			
0620	Finish Carpentry Labor			
0622	Interior Millwork and Trim			
0640	Cabinetry			
0643	Wood Stairs and Railings			
0700	**THERMAL AND MOISTURE PROTECTION**			
0710	Waterproofing			
0720	Insulation			
0731	Roof—Composition Shingle			
0732	Roof—Tile			
0740	Exterior Siding			
0750	Roof—Membrane Build-up			
0760	Flashing and Sheetmetal			
0780	Skylights			
0800	**DOORS, WINDOWS, AND GLASS**			
0820	Exterior Doors			
0821	Interior Doors			
0830	Sliding Glass Doors			
0850	Metal Windows			
0860	Wood Windows			
0870	Finish Hardware			
0872	Garage Doors and Operators			
0883	Mirrors			

FIGURE 7.1 *(Continued)*

0900	**FINISHES**			
0925	Gypsum Drywall			
0930	Tile Countertop			
0934	Marble Countertop			
0938	Formica Countertop			
0960	Wood Flooring			
0961	Wood Mantels			
0965	Linoleum			
0968	Carpet			
0970	Tile Floor			
0990	Exterior Paint			
0991	Interior Paint			
0995	Wallpaper			
1000	**SPECIALTIES**			
1017	Tub and Shower Enclosures			
1030	Prefabricated Fireplace			
1040	Signage			
1080	Toilet and Bath Accessories			
1100	**EQUIPMENT**			
1105	Vacuum System			
1140	Kitchen Appliances			
1200	**FURNISHINGS**			
1250	Window Coverings			
1400	**CONVEYING SYSTEMS**			
1420	Elevators			
1500	**MECHANICAL**			
1540	Plumbing—Rough			
1545	Plumbing—Finish			
1546	Plumbing Fixtures			
1550	Fire Protection			
1580	Heating and Air Conditioning			
1600	**ELECTRIC**			
1610	Rough—Electrical			
1614	Finish—Electrical			
1650	Lighting Fixtures			
1670	Telephone and Prewire			
1676	Intercom and Prewire			
1678	Television and Prewire			

FIGURE 7.1   *(Continued)*

2000	MISCELLANEOUS			
	Clean-Up			
3000	OVERHEAD AND PROFITS			
	Overhead			
	Profits			
	SUB-TOTAL	$	$	
	Loan Cost			
	Insurance, Compensation, P.L. & P.D., Social Security & Unemployment Insurance			
	Builders Overhead and Profit _____%			
	TOTAL COST OF CONSTRUCTION	$	$	
	LAND COST			
	TOTAL	$	$	$

I certify that to the best of my knowledge the above is a true and correct statement of the estimate cost of this job.

Signed: _____ Date: _____

FIGURE 7.1   *(Continued)*

Budget development is particularly important for obtaining a construction loan. Before calculating the construction loan amount, a basic budget is required, the main components of which are:

1. Hard costs: Direct costs associated with the labor and materials used for the actual physical construction of the project.
2. Soft costs: Indirect or "off-site" costs not directly related to labor or materials for construction (architectural plans, engineering, and permit fees).
3. Closing costs: All costs associated with origination and closing the construction loan such as title cost, loan fees, discount fees, insurance, appraisals, and closing fees.
4. Land acquisition costs.
5. Inspection fees.
6. Reserves: Consisting of estimated interest on the loan during construction and contingency reserve for unforeseen expenses and cost overruns.
7. Possible equipment, furnishings, etc.

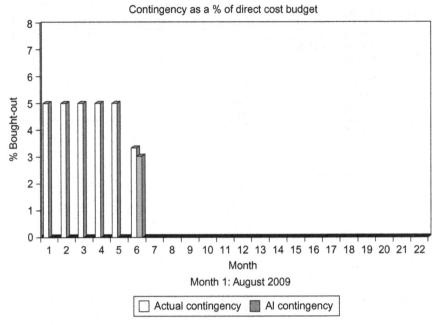

FIGURE 7.2 A graph illustrating the relationship of contingency as a percentage of direct cost budget as the construction process proceeds.

As previously mentioned, the total cost of construction is obtained by adding all the costs incurred in the project including soft and hard costs. Most projects are constrained by limited monetary funding resources. Consequently, they need to have a budget in place to initially define their funding requirement. The project manager develops the budget based on the cost estimates calculated at the beginning of each project phase and refined once there is more accurate information defining the project's scope. Refining the budget occurs through studies and analysis in the design development process. When owners try to fix the budget too early in the project life cycle, they are surprised by the significant increases in the budget over what was set forth. Randy White says, "This reoccurring problem is often caused by the nature of the design process. Design proceeds from general to specific and from conceptual to detailed. Accordingly, there is limited ability to accurately predict construction costs at the onset when initial project planning takes place and accurate costs are needed as part of the business plan to secure financing." With respect to project cost overruns, White says, "Cost overruns are also caused by the traditional design-bid-build process. First the project is designed, and then a contractor is selected by either competitive bid or negotiation to build the project. This process precludes value engineering until the project is already designed. So by the time the bid comes in over budget, the only way to reduce costs is to make major compromises in finishes, quality, or components."

One of the critical tasks and assignments of the project manager (PM) is the development and tracking of the project budget. The PM first develops a project budget in the early feasibility phase and continues to refine it throughout the different project phases until the project is "bought out" by the general contractor prior to the start of the construction phase. All the elements of the budget should be clearly defined, and fine tuned throughout each phase. Specialized estimating software is often used to create, develop, monitor and track budgets. When developing a budget one must follow certain logical steps, such as:

Step 1: It is important to know precisely how much money is available for the project when attempting to develop a project budget. This should include all costs from project initiation as a concept through the award of a construction contract to completion. At various points within the different stages, more detail, specificity, and definition are developed and these estimates become more certain and realistic.

Step 2: Determine the mandatory or vital expenses for the project to succeed. Each project has certain requirements that are essential to the project's success and these should be given priority in the project's budget. For example, incorporating certain green features to achieve LEED certification is very important, which means that this should be high on the project budget's priority list.

Step 3: Collect preliminary estimates from several companies and contractors. Upon having identified the key elements of the project budget, you can start requesting estimates from area businesses to determine which offers the best price or value for building the project. In any case, preconceived cost indices can often be unrealistic and misleading. Building a new building project in an urban setting for example is far more costly than on a green field site. Construction managers usually have a good understanding of true market conditions and pricing in specific regions because their livelihood depends on it. Should your project budget not align with the project's expectations, then one or both will need realignment.

Step 4: Once the final budget is determined, it should be clearly spelled out and written down on paper and distributed to all the team members to ensure that everyone is on the same page. In the final analysis, it will also depend largely on the type of contract entered into between the owner and the contractor, e.g., whether it is a design-bid-build, design-build, cost-plus, etc.

Furthermore, total reliance on "program" and/or "preliminary" level estimates for setting a final project construction budget is inappropriate because it is far too early in the project's design/construction process.

Until the time when a fairly accurate budget estimate can be made, the project remains too conceptual in terms of scope and program size to accurately estimate final costs. After the architect completes the schematic design phase, the project's scope of work is more clearly defined to the extent that a realistic budget estimate can be determined to provide effective discipline and direction on the project. However, this is still insufficient to bid the project, which isn't possible until the contract documents are completed, including all plans and specifications. The different phases or milestones a project must go through from design concept to completion and occupancy are:

1. Project initiation program budget estimate.
2. Planning/programming preliminary project estimate.
3. Design (conceptual design, schematic design budget estimate).
4. Contract documents (drawings, project manual, etc.).
5. Bidding of contract, awarding of contract—project estimate.
6. Construction phase, commissioning.
7. Occupancy phase.

Figure 7.3 shows an example of preliminary construction budget for a typical project (new construction).

Another example of a simple project budget is shown in Figure 7.4, adopted from Charter School Facilities—A Resource Guide on Development and Financing, which shows the typical components of a project budget (it includes the purchase and renovation of a building as opposed to new construction).

7.3 BUYOUTS

The term buyout as it relates to construction project mobilization basically refers to procuring materials and equipment that will be employed in the project and arranging subcontracts. Buyout is the time interval between the preconstruction and the construction phases of a project and is among the most critical first steps in the overall profitability of a construction project. Even before breaking ground, making or losing money could be predetermined based on how well the project is bought out. Also, it is during buyout that purchase orders and subcontracts are issued. This includes selection of both suppliers and subcontractors and finalizing their purchase orders or subcontracts. Unfortunately, buyouts are necessary because often due to time constraints during the bidding phase of a project, complete, meticulous analyses of bids by subcontractors may not have been possible. Figure 7.5 includes a diagram showing the percentage of buyouts of a budget's trade contract cost versus time.

SAMPLE CONSTRUCTION BUDGET

This is an example of a project budget for a government agency that proposes to construct a new facility in which to expand its activities. It should be noted that these costs will vary depending on what region of the country the project is located.

Expenses
Hard Construction Costs (8000 SF @ $97/SF)

Foundation, framing, drywall, flooring, roofing	$581,000
Plumbing, electrical, security system	80,000
Fixtures, furnishings, and equipment	50,000
HVAC	27,000
Landscaping	19,000
Site work	18,000
Sub-total Hard Costs	**$775,000**
Land Acquisitions	175,000

Soft Construction Costs

Architect and Engineers	31,000
Fees	4,000
Sub-total Soft Construction Costs	**$35,000**
Contingency	10,000
Total Expenses	**$995,000**

Revenues

Individual contributions	$295,000
ABC foundation	125,000
Government grant	75,000
Corporate donations	100,000
DEF foundation	85,000
Other foundations	65,000
XYZ corporation (in-kind)	50,000
Other corporate donations	25,000
Fundraising events	15,000
ABC corporation	50,000
XYZ foundation (pending)	50,000
To be raised from other sources	60,000
Total Revenues	**$995,000**

Notes:
1. Hard construction costs include any costs that cannot be physically moved, in other words site work, renovations or construction work, plumbing, electrical, landscaping, parking lot, demolition, flooring, roofing, HVAC, wiring, fire and security alarms, playgrounds, fixtures, appliances, etc. that become a permanent part of the site.
2. Soft construction costs include fees, surveys, permits, architect and engineer fees, etc.
3. Contingencies are usually between 5 and 10 percent of construction costs, depending on the size and complexity of the project.

FIGURE 7.3 A sample construction budget cost estimate for a government agency that is considering constructing a new facility to allow expansion of its activities. It should be noted that these costs can vary depending on what region of the country the project is located.

USES OF FUNDS:

Acquisition of building	**$250,000**

Construction/Renovation Costs (Hard Costs)

Demolition of old walls	75,000
Electrical	65,000
Plumbing	80,000
Heating/ventilation	40,000
Roof	50,000
Drywall and painting	140,000
Carpet	35,000
Windows	40,000
Fixtures and fit-out	55,000
Site work	20,000
Total Construction:	**$600,000**

Hard Cost Contingency (15%)		90,000
	Total Acquisition & Construction:	**$940,000**

Legal Fees	10,000
Appraisal	5,000
Architect	30,000
Project manager	10,000
Engineering	5,000
Insurance during construction	3,000
Closing costs	5,000
Financing fees (loan origination fee etc.)	7,000
Interest during construction	35,000
Inspection fees	5,000
Environmental studies	12,500
Accountant	5,000
Security	8,000
Bonding	6,000
Total:	146,500

Soft Cost Contingency (5%)	**7,325**
Grand Total:	**$1,093,825**

SOURCES OF FUNDS:

Startup grant	$150,000
Donations	238,325
Loan	705,500
Grand Total:	**$1,093,825**

Notes:
1. Hard cost contingencies are anything related to the building structure or its materials. The large contingency budgeted is due to the extensive nature of renovations. On average renovation projects have a 7–10 percent contingency and for new construction it is usually about 5 percent.
2. Soft costs are related to all other costs, including architectural, financing, inspection and legal fees. These costs are normally about 5 percent of the total project costs.

FIGURE 7.4 An example of a simple project budget adopted from Charter School Facilities—A Resource Guide on Development and Financing. Again it shows the typical components of a project budget and includes the purchase and renovation of a building as opposed to new construction.

Most construction literature ignores the issue of buyouts and concentrates on addressing either estimating or project management. The process starts during the tender preparation stage, as the contractor solicits and assesses offers in the process of assembling the cost estimate. If the contractor is awarded the contract, the next step is to attempt to contract with the firm that submitted the best offer. Material procurement and subcontracting are typically the two distinct parts of the buyout process discussed, even though both may be the responsibility of a single department or individual.

Most construction professionals are familiar with the concepts of project buyout and bid shopping. Project buyout is an ethical and necessary practice conducted during the preconstruction process which enables a general contractor to clarify scopes of work and streamline specific activities for the project. As construction professionals more fully understand the ethical issues separating the unacceptable practice of bid shopping from the ethical practice of project buyout, the efficiency and quality of the estimating and subsequent project management processes will be improved. Estimators are normally required to bear the responsibility of obtaining bids and performing project buyouts while maintaining high ethical standards.

A buyout estimate is different from a bid estimate. A bid estimate is detailed in order to bid a project, whereas the purpose of a buyout estimate is to order materials once the project becomes a viable job for the contractor or subcontractor. A typical example of a bid estimate versus a buyout estimate would be metal studs for drywall. During the bid period, it is good enough to know the total linear footage of studs by size

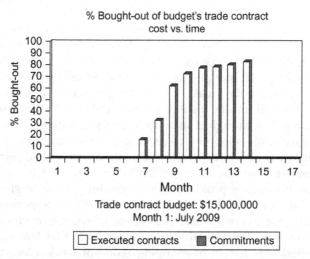

FIGURE 7.5 A diagram showing the percentage of buyouts (executed and committed) of a budget's trade contract cost versus time. The estimated budget in this case is $15,000,000.

and gauge. A buyout however requires greater detail, for example, in addition to the bid information a buyout would require the lengths for each application. However, in mechanical and electrical scopes of work, the bid estimate and buyout estimates are very similar.

Project buyout takes place between the award of a bid to the general contractor and the issuing of subcontracts and purchase orders. While bid shopping is not illegal, it is considered an unethical practice where details of a bid are revealed to a competitor in an effort to solicit an over-all lower bid. A better understanding of ethical versus acceptable construction practices can help construction professionals identify the basic differences between bid shopping and project buyout, while at the same time steering clear of unethical practices and remaining competitive.

According to Cody Andreasen, Mark Lords, and Kevin R. Miller of Brigham Young University, "The justification, for some contractors, when arguing in favor of bid shopping is that if a bid is revealed to another subcontractor, then a lower bid may be forthcoming, which may translate into a lower overall bid on the project, benefit the owner, and thus increase the likelihood of being awarding the project. It can be argued that this is no different than shopping for a car or bartering for goods in a foreign country. However, in the auto industry there is an expectation that pricing will be disclosed to other dealers in the buying process. That expectation does not exist in the construction industry. A construction project is something yet to be built. It is not an existing product, and any changes in the cost typically will affect the quality or schedule of the project. Therefore, the owner is not receiving the same product if bid shopping occurs." A technique that is often used to prevent bid shopping is for a subcontractor to submit a bid at the last minute, thus preventing the general contractor receiving the bid from shopping it.

By and large, bid shopping occurs because most subcontractors being shopped believe that if they do not reduce their price they won't get the job. Additionally, if business is slow subcontractors may be willing to accept lower profit margins just to keep their crews busy even when it means they will only break-even on the project. Sometimes the general contractor will induce a subcontractor additional work upon being awarded the job. Although it may appear that the owner is the principal beneficiary of bid shopping by receiving a lower price for the project, usually the owner also receives a lower quality project in addition to running greater risks and warranty problems down the road.

The main benefits of project buyout according to the Brigham Young University authors are that "Project buyout allows a period of time for the contractor to ensure that each scope of work is covered by only one subcontractor. Occasionally a contractor finds that two subcontractors submitted bids for an overlapping scope of work. Since both subcontractors do not need to perform the work, the general contractor will determine which subcontractor will perform the work for the overlapping scopes. The subcontractor that doesn't perform the overlapping work will generally provide a credit to the general contractor for the reduction in their scope of work. If the opposite is found and there has been work that was assumed to be included in a subcontractor's scope of work but was not included in the bid, the general contractor generally negotiates with a subcontractor to have the work included in their subcontract and that negotiation may increase the contract amount for the subcontractor.

Another instance where changes could be made during the project buyout process is when a subcontractor anticipated a different project schedule than the general contractor. As a result, the subcontractor may not have sufficient crews to complete the job in the time frame or manner desired by the general contractor. In this case, the general contractor may elect to use a different subcontractor in order to maintain the project schedule."

Darin C. Zwick and Kevin R. Miller, authors of an article entitled *Project Buyout*, emphasize the importance of completing the buyout process as early as possible and say, "By completing buyout early, future delays are avoided in the event that a given scope of work is difficult to buyout due to conflicts with subcontractors or suppliers. It also protects the project from price escalation.

Other tasks that occur during the buyout process by the general contractor include checking the following items to ensure that the subcontractor can perform the work for the project:

1. Insurance and liability coverage.
2. Evidence of state workers' compensation coverage.
3. Evidence of proper local and state subcontractor licenses.
4. Evidence of proper bonding requirements if required.

The expiration dates of the previous items need to be verified to prevent lapses of coverage while the subcontractor is working on the project.

Another consideration that companies need to examine during the buyout process is the financial stability of the subcontractors and vendors. During economic downturns, companies may declare bankruptcy, leaving the general contractors in a precarious situation."

Among the main duties of the project buyout specialist is to focus on awarding scopes of work to subcontractors and to act as a liaison between field operations and subcontractors while keeping the contract amount in budget. This is particularly important because it relates to disputes and problems that often exceed field management's ability to solve in a timely manner. During the pre-construction phase this includes technical support to both the design build and estimating departments. In addition, the buyout specialist is responsible for the acquisition of new subcontractors for all projects including cost control, adherence to corporate and contract compliance, quality control, and customer satisfaction. The project buyout specialist is also responsible for the quality and completeness of project buyout and small business utilization, as well as customer relations and client satisfaction in all areas within the firm.

To achieve maximum support for and from each staff member requires team building, setting and monitoring goals, and collaborating with purchasing and operational goal setting. There are currently several proprietary buyout software packages on the market that can save time and reduce effort by automating the bid solicitation process. Buyout software also provides an important tool for determining where the project is in the buyout process at any particular moment. It can also establish the percent complete, minimize exposure, and rapidly see how the actual prices compare with the estimated costs. Moreover, it can offer access to standard cost codes, categories, and tax groups stored in various applications such as accounts payable applications. One example of such software is from Sage Software, Inc. (www.sagecre.com), which offers a software package called Buyout that reportedly offers the following features (Figure 7.6):

- Build a worksheet of material and subcontract items to be bought out automatically by reading the estimating file
- Combine multiple estimates into a single worksheet, an important feature for contractors who receive price discounts based on volume purchases
- Create one-time items in the Buyout item window
- View items the way you want to see them—by WBS, location, phase, material class, and so forth
- Group materials or subcontract items for ease in obtaining prices. Create quote sheets and assign material items and subcontract items to the quote sheets
- Assign multiple vendors and subcontractors to quote sheets
- Use prices from Buyout's standard price database for items in the quote sheet
- Automatically submit requests for quotes and send purchase orders via email, fax, or hard copy
- Split items out of one quote for the creation of a new quote sheet
- Use the summary quote sheet to organize vendors' or subcontractors' quotes from low to high.
- Save prices from the quote sheet to a Buyout standard price database
- Change prices for any item and update the estimating database with pricing from Buyout
- Generate RFQs and POs directly from Buyout and issue automatically via email, fax, or hard copy
- Update estimating estimates with Buyout prices, revised quantities, and vendor/subcontractor selections.

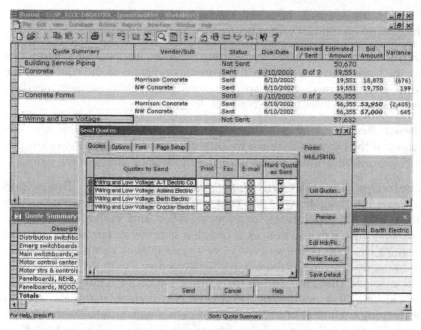

FIGURE 7.6 An illustration of Buyout, computer software that helps sort the items in an estimate into desired groups of materials, produce documents, and perform the tasks necessary to buyout a job. Among other features, it can automatically generate, sort, and send RFQs to suppliers and subcontractors via email, fax, or printed hard copy. It also records, tracks, analyzes, and selects bids received. One of the more important features is the ability to transfer commitments to purchasing for purchase order and subcontract generation.

7.4 GENERAL CONDITIONS AND SUPPLEMENTAL CONDITIONS

The general conditions document is among the most important documents of the project manual because it sets forth and defines the rights and responsibility of the different parties, particularly the owner and contractor, in the construction process, as well as the specific terms of the contract. It also specifies and defines the surety bond provider, the design professional's role, authority, and responsibilities, and the requirements governing the various parties' business and legal relationships. These conditions are "general" and can apply to almost any project. It is vital that the contractor knows exactly what is contained in this section. It should be noted that many trade and professional organizations have developed their own standard documents and general conditions, and the most widely used may be that published by the American Institute of Architects, AIA Document A201. This document has been well tested in the courts and is familiar to most contractors.

Below are some of the main standard clauses that typically appear in the general conditions with a brief description of each.

Definitions and General Provisions   This clause provides definitions for the purpose of the Contract Documents relevant to the contracts, the work, and the drawings and specifications. It also clarifies the ownership, use, and overall intent of the contract documents.

Owner Responsibilities   This section defines the information and services that the owner is required to supply. It also defines the owner's rights and responsibilities and the owner's right to stop or carry out the work.

Contractor General Obligations and Responsibilities   This section lays out the obligations of the contractor regarding construction procedures and site operations, employees, labor and materials, warranty, taxes, permits, fees and notices, schedules, samples and product data, and cleaning up. The contractor is required to execute and complete the works and remedy any defects therein in strict accordance with the contract with due care and diligence and to the satisfaction of the architect, and shall provide all labor, including the supervision thereof, materials, and all other things, whether of a temporary or permanent nature. The contractor shall also take full responsibility for the adequacy, stability, and safety of all site operations and methods of construction, but the contractor shall not be responsible, unless expressly stated otherwise in the contract, for the design or specification of the permanent works or of any temporary works prepared by the architect.

Administration of the Contract   This section delineates the duties, responsibilities, and authority of the architect for the administration of the contract. Specific clauses are included dealing with the architect's responsibility for making periodic site inspections and issuing periodic reports to the owner or lender. This section also deals with issuing modifications in drawings and technical specifications and assisting the contractor in the preparation of change orders and other contract modifications, as well as assisting in inspections, signing certificates of completion, and making recommendations with respect to acceptance of work completed under the contract. The architect is also required to review detailed drawings and shop drawings, price breakdowns, and progress payment estimates, as well as suggest how requests for additional time, claims, and disputes will be handled.

Pre-construction Conference and Notice to Proceed   This section outlines the procedures to conduct a pre-construction conference to acquaint the different parties with one another. For example, within 10 calendar days (or as stated in the contract documents) of contract execution, and prior to the commencement of work, the contractor shall attend a preconstruction conference with representatives of the owner, architect, and other interested parties and stakeholders. This clause also deals with the notice to proceed, in that the contractor can only begin work upon receipt of a written notice to proceed from the owner or designee. The contractor may not begin work prior to receiving such notice.

Availability and Use of Utility Services This section deals with the availability of utility services. Here the project owner shall ensure that all reasonably required amounts of utilities are available to the contractor from existing outlets and supplies, as specified in the contract. Unless otherwise provided in the contract, the amount of each utility service consumed shall be charged to or paid for by the contractor at prevailing rates charged to the owner.

Assignment and Subcontracting This section deals with the assignment and awarding of subcontracts by the general contractor for portions of the work. This clause generally states that the contractor shall not, except after obtaining prior written approval of the project owner, assign, transfer, pledge, or make other disposition of the contract or any part thereof or of any of the contractor's rights, claims, or obligations under the contract. In the event the contractor requires the services of subcontractors, the contractor shall also obtain prior written approval of the owner for all such subcontractors. The approval of the owner does not relieve the contractor of any of his obligations under the contract, and the terms of any subcontract shall be subject to and be in conformity with the provisions of the contract.

Construction by Owner or Others This clause deals with the owner's right to perform some of the construction work with his/her own forces or to award separate contracts to other parties besides the general contractor. The contractor shall in accordance with the requirements of the architect/ project manager and the contract afford all reasonable opportunities for carrying out portions of the work by the owner or to any other contractors employed by the owner and their workmen or the owner's workmen who may be employed in the execution on or near the site of any work not included in the contract or of any contract which the owner may enter into in connection with or ancillary to the works.

Permits and Codes This section basically states that the contractor shall give all notices and comply with all applicable laws, ordinances, codes, rules, and regulations. Before installing the work, the contractor shall examine the drawings and the specifications for compliance with applicable codes and regulations bearing on the work and shall immediately report any discrepancy it may discover to the architect/project manager.

Change Orders This section explains how changes are authorized and processed according to the relevant clauses of the contract. Changes orders are one of the areas of greatest contention between the owner and the contractor. Generally, the architect may instruct the contractor, with the approval of the owner and by means of change orders, all variations in quantity or quality of the works, in whole or in part, that are deemed necessary by the architect. Processing of change orders shall be governed by appropriate clauses of the general conditions.

Construction Progress Schedule   Time is always a pivotal factor on any project. Project schedules depict project startup, progress, and anticipated completion dates. They address issues associated with delays and time extensions to the contract. For example, schedules shall take the form of a progress chart of suitable scale to indicate appropriately the percentage of work scheduled for completion by any given date during the construction period.

Progress Payments   This section specifies how applications for progress payments are to be processed. The owner/lender shall make progress payments approximately every 30 days as the work proceeds, on estimates of work completed which meets the standards of quality established under the contract, as approved by the project manager/architect. Before the first progress payment can be processed under this contract, the contractor shall furnish a breakdown of the total contract price showing the amount included therein for each principal category of the work, which shall substantiate the payment amount requested in order to provide a basis for determining progress payments. This section also deals with the withholding of payments and failure to pay issues.

Protection of Persons and Property   This section is intended to address issues relating to safety of both the project owner's property and the people on the project. It deals with specific issues such as the handling of hazardous materials and emergencies, as well as overall safety programs and requirements. The contractor shall (unless stated otherwise in the contract) indemnify, hold and save harmless, and defend at his own expense the project owner, its officers, agents, and employees from and against all suits, claims, demands, proceedings, and liability of any nature or kind, including costs and expenses, for injuries or damages to any person or any property which may arise out of or in consequence of acts or omissions of the contractor or its agents, employees, or subcontractors in the execution of the contract.

Insurance of Works and Bonds   This section deals with insurance (including liability insurance) and bonding requirements of the various parties which should cover the period stipulated and also cover the defects liability period for loss or damage arising from a cause occurring prior to the commencement of the defects liability period and for any loss or damage experienced by the contractor in the course of any operations carried out for the purpose of complying with the contract obligations.

Examination of Work Before Covering Up   This clause has to do with acceptance of the work by the architect (as agent of the owner). It stipulates how and when the contractor is responsible for uncovering and/or correcting any work deemed unacceptable. No work shall be covered up or put out of view without the prior approval of the project manager/architect. The contractor shall afford full opportunity for the project manager/architect to examine and measure any work that is about to be covered up or put out of view and to examine foundations before permanent work is placed thereon. The contractor shall give due notice to the project manager/architect whenever any such work or foundations are ready for examination and shall without unreasonable delay advise the contractor accordingly to attend for the purpose of examining and measuring such work.

Clearance of Site on Substantial Completion   This section basically states that upon the substantial completion of the works the contractor shall clear away and remove from the site all rubbish, constructional plant, surplus materials, and temporary works so as to leave the whole of the site and works clean and in a workmanlike condition to the satisfaction of the project manager/architect.

As-built Drawings   As-built drawings, as used in this clause, refers to drawings submitted by the contractor or subcontractor at any tier to show the construction of a particular structure or work as actually completed under the contract. As-built drawings shall be synonymous with "record drawings."

Miscellaneous Provisions   This section deals with various matters such as liquidated damages, taxation, disputes, prohibition against liens, warranty of construction, energy efficiency and other green issues, waiver of consequential damages, etc.

Termination/Suspension of the Contract   Either party has the right to terminate the contract under certain conditions. The conditions under which the parties may terminate or suspend the contract are clarified in this clause. For example, the contractor shall on the written order of the architect/project manager suspend the progress of the works or any part thereof for such time or times and in such manner as required by the architect/project manager and shall, during such suspension, properly protect and secure the works as specified by the architect/project manager. The project owner should be notified and written approval sought for any suspension of work in excess of three days.

Supplemental Conditions

These are special conditions also known as supplementary general conditions, special provisions, or particular conditions, that normally deal with matters that are project specific and which are beyond the scope of the standard general conditions. These sections may either add to or amend provisions in the general conditions. Examples of project specific information that may appear in the supplemental conditions are:

- Safety requirements
- Contractor's bond requirements
- Bonus payment information
- Defects liability period
- Cost fluctuation adjustments
- Progress payment retainage
- Services provided by owner
- Temporary facilities provided by owner
- Owner provided materials.

7.5 CONTINGENCIES AND ALLOWANCES

Contingencies are generally necessary to cover unknowns, unforeseen and/or unanticipated conditions, or circumstances that are not possible to adequately evaluate or determine from the information on hand at the time the cost estimate is prepared. Contingency allocations specifically relate to project uncertainties of the current known and defined project scope and that may arise and are not a prediction of future project scope or schedule changes. The amount of contingency allocated relates to the amount of assessed risk and should not be reduced without appropriate supporting justification. Furthermore, inclusion of a contingency amount in the cost estimate mitigates the impact of cost increases inherent in an overly optimistic estimate and provides for the opportunity of an earlier discussion of how to address potentially adverse circumstances.

Contingencies in a project budget represent the degree of risk within the estimate and are traditionally calculated as an across-the-board percentage addition on the base estimate, typically based on initial estimates, previous experience, and historical data. This estimating approach has serious flaws because it is generally illogically arrived at and therefore often not appropriate for the project at hand. Moreover, this method of arbitrary contingency calculation is difficult for an estimator to justify or defend. A percentage addition results in a single-figure prediction of estimated cost which is often unjustified because it does not reflect reality, nor does it encourage creativity in estimating practice.

The following reflect some of contingency types found in typical budgets that should be considered when major projects are involved:

- Construction contingency is basically used to cover cost growth during construction. It is a percentage of construction cost held by the

PM to resolve issues during construction, which is why it should not be used until the project is in the construction phase. This contingency will be higher for renovations in older buildings, buildings with complicated site conditions, or in complex projects. The construction contingency is contained in the contractor's guaranteed maximum price (GMP) but the PM must approve use of these funds prior to being committed by the general contractor.

- Design contingency is for changes or modifications during the design process for such factors as incomplete scope definition and inaccuracy of estimating methods and data. Design contingency amounts are based on the amount of design completed and are a percentage of construction cost held to represent the completeness of the design. The design contingency is understandably higher during the early phase of the project's design. As the design is completed and the scope of work is more defined, this contingency is reduced until it becomes zero in the cost element at the completion of the permit phase.
- Project contingency is a percentage of project cost retained for risks in other project costs such as professional fees, hazardous materials abatement (e.g., asbestos), communications wiring, etc. Money allocated as contingency in the project budget should not be utilized for additional scope or other changes to the project once the design is completed.
- Program contingency is optional and may be employed to cover scope or program changes requested by the user group or owner. An alternative to this contingency is to have the general contractor carry an *allowance* line item in the GMP contract.
- Various other contingencies for areas or items that may show a high potential for risk and change, i.e., environmental mitigation, utilities, highly specialized designs, etc.

7.5.1 Construction Contingencies

A construction contingency is essentially a set percentage of the construction contract amount budgeted for unexpected and unforeseen emergencies or design shortfalls identified after a construction project has commenced. When underwriting a commercial construction loan request, it is prudent to analyze the four major elements of the total construction cost which are the land cost, the hard costs, the soft costs, and the contingency reserve. Since there are always cost overruns in almost any commercial construction process, a contingency reserve is put in place to build in a cushion in the project's construction budget to cover these cost overruns.

While there is no specific formula for computing a contingency reserve, many underwriters feel comfortable using 5 percent of the construction estimate/bid for new construction (although in complicated projects the contingency can be as high as 10 percent) and 7 percent of the construction estimate/bid for remodeling/renovation projects. The land costs are not included because it is usually known in advance and is fixed. It is unlikely that there will be a cost overrun connected to the purchase of the land itself.

A construction contingency is included in the budget to allow the project to proceed with minimal interruption for small or insignificant (non-scope) changes or cost overruns. The typical construction contract will include a specific completion date or specific number of working days to complete, and the contractor can be required to pay liquidated damages if the work is not completed within this specified period. At the same time, the contractor is entitled to proceed with the work without undue interruption. To minimize delays due to external causes, the client must be capable of implementing minor (i.e., non-scope) changes without causing any administrative delay.

Whatever the case, changes always occur on construction projects. The owner must therefore ensure that an appropriate contingency is included to cover the costs of any changes in the scope of a project, such as adding upgrades, additional equipment, or perhaps enlarge the footprint of the building. Moreover, financing costs may change with the market. A small contingency may be sufficient to cover final documentation of drawings by designers, but plan a construction contingency of roughly 3 percent to cover the changes in market conditions and estimate variances.

The objective of contingency planning is to determine a confidence value by means of a percentage in potential cost and schedule growth. The contingency value is an indicator of the level or degree of project development, and typically, the less defined a project, the higher the contingency value. Issues such as scope definition and quality assurance have a significant impact on confidence, risks, and resulting contingency development. In determining a contingency value, consideration must be given to the details and information available at each stage of planning, design, and construction for which a cost estimate is being prepared.

As previously mentioned, most construction budgets contain an allowance for contingencies or unexpected costs occurring during construction. This contingency amount may either be included within each cost item or be included as a single category in the construction contingency. The estimated amount of contingency to be retained is based on historical experience and the anticipated difficulty of a particular construction project. For example, one construction firm places estimates of the anticipated cost into five specific categories. These are:

- Design development changes
- Schedule adjustments

- General administration changes (such as wage rates)
- Differing site conditions from those expected
- Third party requirements imposed during construction, such as new permits.

Any contingent amounts not disbursed during the construction process can be released toward the end of construction to the project owner or alternatively used to add additional project elements. The construction cost process consists of two essential components, *hard costs* and *soft costs*.

Hard costs are considered to be by far the largest portion of the allocated expenses in a construction budget and generally consist of all of the costs for physical items and visible improvements (i.e., actual construction costs incurred to build the project), line items including site preparation (grading/excavating), concrete, framing, electrical, carpentry, roofing, and landscaping. Hard costs have been described as the bricks and mortar expenses. In some cases, it may include the land, but that particular cost is usually separated in order to find out the actual construction expenses.

Soft costs are the non-physical expenses and involve all of the other fees involved in the completion of the project. Soft costs include transfer taxes, origination points, mortgage insurance (if applicable), overhead expenses, attorney fees, professional fees, permits, title insurance, appraisal fee, testing, hazard insurance, marketing, construction insurance, etc. Another primary soft cost category if applicable is that of fixtures, furnishing, and equipment (FF&E). The soft costs are generally estimated as a percentage of the total project budget during the planning stages of a project. And as the planning and design of a project progresses, the soft cost contingency percentage can be increased or decreased.

To arrive at the total cost of a commercial construction project one must include the hard costs, soft costs, the cost of land, as well as the contingency reserve, which for new construction is generally about 5 percent of the total project cost.

Design professionals are able to establish a project cost estimates by employing several methods. For example, one approach is to use estimates whose development is based on project parameters and major cost elements, or which are based on an analysis of historical bid data, actual cost, or a combination of these methods. But whatever method is used, in the final analysis special care must be taken to ensure that the capital cost estimate undertaken is complete and realistic and not overly optimistic. Underestimation of project construction and related costs is one of the more common problems faced in the economic analysis and budgeting of a project. Contingency funding is a fiscal planning tool that is used to help manage the risk of cost escalations and cover potential cost estimate shortfalls. Inclusion of a contingency amount in the cost estimate will mitigate the adverse impact of cost increases inherent in an overly optimistic estimate and provide an opportunity for an earlier discussion of how potential circumstances can be addressed.

Having an overall management contingency is strongly advised for large projects. This contingency is usually a "stand alone" amount of the cost estimate that is managed by an executive and used for a broad array of uncertainties and potential risks. Some of the project oversight management contingency allowance will be disbursed to manage costs, manage the approved budget and schedule deviations, address adverse impacts caused by modifications, and for initiatives being analyzed or implemented in order to address or mitigate potential cost overruns or schedule delays.

Management of the transfer of costs to and from contingency and allowance line items needs to be administered and tracked carefully to allow decision makers to take appropriate action. Cost transfers should correspond to the major component type of cost escalation. Thus, if a proposed work is clearly outside of a well-defined scope but is found to be essential to the well-being of the project and can be readily justified, then a management decision can be made to disburse payment for the added work or change order from either the management contingency or another appropriate contingency. On the other hand, if there are distinct fees or FF&E issues that have a fee or FF&E contingency, careful tracking of these particular contingencies can help the PM and management to better analyze potential cost overruns.

Rationale for supporting contingency transfers should be noted and incorporated into all relevant reporting. This is to allow a periodic comparison analysis of available contingency amounts to establish contingency usage rates. This analysis will alert project managers if potential problems exist as well as confirm if a reasonable and sufficient amount of contingency remains to keep the project within the latest approved budget.

Construction cost estimates should not be presented as a lump sum total, but rather as the sum of costs for each major element of the project. The contingency allowances can be clearly identified as individual line items associated with each major element. This allows the PM and reviewers of future updates to track where and how project costs are changing and how they may impact the completion of the project. This may be achieved by providing information on reoccurring patterns and reasons behind cost escalation. Contingencies are normally disclosed as a dollar value or a percent of the major element cost.

7.5.2 Budget Allowances

Budget Allowances are generally similar to contingencies in that their purpose is to reserve funds for circumstances that are ill defined and thus more prevalent in the earlier design phases of a project when the uncertainties are most evident. However, unlike contingencies, allowances are usually identifiable single items/issues and are placed in budgets as individual line items. Certain Allowances may also be carried by the general contractor upon attaining the approval of the PM and provided they do not exceed their budgets or estimates to cover such items that they believe may arise (based on prior experience). Furthermore, as an optional allowance, the general contractor may also carry a contingency to cover scope or program changes that the user group/owner may request during construction. This allowance is an agreed upon amount between the contractor and the PM and must be approved by the PM prior to being committed by the contractor. The PM can carry Allowances in any of the cost and time summary (CATS) categories for questionable or additive alternate construction and non-construction items as necessary.

In discussing allowances, Sabo & Zahn of Sabo & Zahn, Attorneys at Law state that "An allowance is a line item in a construction budget that serves as a placeholder during the bidding and initial construction contract phase. It is used when a particular item to be used in the construction has not been selected or completely specified. For example, if the carpeting has not been selected at the time of bidding, rather than delay the bidding, an allowance for the carpeting can be used. Normally in this situation the total amount of carpeting to be used is known. If, for instance, the house will have 300 yards of carpeting, with a $50 per square yard allowance, the contractor will include a carpet allowance of $15,000 in the bid. This allowance will cover the cost of the materials as well as the cost of installation. The contractor will also have its overhead and profit included in the proper category. At some later date, the owner will pick the actual carpeting. If the actual cost for that carpeting is $60 per yard, then the contractor will be entitled to a change order for the increased cost—in this example, $3,000. On the other hand, if the actual cost of the carpeting turns out to be only $40, then the change order will reflect a deduct of $3,000.

The key to properly administering allowances is to account for them by proper change orders. At the time that the actual material is selected and approved by the owner, a change order must be issued and signed. This change order must indicate that the allowance for that item is being deleted, with a credit to the contract for the allowance amount, with a corresponding increase in the construction cost in the amount of the actual cost. In our example with a $60 carpet cost, the allowance of $15,000 would be credited to the owner and the $18,000 actual cost would be added to the contract, for a net increase of $3,000." However, the best practice is not to provide any allowances if possible, but instead ensure that everything is clearly identified and specified prior to solicitation of bids.

In this respect, the allowance section of AIA Document A201-1997 states that, "The contractor shall include in the contract sum all allowances stated in the contract documents. Items covered by allowances shall be supplied for such amounts and by such persons or entities as the owner may direct, but the contractor shall not be required to employ persons or entities to whom the contractor has reasonable objection."

The AIA Document A201-1997 also states that, "Unless otherwise provided in the contract documents:

1. Allowances shall cover the cost to the contractor of materials and equipment delivered at the site and all required taxes, less applicable trade discounts.
2. The contractor's costs for unloading and handling at the site, labor, installation costs, overhead, profit, and other expenses contemplated for stated allowance amounts shall be included in the contract sum but not in the allowances.
3. Whenever costs are more than or less than allowances, the contract sum shall be adjusted accordingly by change order. The amount of the change order shall reflect (1) the difference between actual costs and the allowances under Clause 3.8.2.1 and (2) changes in the contractor's costs under Clause 3.8.2.2."

The AIA document further stipulates that materials and equipment under an allowance shall be selected by the owner within sufficient time to avoid causing delay to the work. This means that if the additional time caused by the delay is sufficiently significant, the contractor may be entitled to additional compensation due to that delay.

7.6 PROJECT COST MANAGEMENT

Optimum results are most often achieved when the project's activities are integrated and costs are managed collaboratively. The integrated project team should always be engaged at the earliest phases of design, using target costing, value management, and risk management. Owners/developers are sometimes tempted to put in place a guaranteed maximum price on the project before the design stage is complete, but this should be resisted to ensure quality and functionality for the building owner or stakeholder. If the project owner comes under pressure to seek a fixed price at an earlier stage of the process, it would be prudent to agree on an incentive scheme for the sharing of benefits. It goes without saying that the owner should have a clear understanding of actual construction costs, both hard and soft costs. Likewise, the owner and project manager must be able to identify and differentiate between underlying costs and risk allowances in addition to being able to distinguish between profit and overhead margins.

7.6.1 Successful Cost Management Procedures

The project manager (PM) is generally responsible for managing the operation and overall cost of the project and who in turn reports regularly to the owner (or lender, depending on the contract documents). One of the project manager's responsibilities is to maintain ongoing reviews of designs as they develop and provide advice on costs to the integrated project team as well as receive feedback from the project team. This continuous cost oversight is of particular benefit in assessing individual decisions and is especially useful on large and complex schemes. It may also prove useful to schedule in periodic formal assessments of the whole scheme, as budgetary estimates, at each phase of the project (Figure 7.7). The roles and responsibilities and limits of authority for the project manager should be clearly agreed upon at the start of the project so that everyone knows exactly what the PM is empowered to do in managing project costs and cost overruns.

According to the UK Office of Government Commerce the main ingredients for successful project cost management are:

- To manage the base estimate and risk allowance
- To operate change control procedures
- To produce cost reports, estimates, and forecasts (The project manager is directly responsible for understanding and reporting the cost

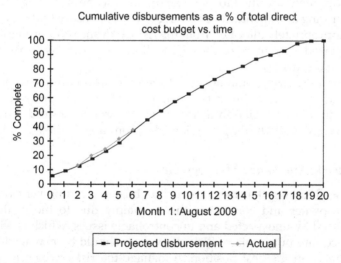

FIGURE 7.7 An example of a cumulative disbursements schedule as a percentage of total direct cost budget versus time and based on the current project budget and a 20.5-month construction period. The CM's projection of disbursements is shown plotted in the graph, and generally follows a realistic "S" curve. The project's cumulative net direct cost disbursements to date indicate being roughly in line with initial budget estimates.

consequences of any decisions and for initiating corrective actions if necessary)

- To maintain an up-to-date estimated outturn cost and cash flow
- To manage expenditure of the risk allowance
- To initiate action to avoid overspending
- To issue a monthly financial status report.

Additionally, the cost management objectives during the construction phase include delivery of the project at the appropriate capital cost using the value criteria established at the project's inception and ensuring that throughout the project, comprehensive and accurate accounts are kept of all transactions, payments, and changes.

The UK Office of Government Commerce also believes that the chief areas that cost management teams should consider during the design and execution of a construction project are:

- Identifying elements and components to be included in the project and constricting expenditures accordingly.
- Defining the project program from inception to completion.
- Making sure that designs meet the scope and budget of the project and that appropriate quality is delivered and conforms to the brief.
- Checking that orders are properly authorized.
- Certifying that the contracts provide full and proper control and that all incurred costs are as authorized. All materials are to be appropriately specified to meet the project's scope and design criteria.
- Monitoring all expenditures relating to risks to ensure that they are appropriately allocated from the risk allowance and properly authorized. Also monitor use of risk allowance to assess impact on overall outturn cost.
- Maintaining strict planning and control of both commitments and expenditures within budgets to help prevent any unexpected cost over/under runs. All transactions are to be properly recorded and authorized and where appropriate, decisions are justified.

7.6.2 Risk Allowance Management

It is well known that the construction business can be very risky for both the owner and contractor; this is mainly due to the plethora of risks caused by unexpected and uncontrollable issues, which is why risk allowances are put in place. Risk allowances should be managed by the party that is in the best position to manage the risks, which is usually the project owner or someone representing the owner, with the advice and support of the project manager. What risk allowance management essentially consists of is a procedure to move costs out of the risk allowance column into the base estimate for the project work either as risks materialize or actions are taken to manage the risks. Formal procedures are required to be put in place for controlling quality, cost overruns, project delays, and change orders. Risk allowances should not be disbursed unless the identified risks to which they relate actually occur. When risks occur that have not previously been identified, they should be treated as change orders to the project. Likewise, risks that materialize but have insufficient risk allowance allocated for them should also be treated as change orders (variation orders).

Figure 7.8 is a graph that is designed to assist the PM and owner in monitoring the key elements of the project as well as present an overall picture of how the project is progressing. It should be noted that the graph is to be supplemented with notes for each key element. For example, "Site Work: The site has been cleared. The north waterline connection, and two (2) north and two (2) south sanitary connections have been installed and stubbed through the foundation wall. Electrical connections to the temporary switchgear have been made. A temporary concrete sidewalk has been placed along Washington Avenue."

It is always preferable to define potential risks by allocating specific costs to them, as opposed to just inflating the total cost to compensate for inadequate early planning. A risk allocated cost contingency is normally needed and included in the total project cost estimate to help mitigate potentially significant risks. Risk management and contingency funding is particularly useful for mitigating those risks that cause cost escalations and project cost overruns during the course of a project's execution.

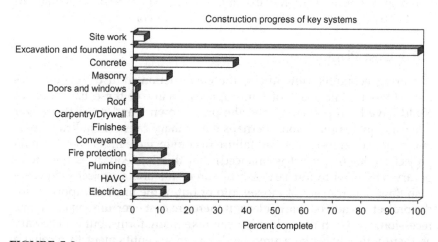

FIGURE 7.8 A graph designed to assist the project manager and owner in monitoring the key elements of the project as well as present an overall picture of how the project is progressing.

In the process of preparing the initial project budget, it is strongly recommended to perform a risk assessment on the entire project in order to identify and quantify the potential risk areas and types. This will help mitigate the uncertainties and help create a conservative cost expectation. Risk assessments should also be performed on a regular basis throughout the project's execution and to update contingency amounts. Examples of risk assessment areas that may cause concern include failure to perform, analysis of heterogeneous or irregular site conditions, utility impacts, hazardous materials, environmental impacts, third-party concerns, etc.

When quantifying risk as a contingency amount, expectation of occurrence, severity, and anticipated dollar value are variables that may be considered and utilized. After all known risk mitigation, the budget's cost estimate contingency allowance levels should reflect the actual amount of remaining risk associated with the project's major cost elements. An overall management contingency can also be included to cover unknown, unanticipated risks.

Risks and risk allowances should be reviewed and evaluated on a regular basis, particularly when formal estimates are prepared, from the design, construction phases and through substantial completion and occupancy. The introduction of changes after the briefing and outline design stages are complete should be avoided as much as possible. Change orders can be minimized by ensuring that from the start of the project the contract documents are as clear, complete, and comprehensive as possible and have been approved by the stakeholders. This may require early meetings with planning authorities to discuss their requirements and to ensure that the designs are adequately developed and coordinated before construction begins. For renovation of existing buildings the type of risks may differ slightly and may require site investigations or condition surveys.

7.6.3 Cost Planning

During economic downturns, the construction industry experiences more than its fair share of bankruptcies. Many of these bankruptcies could have been prevented had the project owner and project manager taken adequate precautions. Perhaps the primary cause of bankruptcies is inadequate cash resources and failure to convince creditors and the main project lender (if project is financed) that this inadequacy is only temporary. The need to forecast cash requirements and a project's expected cash flow (i.e., transfer of money into or out of the firm) is important for the project to succeed, particularly if there are cost overruns, an economic recession, etc. Cash flow planning can take many forms, but is necessary as there will always be a time lag between an entitlement to receive a payment for work executed and actually receiving it.

In an elemental designed cost plan the estimate is broken down into a number of components that can then be compared with later estimates, or with actual costs as the project progresses. In applying this approach each element or item is treated as a distinct cost center, although money can be still be transferred between elements as long as a reasonable balance between elements is maintained and the overall target budget is maintained. It must be emphasized that green buildings require intensive planning to ensure optimal results, but any additional effort is usually worth it if you consider that operating costs will be substantially reduced over the life of the facility.

Often the initial cost plan is based on unsubstantiated estimates, which nevertheless provide a fair basis for determining the validity of future assessments. The project manager is able to control costs by instituting ongoing reviews of estimates for each cost center against its target budget. Figure 7.9 is a practical of a detailed cost estimate with comments included. It is interesting to note some of the buyouts and line items which appear low or acceptable. As the project design continues to develop and is priced, variances in cost from the initial cost plan are noted and recorded. A decision must then be made as to whether that item can be authorized with a corresponding increase in cost, which would then require an equal reduction elsewhere, or whether the element in question needs to be redesigned in order to keep within the proposed budget. Furthermore, if a lender is involved, the consultant or PM needs to check the lender's policy to determine if funding of contract deposits is permitted on such subcontracts as structural steel, precast panels, curtain wall systems, etc. In most cases, the Lender's policy is to refrain from funding deposits.

The payment process is normally managed in the same manner as the design/construction process. All payments should be made as per the contract agreement and on time. Payments for change orders, provisional sums, etc should be discharged after formal approval is given and as the work is carried out.

With the majority of projects it is the owner's designated representative who has overall responsibility for general management of the project, including the estimated cost, and therefore must be satisfied that suitable methods are in place for controlling the project's cost. Where the design process requires the allocation of a significant amount of money, such costs should be appropriately assessed against the budget amount and properly authorized. To facilitate matters, the owner's representative will frequently delegate an appropriate degree of financial authority for design development decisions to the integrated project team. For particularly large or complex projects the owner may decide to change the delegated levels for each cost center.

However, once the construction process begins, any instructions issued to the integrated project team requesting a change through a formal change order procedure can have a pronounced impact on the project's cost and possibly other impacts such as time delay. This is why the project team should have specific procedures and protocols in place for issuance

Budget Review

XYZ Office Building - Leesburg, Virginia
ABC Project No. A708
November 3rd 2009

Site Area (Acres):   26310   Building Area (SFG):   358,000

Item of Work	Borrower's Budget	$/SFG	$/ACRE	% Total	Comments
Mass Grading	0	0.00	0	0.00%	By previous land owner
Fine Grade and Spoil Removal	240,000	0.67	9,122	0.95%	OK-Bought Out
Site Concrete	230,000	0.64	8,742	0.91%	OK-Bought Out
Asphalt Paving, Striping and Signage	460,000	1.28	17,483	1.83%	Seems Low
Landscaping Allowance	75,000	0.21	2,851	0.30%	Seems Low-Allowance
Site Irrigation Allowance	25,000	0.07	950	0.10%	Allowance
Parking Equipment	60,000	0.17	2,280	0.24%	Seems High
Site Plumbing	334,500	0.93	12,713	1.33%	OK-Bought Out
Site Electrical	93,250	0.26	3,544	0.37%	Low-Buy Out Loss
Fencing with Gate for Secure Parking	15,000	0.04	570	0.06%	Acceptable
Total Site Work	**$1,532,750**	**$4.28**	**58,255**	**6.09%**	**Acceptable**
Building Excavation, Stone for Fill	263,000	0.73	N/A	1.04%	OK-Bought Out
Building Concrete	1,841,000	5.14	N/A	7.31%	OK-Bought Out
Precast Concrete	800,000	2.23	N/A	3.18%	Low/OK-Bought Out
Caulking of Precast and Windows	70,000	0.20	N/A	0.28%	Acceptable
Masonry	20,000	0.06	N/A	0.08%	Acceptable
Structural Steel and Metal Decking	2,585,000	7.22	N/A	10.26%	Low/OK-Bought Out
Utility Court Steel Doors	10,000	0.03	N/A	0.04%	Acceptable
Steel Stairs and Misc. Metals	300,000	0.84	N/A	1.19%	OK-Bought Out
Steel Precast Support at Tall Entries	50,000	0.14	N/A	0.20%	Unknown Scope
Spray on Fireproofing	362,000	1.01	N/A	1.44%	Acceptable
Foundation and Basement Waterproofing	35,000	0.10	N/A	0.14%	Acceptable
Misc. and Sanitary Caulking	9,150	0.03	N/A	0.04%	Acceptable
Metal Penthouse Siding and Louvers	255,100	0.71	N/A	1.01%	Seems High
60 Mil EPDM Roofing	249,432	0.70	N/A	0.99%	Acceptable
Windows, Entrances, Glass and Glazing	864,000	2.41	N/A	3.43%	OK-Bought Out
Mirrors	5,000	0.01	N/A	0.02%	OK-Bought Out
Overhead Dock Doors	10,000	0.03	N/A	0.04%	Acceptable
Drywall and Metal Studs	1,433,583	4.00	N/A	5.69%	Acceptable
Acoustical Ceilings	469,063	1.31	N/A	1.86%	Acceptable
Carpeting	470,000	1.31	N/A	1.87%	Seems Low
Lobby Floors Allowance	42,000	0.12	N/A	0.17%	Allowance
Ceramic Floor and Wall tile	75,000	0.21	N/A	0.30%	Seems low
VCT	90,000	0.25	N/A	0.36%	Acceptable
High Pressure P-Lam Flooring	25,000	0.07	N/A	0.10%	Acceptable
Painting and Vinyl Wall Covering	360,000	1.01	N/A	1.43%	Acceptable
Toilet Partitions	30,000	0.08	N/A	0.12%	Seems Low
Flag Pole	3,500	0.01	N/A	0.01%	Acceptable
Install Interior Signage	10,000	0.03	N/A	0.04%	Acceptable
Fire Extinguishers and Cabinets Allowance	20,000	0.06	N/A	0.08%	Allowance
Dock Equipment	15,000	0.04	N/A	0.06%	Acceptable
Access Flooring	145,000	0.41	N/A	0.58%	Acceptable
Window Blinds and Draperies	64,000	0.18	N/A	0.25%	Unknown Scope
Operable Walls	70,000	0.20	N/A	0.28%	Acceptable
Carpentry	124,500	0.35	N/A	0.49%	Acceptable
P-Lam Vanity Tops	24,000	0.07	N/A	0.10%	Acceptable
Hollow Metal Doors & Frames Allowance	79,000	0.22	N/A	0.31%	Allowance
Wood Doors Allowance	86,000	0.24	N/A	0.34%	Allowance
Hardware Allowance	149,500	0.42	N/A	0.59%	Allowance
Lobby Features Allowance	20,000	0.06	N/A	0.08%	Allowance
Toilet Accessories	35,957	0.10	N/A	0.14%	Acceptable
Building Directory and Floor Directories	7,500	0.02	N/A	0.03%	Acceptable
Misc. Counter Tops	500	0.00	N/A	0.00%	Unknown Scope
Closet Shelves and Rods	1,500	0.00	N/A	0.01%	Acceptable
Chair Rail	500	0.00	N/A	0.00%	Acceptable
T.V. Brackets	500	0.00	N/A	0.00%	Acceptable
Sound Absorbing Wall Panels	35,000	0.10	N/A	0.14%	Acceptable
NovaWall Wall System	21,000	0.06	N/A	0.08%	Unknown Scope
Elevators	795,000	2.22	N/A	3.16%	Acceptable
Plumbing	865,000	2.42	N/A	3.43%	OK-Bought Out

FIGURE 7.9 A project budget review for an office building with a total building area of 358,000 square feet and a total cost budget of $25,286,000. It is interesting to note the various line item comments in the end column.

Fire Protection (no fire pump)	392,000	1.09	N/A	1.56%	OK-Bought Out
HVAC	3,310,000	9.25	N/A	13.14%	OK-Bought Out
Electrical	3,555,750	9.93	N/A	14.12%	OK-Bought Out
Total Building	**$20,554,035**	**$57.41**	**N/A**	**81.61%**	**Compares Low**
Total Trade Costs Site and Building	**$22,086,785**	**$61.69**	**N/A**	**87.69%**	**Compares Low**
General Conditions	1,618,000	4.52	N/A	6.42%	Acceptable
General Liability Insurance	150,215	0.42	N/A	0.60%	Acceptable
Subcontractor Bond Costs	221,000	0.62	N/A	0.88%	Acceptable
GC's Fee	710,000	1.98	N/A	2.82%	Acceptable
GC's Construction Contingency	400,000	1.12	N/A	1.59%	Acceptable
Total General	**$3,099,215**	**8.66**	**N/A**	**12.31%**	**Acceptable**
Total Direct Cost Budget	**$25,186,000**	**$70.35**	**N/A**	**100.00%**	**Needs Borrower Contingency**

FIGURE 7.9 (*Continued*)

of instructions and information to ensure that any issued instructions are within the assigned authority. Also, before being issued, the cost of proposed change orders should be properly estimated and their impact fully evaluated. Any issued change order instruction should be fully sustainable in terms of value for money and overall positive impact on the project. Furthermore, adequate and continuous monitoring of the total costs of all issued instructions is necessary, and where costs are determined to be outside the delegated authority, specific approval is required.

Payments are discussed in Chapter 4. The normal payment procedure requires the client, as the contracting party, to make all payments to the integrated project team including the interim and final payments as per the contract. These payments usually take the form of payments made at various stages of the work in progress or upon application for payment by the general contractor (usually at monthly intervals) upon inspecting and assessing the value of work in place. The client's lender (e.g., bank) should be updated constantly on the project's progress—whether satisfactory or otherwise—by means of regular reports, memos, and cash flow and budget forecasts. It should be noted that some lenders have a policy of not permitting funding for mobilization, only for actual work-in-place, while others do permit funding for certain trades. The terms of the contract should be verified prior to commencement of the construction process.

The contract may also include clauses that allow the project manager in certain circumstances to claim additional payments as specified in the contract's general conditions. Any justification for additional payment claims may be the result of or caused by occurring risks that are essentially considered to be client risks under the contract. Examples include requesting a change in scope or additional work, or by the client's failure to comply with its contract obligations which may be caused by a disruption to the project's scheduled program because of modifications, delivery delays, or other reason.

8

Green Design and Construction Economics

304

Doi: 10.1016/B978-1-85617-676-7.00008-7

8.1 GENERAL

Green building systems are changing so expeditiously that each new project can involve employing materials and systems that designers and officials have not previously worked with. However, the current financial crisis and global economic recession have put an enormous strain on the nation's construction industry. Peter Morris, principal of the global construction consultancy Davis Langdon, believes that the dramatic reduction in construction activity is encouraging increased competition among bidders and lower escalation pressure on projects to the extent that in many projects, cost trends have become negative, leading to moderate construction price deflation. But one of the biggest causes of concern according to Morris is the issue of contractor financing and working capital. Many contractors are finding it increasingly difficult to maintain adequate cash flow for their operations, and none have the resources to manage significant expansion of working capital. This has caused considerable concern and obliged bidders to be more cautious and judicious in project selection, with sound cash flows being a major consideration in project selection.

There has emerged an increasingly broad awareness of the significant benefits that green buildings have to offer. For example, the U.S. military, including the U.S. Air Force and Navy, now require that their new buildings be LEED™ green buildings. This may be in part because they recognize the linkage between wasteful energy consumption and the exposure of U.S. military forces to military confrontation related to oil resources. As Boston Mayor Thomas M. Menino put it, "High performance green building is good for your wallet. It is good for the environment. And it is good for people." As costs fall, the appealing financial performance of existing green buildings becomes clearer. In a 2006 survey of developers by McGraw-Hill Construction, respondents reported they expected to see occupancy rates for green buildings 3.5 percent higher than market norms, and rent levels to increase by 3 percent. Operating costs are estimated to be 8 to 9 percent lower as well. These numbers are getting the attention of developers and investors, which is driving the growth of today's eco-construction.

The rapid emergence of eco-construction reflects the building industry's growing confidence that the extra costs of building green are a good investment. Although the upfront costs of building green may be higher than using conventional materials, that premium is shrinking. Precise benefits such as reduced energy bills and reduced potable water consumption can easily be computed, while other benefits such as green design's impact on occupant health or security are usually much more difficult to quantify. Incisive Media's "2008 Green Survey: Existing Buildings" found that nearly 70 percent of commercial building projects in the United States have already incorporated some kind of energy monitoring system. The survey also found that energy conservation is the most widely implemented green program in commercial buildings; this is followed by recycling and water conservation. Moreover, approximately 65 percent of building owners who have implemented green building features claim their investments have already resulted in a positive return on their investment. The return on investment is expected to improve even further as the market for green materials and design expertise grows and matures. And according to Taryn Holowka, director of marketing and communications for USGBC, "The supply of materials and services is going up and the price is coming down."

Turner Construction Company's 2008 Green Building Barometer notes that approximately 84 percent of respondents maintain that their green buildings have resulted in lower energy costs, and 68 percent recorded lower overall operating costs. Likewise, "75 percent of executives said that recent developments in the credit markets would not make their companies less likely to construct green buildings." In fact the survey contends that 83 percent would be "extremely" or "very" likely to seek LEED certification for buildings they are planning to build within the next three years. In the same survey executives reported that green buildings generally have better financial performance than non-green buildings, especially in the following areas:

- Higher building values (72 percent)
- Higher asking rents (65 percent)
- Greater return on investment (52 percent)
- Higher occupancy rates (49 percent).

Another similar study contends that 60 percent of commercial building owners offer education programs to assist tenants in carrying out green programs in their space, reflecting a growing understanding of the significance of environmental awareness among employees and customers in addition to the use of green materials and systems application.

Davis Langdon conducted a comparative study in 2006 in which the construction costs of 221 buildings were analyzed, and it was found that 83 buildings were constructed with the intent of achieving LEED™ certification and 138 lacked any sustainable design intentions. The study found that a majority of the buildings analyzed were able to achieve LEED™ certification without increased funding. In another investigation conducted by Davis Langdon of a wide and diverse range of studies by other organizations found that the average construction cost premium required to achieve a moderate level of green features, equivalent to a Silver LEED™ certification, was roughly between 1 percent and 2 percent. However, what is particularly interesting is that it was also found that half or more of the green projects in these studies often revealed a zero increase in construction costs.

Today, even with the increased awareness of the benefits for sustainable design, property owners and developers are not always quick off the mark to embrace green building practices. This appears to be the prevailing sentiment of CB Richard Ellis's "Green Downtown Office Markets: A Future Reality," a report depicting the general progress of the green building movement. The report scrutinizes the obstacles preventing a broad-based acceptance of sustainable design in office construction. Perhaps the main obstacle to embracing design sustainability is the perception of initial outlay compared to long-term benefits; even though an increasing number of studies similar to the one conducted by Davis Langdon in 2006 clearly conclude that there is no significant difference in average costs for green buildings as compared to conventionally constructed buildings. Another hurdle that requires addressing is the lack of sufficient data on development, construction costs, and time needed to recoup costs.

The amount of research undertaken relating to green building in the United States is dismal, constituting an estimated $193 million per year, or roughly 0.2 percent of federally funded research. This approximates only 0.02 percent of the estimated $1 trillion value of annual U.S. building construction, despite the fact that the building construction industry represents approximately 9 percent of the U.S. GDP. It is unfortunate that the construction industry can currently manage to reinvest only 0.6 percent of sales back into research. This is markedly less than the average for other U.S. industries and private sector construction research investments in other industrialized countries around the world.

Some green organizations strongly suggest that unless we move decisively toward increasing and improving green building practices, we are likely to soon be confronted with a dramatic backlash in adverse impact of the built environment on human and environmental health. Building operations today are estimated to account for 38 percent of U.S. carbon dioxide emissions, 71 percent of electricity use, and 40 percent of total energy use. If the energy required in the manufacture of building materials and constructing buildings is included, this number then goes up to an estimated 48 percent. Buildings also consume roughly 12 percent of the country's water, in addition to rapidly increasing amounts of land. Moreover, construction and remodeling of buildings accounts for 3 billion tons, or 40 percent, of raw material used globally each year, which in turn has a negative impact on human health; in fact, up to 30 percent of new and remodeled buildings may experience acute indoor air quality problems such as Sick Building Syndrome (SBS).

Like with most projects, determining building strategies early in the design process and sticking with those decisions can result in the most efficient cost models for building. Implementing a goal-setting session at the beginning of each project to determine appropriate strategies and levels of cost and time investments can result in lower sustainable design construction costs. As mentioned above, many of these strategies come with little or no additional costs, which helps to make green design an easy sell.

An analysis of green construction costs can be achieved through the application of several methods. For example, it is possible to use the LEED® Rating System or the Green Globes rating system as benchmarks for success. Higher levels of certification may carry increased costs, but empirical market data suggests that the "certified" and "silver" levels with the LEED system and one or two "globes" with the Green Globes system carry little or no premium over traditional building costs for most building types. Specialty project types, such as healthcare or research, often have program criteria and specific needs that are at odds with the principles of sustainable design, primarily in environmental constraints and energy usage.

8.2 THE ECONOMICS OF GREEN DESIGN—COSTS AND BENEFITS

Most people would obviously support policies that protect the environment, but what developers and investors really want to know is, "at what cost," and how will building green benefit the financial viability of their investment? In this respect, Peter Morris opines that, "Clearly there can be no single, across the-board answer to the question, 'what does green cost?' On the other hand, any astute design or construction professional recognizes that it is not difficult to estimate the costs to go green for a specific project. Furthermore, when green building concepts and features are incorporated early in the design process, it greatly increases the ability to construct a certified green building at a cost comparable to a code compliant one. This means that it is possible today to construct green buildings or buildings that meet the U.S. Green Building Council's (USGBC) Leadership in Energy and Environmental Design (LEED®) third party certification process with minimal increase in initial costs." Figure 8.1 clearly reflects this point.

Also with respect to LEED certification, there are studies that suggest that conventionally constructed buildings can often qualify for 12 or more LEED® points by virtue of current building standards and inherent design qualities. In many cases, between 15 and 20 additional points can be achieved with little to no additional costs, qualifying most buildings for the minimum rating classification. Some studies concluded that the cost of achieving Silver certification varies between 2 and 6 percent above traditional construction. To achieve the higher levels of certification, Gold and Platinum, can add significantly to the cost of the project, primarily

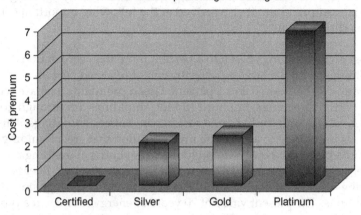

FIGURE 8.1 A diagram showing one study's estimated cost of building green for the various LEED™ Rating Systems. (Source: USGBC.)

due to the costs of applying more efficient technologies for water conservation and energy performance.

Davis Langdon suggests that to be successful in building green and to keep the costs of sustainable design under control, three critical factors must be understood and implemented, and they are:

- Clear goals are critical for managing the cost. It is insufficient to simply state, "We want our project to be green"; the values should be determined and articulated as early in the design process as possible.
- Once the sustainability goals have been defined, it is essential to integrate them into the design and to integrate the design team so that the building elements can work together to achieve those goals. Buildings can no longer be broken down and designed as an assemblage of isolated components. This is the major difference between traditional building techniques and the new sustainable design process.
- Integrating the construction team into the project team is critical. Many sustainable design features can be defeated or diminished by poor construction practices. Such problems can be eliminated by engaging the construction team, including subcontractors and site operatives, in the design and procurement process.

Another important study reflecting the largest international research of its kind, "Greening Buildings and Communities: Costs and Benefits," is based on extensive financial and technical analysis of 150 green buildings across the U.S. and in 10 other countries and provides the most detailed findings to date on the costs and financial benefits of building green. This study found that benefits of building green consistently outweigh any potential cost premium. It also arrived at the following important conclusions as shown below:

1. Most green buildings cost 0 to 4 percent more than conventional buildings, with the largest concentration of reported "green premiums" between 0 and 1 percent. Green premiums increase with the level of greenness, but most LEED™ buildings, up through gold level, can be built for the same cost as conventional buildings. This stands in contrast to a common misperception that green buildings are much more expensive to build than conventional buildings.
2. Energy savings alone make green building cost effective. Energy savings alone outweigh the initial cost premium in most green buildings. The present value of 20 years of energy savings in a typical green office ranges from $7 per square foot (certified) to $14 per square foot (platinum), more than the average additional cost of $3 to $8 per square foot for building green.
3. Green building design goals are associated with improved health and enhanced student and worker performance. Health and productivity benefits remain a major motivating factor for green building owners, but are difficult to quantify. Occupant surveys generally demonstrate greater comfort and productivity in green buildings.
4. Green buildings create jobs by shifting spending from fossil fuel-based energy to domestic energy efficiency, construction, renewable energy, and other green jobs. A typical green office creates roughly one-third of a permanent job per year, equal to $1 per square foot of value in increased employment, compared to a similar non-green building.
5. Green buildings are seeing increased market value (higher sales/rental rates, increased occupancy, and lower turnover) compared to comparable conventional buildings. CoStar, for example, reports an average increased sales price from building green of more than $20 per square foot, providing a strong incentive to build green even for speculative builders.
6. Roughly 50 percent of green buildings in the study's data set see the initial "green premium" paid back by energy and water savings in five years or less. Significant health and productivity benefits mean that over 90 percent of green buildings pay back an initial investment in five years or less.
7. Green community design (e.g., LEED™-ND) provides a distinct set of benefits to owners, residents, and municipalities, including reduced infrastructure costs, transportation and health savings, and increased property value. Green communities and neighborhoods have a greater diversity of uses, housing types, job types, and transportation options

    and appear to better retain value in the market downturn than conventional sprawl.

8. Annual gas savings in walkable communities can be as much as $1000 per household. Annual health savings (from increased physical activity) can be more than $200 per household. CO_2 emissions can be reduced by 10 to 25 percent.

9. Upfront infrastructure development costs in conservation developments can be reduced by 25 percent, approximately $10,000 per home.

10. Religious and faith groups build green for ethical and moral reasons. Financial benefits are not the main motivating factor for many places of worship, religious educational institutions, and faith-based nonprofits. A survey of faith groups building green found that financial cost effectiveness of green building makes it a practical way to enact the ethical/moral imperative to care for the Earth and communities. Building green has also been found to energize and galvanize faith communities.

Even when green building up front costs are in excess of what was originally estimated due primarily to inefficient planning and execution, these costs can be quickly recouped through lower operating costs over the life of the building.

8.2.1 The Economic Benefits of Green Buildings

Maximum cost savings can only be achieved when the green design strategies are incorporated at the project's conceptual design phase in collaboration with an integrated team of professionals. Using an integrated systems approach ensures that the building is designed in a holistic manner as one system rather than a number of stand-alone systems as is normal with conventional methods. The challenge here is that not all green building benefits are easy to quantify; for example, how do you measure improving occupant health, comfort, productivity, or pollution reduction? This is why they are excluded from being adequately considered in cost analysis. It would appear to be prudent therefore to consider setting aside a small portion of the building budget (e.g., as a contingency) to cover differential costs associated with less tangible green building benefits or to cover the cost of researching and analyzing green building options. Even when experiencing difficult times, many green building measures can be incorporated into a project with minimal or zero increased up front costs. Yet this would be capable of yielding substantial savings and other benefits (Figure 8.2).

One of the first questions often raised by owners and developers regarding sustainable design is what does "green" cost?; typical translation: does

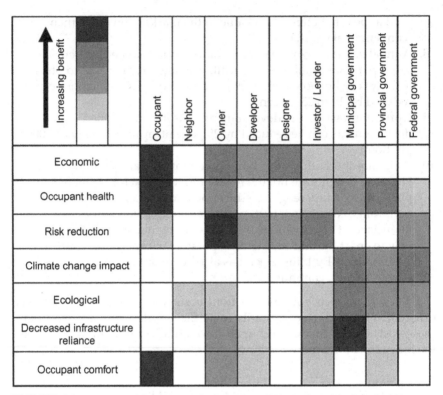

FIGURE 8.2   A matrix illustrating various green building stakeholder benefits. (Source: Report—A Business Case for Green Buildings in Canada.)

it cost more? The answer to this question has thus far been largely elusive due to the lack of hard data that was available until recently. Over recent years there has been considerable research undertaken, and we now have a substantial databank on building costs that allows us to compare the costs of green buildings and non-sustainable buildings with comparable characteristics.

When assessing sustainable design and how it relates to construction costs, it is imperative to analyze the costs and benefits using a holistic approach. This basically means including evaluation of operations and maintenance costs, user productivity and health, and design and documentation fees, among other financial measurements. This is largely because empirical experience continues to demonstrate that it is the construction cost implications that have the greatest impact in determining decisions about sustainable design. Helping teams to really understand the actual construction costs of real green projects and providing a methodology that allows teams to viably manage these construction costs

Category	20-years Net Present Value
Energy Savings	$5.80
Emissions Savings	$1.20
Water Savings	$1.50
Operations and Maintenance Savings	$8.50
Productivity and Health Value	$36.90 to $55.30
Subtotal	**$52.90 to $71.30**
Average Extra Cost of Building Green	(–3.00 to –$5.00)
Total 20-year Net Benefit	**$50 to $65**

FIGURE 8.3   Financial benefits of green buildings—summary of findings (per ft^2). The financial benefits of going green are related mainly to productivity. (Source: Capital E. Analysis.)

can go a long way to facilitate a team's ability to get past the question of whether or not green is the answer. Green construction is helped by the fact that the cost of green design has dropped significantly in the last few years as the number of green buildings has increased. The trend of declining costs associated with increased experience in green building construction has manifested itself in a number of states throughout the country.

From such an analysis it can be concluded that many projects are able to achieve sustainable design within the initial budget, or with minimal supplemental funding. This suggests that developers continue to find ways to incorporate project goals and values, regardless of budget, by making choices. However, every building project is unique and should be considered as such as there is no one-size-fits-all answer, and benchmarking with other comparable projects can be valuable and informative, but not predictive. Any estimate of cost relating to sustainable design for a specific building must be made with reference to that building, its objectives and particular circumstances.

Figure 8.3 provides a summary of financial benefits from going green, as depicted in a recent study by Greg Kats of Capital E Analysis. The report concludes that financial benefits for building green are estimated to be between $50 and $70 per square foot in a LEED™ certified building; this represents more than 10 times the additional cost associated with building green. These financial benefits come in the form of lower energy costs, waste and water costs, lower environmental and emissions costs, and lower operational and maintenance costs, lower absenteeism, increased productivity and health, greater retail sales, and easier re-configuration of space resulting in less downtime and lower costs. Cost estimates, which are based on a sample of 33 office and school buildings, suggested only

.6 percent greater costs for LEED™ certification, 1.9 percent for silver, 2.2 percent for gold, and 6.8 percent for platinum certification. Although these estimates are direct costs, they nevertheless closely reflect those provided by the USGBC. What is perhaps surprising is that not many studies have been undertaken with regard to rating systems other than LEED, such as Green Globes.

The principal motivators that can impact the long-term value of green building appear to be:

1. Increasing energy costs (75 percent).
2. Government regulations/tax incentives (40 percent).
3. Global influences (26 percent).

The lack of adequate clear data pertaining to development, construction costs, and time required to recoup costs is the main challenge/obstacle to the industry's acceptance of green construction, which is why education has become the most important tool in promoting green construction strategies. The main obstacles measured in the Kats study include:

- Too multi-disciplinary—41 percent
- Not convinced of increased return on investment (ROI)—37 percent
- Lack of understanding benefits—26 percent
- Lack of service providers—20 percent
- Too difficult—17 percent
- Greenwashing—16 percent
- Lack of shareholder support—10 percent.

The main objective of most building owners and developers is to build a project and then sell it, in the sense that they construct or revamp an office building, lease its space, with the hope or calculation of selling the asset within a 3 to 5 year time frame in order to repay debts and ensure a profit. The speed with which the process is completed impacts the amount of profit generated upon execution of the sale of the building. Some uninformed developers are under the false perception that green construction costs more, which triggers fear since they are already concerned about the cost of short-term debt and conventional building materials. Typical benefits for green building owners and tenants alike, according to Davis Langdon, include:

- Reduced risk of building obsolescence
- Ability to command higher lease rates
- Potential higher occupancy rates
- Higher future capital value
- Less need for refurbishment in the future
- Higher demand from institutional investors
- Lower operating costs

- Mandatory for government tenants
- Lower tenant turnover
- Enhanced occupant comfort and health
- Improved interior air quality (IAQ); increased employee productivity and satisfaction.

8.2.2 Cost Considerations of Green Design

With respect to green buildings, there are basically two categories of costs: direct capital costs and direct operating costs.

Direct Capital Costs are costs associated with the original design and construction of the building and generally include interest during construction (IDC). There is a general misperception by some building stakeholders that the capital costs of constructing green buildings are significantly higher than those of conventional buildings, whereas many others within the green building field believe that green buildings actually cost less or no more than conventional buildings. Empirical evidence shows that savings are achieved by the downsizing of systems through better design and the elimination of unnecessary systems which will offset any increased costs caused by implementing more advanced systems. Capital and operational costs are normally relatively easy to measure because the required data is readily available and quantifiable. Productivity effects on the other hand are difficult to quantify, yet are nevertheless important to consider due to their potential impact. There are other indirect and external effects that can be wide reaching, and which quantifying may prove difficult.

Direct operating costs include all applicable expenditures required to operate and maintain a building over its full life. Included are the total costs related to building operation, such as energy use, water use, insurance, maintenance, waste, property taxes, etc. over the entire building life. The primary costs are those associated with heating and cooling and maintenance activities such as painting, roof repairs and replacement. Included in this cost category are less obvious items such as churn (the costs of reconfiguring space and services to accommodate occupant moves). All costs relating to major renovations, cyclical renewal and residual value, or demolitions costs are excluded from this category.

The question of insurance is discussed in greater detail in Chapter 11. Insurance is essentially a direct operating cost and green buildings have many tangible benefits that reduce or mitigate a variety of risks, and which should be reflected in the insurance rates for the building. Likewise, the fact that green buildings generally provide a healthier environment for occupants should be reflected in health insurance premiums. Indeed, the general attributes of green buildings (e.g., the incorporation of natural light, off grid electricity and commissioning) should reduce a broad range of liabilities, and the general site locations also potentially reduce risks of property loss due to natural disasters. Furthermore, a fully integrated design of a building will typically reduce the risk of inappropriate systems or materials being employed, which could have a positive impact on other insurable risks. Insurance companies sometimes offer premium reductions for certain green features, such as commissioning or reduced reliance on fossil fuel-based heating systems. The list of premium reductions will undoubtedly increase with further education and awareness and as the broad range of benefits are more fully recognized and understood. In any case, it is always advisable to consult an insurance agent or attorney prior to taking out a policy.

Churn rate reflects the frequency with which building occupants are moved, either internally or externally, including occupants who move but remain within a company, and those who leave a company and are replaced. It has been found that because of increased occupant comfort and satisfaction, green buildings typically have lower churn rates than conventional buildings.

8.2.3 Increased Productivity

Green buildings offer many benefits, one of which is the positive effect on productivity. There are numerous studies that clearly illustrate that green buildings dramatically affect productivity. However, because these studies are often broad in nature and rarely focus on unique green building attributes, they need to be supplemented by other thorough, accurate, and statistically sound research to fully comprehend the effects of green buildings on occupant productivity, performance, and sales. It seems prudent, however, that any productivity gains attributable to a green building should be included in the life-cycle cost analysis, particularly for an owner-occupied building.

Key features of green buildings relating to increased productivity are systems relating to ventilation, temperature, and lighting; daylighting and views; natural and mechanical ventilation; pollution-free environments; and vegetation. It is not always clear why these features produce improved productivity, although studies show healthier employees typically means happier employees, which in turn creates increased worker satisfaction, improved morale, increased productivity, and reduced absenteeism.

A recent study undertaken by Lawrence Berkeley National Laboratory concluded that improvements to indoor environments such as commonly found in green buildings could reduce health care costs and work losses as follows:

- From communicable respiratory diseases by 9 to 20 percent
- From allergies and asthma by 18 to 25 percent
- From non-specific health and discomfort effects by 20 to 50 percent

Hannah Carmalt, a project analyst with Energy Market Innovations, notes that, "The most intuitive explanation is that productivity increases due to better occupant health and therefore decreased absenteeism. When workers are less stressed, less congested, or do not have headaches, they are more likely to perform better." High performance buildings have many potential benefits including increased market value, lower operating and maintenance costs, improved occupancy for commercial buildings, and increased employee satisfaction and productivity for owner-occupied buildings.

A William Fisk study concluded that green buildings add $20 to $160 billion in increased worker productivity annually. This is due to the fact that LEED^{TM}-certified buildings were found to yield significant productivity and health benefits, such as heightened employee productivity and satisfaction, fewer sick days, and fewer turnovers. Moreover, other independent studies have shown that better climate control and improved air quality can increase employee productivity by an average of 11 to 15 percent annually. However, it should be noted that in commercial and institutional buildings, payroll costs generally significantly overshadow all other costs, including those involved in a building's design, construction, and operation.

It is necessary to define the particular elements of green buildings that are directly related to productivity. Sound control for example, while recognized as increasing productivity, is often excluded from green related studies, mainly because it is not considered to be particularly a green building feature. Likewise, the presence of biological pollutants, such as molds, is also associated with decreased productivity, yet these are also excluded because typical green buildings do not automatically eliminate the presence of such pollutants even though their presence is reduced because of improved ventilation in green buildings.

Occupant control is one of the most significant elements of green buildings that affect productivity and thermal comfort; control is required over temperature, ventilation, and lighting. Green buildings usually try to incorporate this feature because it can noticeably decrease energy use by ensuring areas are not heated, cooled, or lit more than is necessary. These measures are decisive to maintaining energy efficiency and occupant satisfaction within a building.

Increased productivity has been associated with increased emotional well-being through various studies. A study conducted by the Heschong Mahone Group (HMG) found that higher test scores in daylight classrooms were achieved due to students being happier. HMG also found that when teachers were able to control the amount of daylighting in classrooms, students appeared to progress 19 to 20 percent faster than students in classrooms that lacked controllability. Similar studies performed in office settings clearly showed that there was a significant rise in productivity when there was individual control over temperature, ventilation, and lighting.

A view to the outdoors is another common feature of green buildings that is associated with productivity in office space. The 2003 HMG office study mentioned previously confirmed correlations between productivity and access to outdoor views: test scores were generally 10 to 15 percent higher and calling performance increased by 7 to 12 percent. This reinforces HMG's earlier 1999 study findings of schools where children in classrooms of the Capistrano School District progressed 15 percent faster in math and 23 percent faster in reading when they were located in the classrooms with the largest windows.

The importance of ventilation is that it facilitates the introduction of fresh air to cycle through the building and removes stale or pollutant air from the interior. Germs, molds, and various VOCs, such as those emitted by paints, carpets and adhesives, can often be found within buildings lacking adequate ventilation, causing Sick Building Syndrome (SBS). Typical symptoms include inflammation, asthma, and allergic reaction. Ventilation can also play a critical role in worker productivity, as evidenced by the extensive research that has been conducted to address these issues. This is also why it is imperative to minimize the use of toxic materials inside the building. Many of the products used in conventional office buildings, such as carpets and copying machines, contain toxic materials, and minimizing them decreases the potential hazards associated with them and their disposal. Furthermore, because these materials are known to leak pollutants into the indoor air, proper ventilation is required to avoid an adverse impact on worker productivity. Thus, from the above it becomes evident that productivity can be influenced by many factors.

Job satisfaction is known to generate less staff turnover, thereby improving the overall productivity of a firm. Less time spent on job training allows more time to be spent on productive work. Staff retention is a major factor for many firms that make the decision to green their office buildings in today's competitive world.

Committing to a market value for occupant productivity gains and accurately reflecting them is not easy for speculative or leased facilities. However, there is now adequate data and evidence quantifying the effects to support taking them into account on some basis. And while an owner of a leased facility may not financially benefit directly from increased user productivity, there could be indirect benefits in the form of increased rental fees and occupancy rates. For the majority of commercial buildings, the use of a conservative estimate for the potential reduction in salary costs and productivity gains will loom large in any calculation.

8.2.4 Improved Tenant/Employee Health

The principal features of green buildings normally include superior air quality, abundant natural light, access to views, and effective noise control. Each of these qualities is for the benefit of building occupants, making these buildings better places to work and live. Building occupants are increasingly seeking many green building features, such as superior air quality, control of air temperatures, and views. The Urban Land Institute (ULI) and the Building Owners and Managers Association (BOMA) conducted a survey which found that occupants rated air temperature (95 percent) and air quality (94 percent) most crucial in terms of tenant comfort. The study also determined that 75 percent of buildings did not have the option or capability to adjust features and that many individuals were willing to pay higher rents in order to obtain such features. These features were the only ones that were considered "most important" and yet were also what tenants were least satisfied with. This study also determined that heating or cooling problems are the principal reasons why tenants move out.

Most building occupants understand that natural light, clean air, and thermal comfort are required elements to stay healthy and productive, in addition to providing an enjoyable living and work place. The number of credible studies demonstrating an intrinsic connection between green building strategies and occupant health and well-being is endless. William Fisk in a 2002 study, "How IEQ Affects Health, Productivity," estimated that 16 to 37 million cases of colds and flu could be avoided by improving indoor environmental quality. This translates into a $6 to $14 billion annual savings in the United States while at the same time reducing Sick Building Syndrome (SBS) symptoms (a condition whereby occupants become temporarily ill), by 20 to 50 percent, resulting in $10 to $30 billion annual savings.

8.2.5 Increased Recruitment and Retention

A recent national survey by Harris Interactive reported that more than a third of U.S. workers would be further inclined to work for companies with strong green credentials and highlighted the growing influence of environmental issues over staff recruitment and retention policies.

This is further confirmed by Timothy R. Johnson, a principal at GCA International, who notes, "Firms that focus on the growing market sector of green building and sustainable design will be more attractive to top-notch candidates for reasons of recognized workload availability, progressive growth, and employment stability." This is clear evidence that providing a healthy and pleasant work environment increases employee satisfaction, productivity, and retention. It also clearly increases the ability to compete for the most qualified employees, as well as for business.

Statistical data and other evidence leaves no doubt that high performance green buildings can increase a company's ability to recruit and retain employees due to many factors such as good air quality, abundant

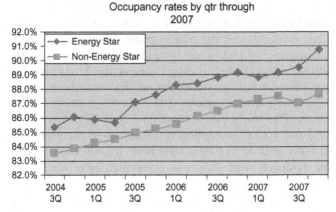

FIGURE 8.4   A diagram comparing the occupancy rates of two types of buildings—Energy Star® and non-Energy Star. (Source: Does Green Pay Off? by Norm Miller, Jay Spivey, and Andy Florance.)

amounts of natural light, and better circulated heat and air conditioning, all of which help provide more pleasant, healthier, and more productive places to work. With this in mind, it is surprising that the willingness to join and remain with an organization are aspects often overlooked when considering how green buildings affect employees. The economics of employee retention is important to seriously consider, as one estimate puts the cost of losing a single good employee at between $50,000 and $150,000. Many organizations experience a 10 to 20 percent annual turnover, some of it from persons they would have really liked to retain. In a workforce of 100 people, turnover at this level implies 10–20 people leaving per year. In some cases, people decide to leave due to poor physical and working environments. Figure 8.4 compares the difference in occupancy rates between Energy Star® and non-Energy Star buildings.

8.2.6 Enhancement of Property Value and Marketability

It is interesting to note that to date few reliable studies have been conducted on the intrinsic relationship between property values and green buildings, even though this is an important aspect that should be quantified and included in economic calculations. There are many factors that will or could increase property values for green buildings. Indeed, enhancement of property value is a key factor for speculative developers who fail to directly achieve operating cost and productivity savings. It is an element of particular relevance to speculative developers who intend to either sell or lease a new building, although it can also have a bearing on the decision process in general, including developers who intend to occupy a building while keeping an eye on the market value of the asset. Much of the real estate industry unfortunately still does not fully comprehend the benefits of green buildings, and therefore are unable to correctly convey these benefits to prospective purchasers.

A well-known green scholar, Jerry Yudelson, maintains that increased annual energy savings have been found to promote higher building values, and cites as an example a 75,000-sq-ft building that saves $37,500 per year in energy costs versus a comparable building built to code (this savings might result from saving 50 cents per square foot per year). At capitalization rates of 6 percent, typical today in commercial real estate, green-building standards would add $625,000 ($8.33 per square foot) to the value of the building. This means that for a small upfront investment, an owner can reap benefits that typically offer a rate of return exceeding 20 percent with a payback of three years or less.

The fact that high performance buildings can offer building owners many important benefits ranging from higher market value to more satisfied and productive employee occupants is not always apparent. The primary reason for this is that the majority of the benefits accrue to tenants, and tenants usually need proof before they are willing to pay for investments that may help them be more productive or save money. It is only very recently, due mainly to increased awareness, that tenants have started to fully appreciate the benefits of cleaner air, more natural lighting, and flexible spaces that can be modified as needed.

8.2.7 Resale Values and Rental Rates

There is considerable evidence that green buildings, particularly green buildings with good quality natural lighting, can have a dramatic effect on sales with respect to commercial buildings. Furthermore, without exception, several American studies report that there is a sound economic basis for green buildings, but only when operational costs are included in the equation. More specifically, whole building studies conclude that the net present values (NET) for pursuing green buildings as opposed to conventional buildings range from $50 to $400 per square foot ($540 to $4300 per square meter). The NET depends on a building's length of time analyzed (e.g., 20 to 60 years) and the degree to which the buildings implement green strategies. One of the main conclusions from these studies is that generally the NET increases as the greenness of the building increases.

A CoStar study found that with regard to rental sales, LEED™ buildings can command rent premiums of about $11.24 per square foot, in addition to a 3.8 percent increase in occupancy rate. The study also found that rental rates in ENERGY STAR buildings can boast a $2.38 per square foot premium versus comparable non-ENERGY STAR buildings in addition to a 3.6 percent greater occupancy rate. However, what is perhaps more remarkable and what may prove to be a trend that could signal greater attention from institutional investors, is that LEED™ buildings are commanding a surprising $171 more per square foot than their conventional counterparts, and ENERGY STAR buildings are commanding an average of $61 more per square foot. This is quite extraordinary since most leasing arrangements, particularly in the office/commercial sectors, provide little incentive to undertake changes that might be construed as being beneficial to the environment. For example, leases often have fixed rates with no regard to energy or water consumption, even though the lessees have control over most energy and water consuming devices.

8.2.8 Other Indirect Benefits

The indirect benefits of green building are many, including such features as improved image, risk reduction, future proofing, and self-reliance. These and other similar benefits may be captured by investors and should not be discarded in decision economics considerations. Although they may be difficult to quantify and in some cases may even be intangible, they should nevertheless be factored into the business case because they are intrinsically connected to sustainable design and they can significantly impact a green building's value.

Improved Image

One of the key messages conveyed by sustainable buildings is concern for the environment. Moreover, even if we disregard the financial benefits attributed to green buildings, they are generally perceived by the public as modern, dynamic, and altruistic. Green buildings can therefore provide a strong symbolic message of an owner's commitment to sustainability. Some of the benefits that companies can enjoy from these perceptions include employee pride, satisfaction, and well-being, which often translate into reduced turnover, advantages in employee recruitment, and improved morale. These powerful images can be an important factor in a company's decision to pursue occupancy in a green building.

Risk Reduction

Commenting on the recent downturn in the U.S. economy, Peter Morris, principal with Davis Langdon tells us, "Risk remains a serious concern for construction projects. Delay and cancellation of projects, even projects under construction, is a growing trend." Morris also proposes a key theme for project owners in the current market turmoil which is that, "The successful adoption of a competitive procurement strategy, in order to secure lower costs, will depend on active steps by the owner and the project team to ensure that the contractor is in a position to provide a realistic and binding bid and that contractors' bidding costs are minimized."

The application of green building principles can mitigate many of the potential perceived risks. In this regard, the U.S. Environmental Protection Agency currently classifies indoor air quality as one of the top five environmental health risks. Sick Building Syndrome (SBS) and Building Related Illness (BRI) are among the issues of major concern and which often end up being resolved in the courts. Business owners and operators are increasingly facing legal action from building tenants blaming buildings for their health problems. The main cause of both SBS and BRI is poor building design and/or construction, particularly with respect to the building envelope and mechanical systems. Green buildings should emphasize and promote safe, exceptional air quality, and no recognized green building should ever contribute to SBS or BRI. Litigation involving mold-related issues has also increased.

Future Proofing

Green buildings are inherently efficient and safe, and as such help ensure that they will not be at a competitive disadvantage in the future. Davis Langdon sums it very well by saying, "Going green is 'future-proofing' your asset." This is largely because there are a number of important potential risks that are significantly mitigated in green buildings such as:

- Energy conservation protects against future energy price increases.
- Occupants of green buildings are generally more comfortable and contented, so it can be assumed that they will generally be less likely to be litigious.
- Water conservation shields against water fee increases.
- A documented effort to build or occupy a healthy green building demonstrates a level of due diligence that could stand as an important defense against future lawsuits or changes in legislation, even when faced with currently unknown problems.

Self Reliance

The fact that green buildings often incorporate natural lighting and ventilation, and internal energy and water generation, makes them less likely to rely on external grids, and less likely to be affected by grid-related problems or failures such as blackouts, water shortages, or contaminated water. This element is acquiring increasing importance globally because of the increased risk of terrorism. Local self-reliance is steadily moving to primetime and the Institute for Local Self Reliance (ILSR) continues to develop cutting edge solutions to the problems facing communities around the globe.

Examples of five American communities that have started to take steps toward energy independence are outlined below. It shows that at the local level, energy independence is now a realistic possibility as numerous communities around the United States explore available renewable resources, and the technology necessary to harness them. Below are five of the many U.S. towns creating models for clean energy production and self reliance:

- Greensburg, Kansas is rebuilding itself as a "model green town" after being hit by a disastrous tornado, and is expecting to provide enough power to meet all the energy needs for the town in the foreseeable future.
- Rock Port, Missouri is a small town with a population of 1400 that has become the first community in the nation to be completely powered by wind.
- Reynolds, Indiana is another small Midwestern town with a population of 540 that was chosen as the community to execute the state government's "Biotown USA" experiment. The plan is to power the town with a range of locally-available biomass.
- San Jose, California's city council recently gave the city manager the authority to negotiate the terms of "an organics-to-energy bio-gas facility."
- Warrenton, Virginia, like San Jose, is taking the "trash to treasure" approach. Mayor George Fitch has spearheaded an effort to build a "biorefinery," and reduce the town's greenhouse gas emissions by 25 percent by 2015.

8.2.9 External Economic Effects

It is not easy to give a precise definition of external effects; they generally consist of costs or benefits of a project that accrue to society and are not readily captured by the private investor. Examples of this are reduced reliance on infrastructure such as sewers and roads, reduced green house gases, and reduced health costs, etc. The extent to which these benefits can be factored to a business case relies on the extent to which they can be converted from the external to the internal sides of the ledger. This constitutes a vital factor in any assessment of the costs and benefits of green buildings. Thus the costs of green vegetated roofs are borne by the developer or investor, while much of the benefit accrues at a broader societal level such as reduced heat island effects and reduced storm water runoff. Where the investor is a government agency, or where a private developer is compensated for including features that produce benefits at a societal level, then the business case can encompass the much broader range of effects. For example, there are jurisdictions, such as in Arlington, Virginia, which allow higher floor space to land coverage ratios for green buildings. Another example is the State of Oregon that offers tax incentives for green building, thereby providing a direct business case payoff to the investor.

Buildings can be singled out to have the largest indirect environmental impact on human health. Other perhaps less critical impacts, such as damage to ecosystems, crops, structures/monuments, and resource depletion, should also be considered even though they do not have a large associated indirect cost relative to human health. Infrastructure costs such as water use and disposal are typically provided by governments and are rarely cost effective or even cost neutral, and in many instances, governments are required to heavily subsidize water use and treatment. On the other hand external environmental costs consist mainly of pollutants in the form of emissions to air, water, and land and the general degradation of the ambient environment.

"Green" has become a common buzzword and journalists everywhere are writing about a "green economy," "green technology," and even "green jobs." Many manufacturers are clinging to the green bandwagon and increasingly claiming to be "green," while many others try to measure the effect of "green" technology on the job market. "Green" encourages job creation, partly because green building attributes are often labor intensive, rather than material or technology intensive. For example, there are significant environmental impacts associated with the transportation of materials for the construction industry. And by promoting the use of local and regional materials, local and regional job creation is encouraged and promoted.

It has been found that green building can also have economic ramifications and export opportunities on a much broader scale as a result of increased international recognition and related export sales. The 2005 Environmental Sustainability Index prepared by Yale and Columbia Universities benchmarked the ability of nations to protect the environment by integrating data sets including natural resource endowments, pollution levels, and environmental management efforts into a smaller set of indicators of environmental sustainability. In this study the United States unexpectedly ranked only 45th of countries in the index.

8.3 LIFE CYCLE COSTING (LCC)

This is a technique of combining both capital and operating costs to determine the net economic effect of an investment, and to evaluate the economic performance of additional investments that may be required for green buildings. It is based on discounting future costs and benefits to dollars of a specific reference year that are referred to as present value (PV) dollars. This makes it feasible to intelligibly quantify costs and benefits and compare alternatives based on the same economic criterion or reference dollar.

A recent study by the World Business Council for Sustainable Development (WBCSD suggests that key players in real estate and construction unfortunately often misjudge the costs and benefits of green buildings.

Peter Morris, a principal with Davis Langdon, says that "Perhaps a measure of the success of the LEED™ system, which was developed to provide a common basis for measurement, is the recent proliferation of alternative systems, each seeking to address some perceived imbalance or inadequacy of the LEED™ system, such as the amount of paperwork, the lack of weighting of credits, or the lack of focus on specific issues. Among these alternative measures are broad-based approaches, such as Green Globes, and more narrowly focused measures, such as calculations of a building's carbon footprint or measurements of a building's energy efficiency (the ENERGY STAR® rating). All these systems are valid measures of sustainable design, but each reflects a different mix of environmental values, and each will have a different cost impact."

Total integration of sustainable design elements into a project during its early project development and design phases can reduce building costs. On the other hand, if green design elements are not considered until late in the design process or during construction, and may require modifications to parts of the project or the entire project, then a significant increase in costs can be expected.

8.3.1 Initial/First Cost

Construction projects typically have initial or up front costs, which may include capital investment costs related to land acquisition, construction, or renovation and for the equipment needed to operate a facility. Land acquisition costs are normally included in the initial cost estimate if they differ among design alternatives. A typical example of this would be when comparing the cost of renovating an existing facility with new construction on purchased land.

The most cited reason for not incorporating green elements into building design strategies is the assumed increase in first cost. Some aspects of design have little or no first cost, such as site orientation and window and overhang placement. Other sustainable systems that incorporate additional costs in the design phase, such as an insulated shell, can be offset, for example, by the reduced cost of a smaller mechanical system. Material costs can be reduced during the construction phase of a project by the use of dimensional planning and other material efficiency strategies. Such strategies can reduce the amount of building materials needed and cut construction costs but they require forethought on the part of designers to ensure a building that creates less construction waste solely on its dimensions and structural design. An example of dimensional planning is designing rooms of 4 foot multiples, since wallboard and plywood sheets come in 4- and 8-foot lengths. Moreover, one dimension of a room can be designed in 6- or 12-foot multiples to correspond with the length of carpet and linoleum rolls.

8.3.2 Life-Cycle Cost Analysis

Life-cycle cost analysis (LCCA) is a method for evaluating all relevant costs over time of a project, product, or measure. It takes into consideration all costs including first costs, such as capital investment costs, purchase, and installation costs; future costs, such as energy costs, operating costs, maintenance costs, capital replacement costs, financing costs; and any resale, salvage, or disposal cost, over the lifetime of the project or product. LCCA is thus an engineering economic analysis tool useful for comparing the relative merit of competing project alternatives. George Paul Demos, estimating engineer at CDOT, echoes this and notes that, "The first component in an LCC equation is cost. There are two major cost categories by which projects are to be evaluated in an LCCA: initial expenses and future expenses. Initial expenses are all costs incurred prior to occupation of the facility. Future expenses are all costs incurred after occupation of the facility. Defining the exact costs of each expense category can be somewhat difficult at the time of the LCC study. However, through the use of reasonable, consistent, and well-documented assumptions, a credible LCCA can be prepared." Demos also considers the following major steps to be essential to performing a benefit cost analysis:

1. Establish objectives.
2. Identify constraints and specify assumptions.
3. Define base case and identify alternatives.
4. Set analysis period.
5. Define level of effort for screening alternatives.
6. Analyze traffic effects.
7. Estimate benefits and costs relative to base case.
8. Evaluate risk.
9. Compare net benefits and rank alternatives.
10. Make recommendations.

Sieglinde Fuller of the National Institute of Standards and Technology (NIST) says, "LCCA is especially useful when project alternatives that fulfill the same performance requirements, but differ with respect to initial costs and operating costs, have to be compared in order to select the one that maximizes net savings. For example, LCCA will help determine whether the incorporation of a high-performance HVAC or glazing system, which may increase the initial cost but result in dramatically reduced operating and maintenance costs, is cost-effective or not." However, LCCA is not useful when it comes to budget allocation.

Although there is no general consensus on the valid basis for adopting a life-cycle approach, most building stakeholders prefer to focus on minimizing direct costs or, at best, applying short time frame payback periods. Many developers, building owners, and other stakeholders believe that basing opinions on anything other than a reduced direct cost approach is fiscally irresponsible, when in reality the opposite is often the case. This lack of adoption is largely due to the typical corporate structure that dissociates direct and operating costs, and with most constructers often lacking the mandate to reduce operating costs, although they are mandated to reduce construction cost. This unfortunate reality is also evidenced by owner/developers who oversee construction of buildings for their own use.

The LCCA's main objective is to calculate the overall costs of project alternatives and to select the design that safeguards the ability of the facility to provide the lowest overall cost of ownership in line with its quality and function. The LCCA should be performed early in the design process to allow any needed design refinements or modifications to take place before finalization to optimize the life-cycle costs (LCC). One of the most important and challenging tasks of an LCCA (or any economic evaluation method for that matter) is to evaluate and determine the economic effects of alternative designs of buildings and building systems and to be able to quantify these effects and depict them in dollar amounts. LCCA is especially suited to the evaluation of design alternatives that satisfy a required performance level, but that may have differing investment, operating, maintenance, or repair costs; and possibly different life spans.

Although lowest life-cycle cost (LCC) provides a straightforward and easy-to-interpret measure of economic evaluation, there are other commonly used methods such as net savings (or net benefits), savings-to-investment ratio (or savings benefit-to-cost ratio), internal rate of return, and payback period. Fuller sees them as being consistent with the lowest LCC measure of evaluation if they use the same parameters and length of study period. Almost identical approaches can be used to make cost-effective choices for building-related projects irrespective of whether it is called cost estimating, value engineering, or economic analysis. And after identifying all costs by year and amount and discounting them to present value, they are added to arrive at total life-cycle costs for each alternative. These include:

- Initial design and construction costs
- Operating costs that include energy, water/sewage, waste, recycling, and other utilities
- Maintenance, repair, and replacement costs
- Other environmental or social costs/benefits including but not limited to: impacts on transportation, solid waste, water, energy, infrastructure, worker productivity, and outdoor air emissions, etc.

There should be appropriate adjustments to place all dollar values expended or received over time on a comparable basis as this is necessary for the valid assessment of a project's life-cycle costs and benefits. Time adjustment is required because a dollar today will not have an equivalent value to a dollar in the future.

Supplementary measures are considered to be relative measures, i.e., they are computed for an alternative relative to a base case. Sieglinde Fuller says, "Supplementary measures of economic evaluation are Net Savings (NS), Savings-to-Investment Ratio (SIR), Adjusted Internal Rate of Return (AIRR), and Simple Payback (SPB) or Discounted Payback (DPB). They are sometimes needed to meet specific regulatory requirements. For example, the FEMP LCC rules (10 CFR 436A) require the use of either the SIR or AIRR for ranking independent projects competing for limited funding. Some federal programs require a payback period to be computed as a screening measure in project evaluation. NS, SIR, and AIRR are consistent with the lowest LCC of an alternative if computed and applied correctly, with the same time-adjusted input values and assumptions. Payback measures, either SPB or DPB, are only consistent with LCCA if they are calculated over the entire study period, not only for the years of the payback period."

The most successful green buildings are produced through active, deliberate, and full collaboration among all the players. Building-related investments typically involve a great deal of uncertainty relating to their costs and potential savings. The performing of an LCCA greatly increases the ability and likelihood of deciding on a project that can save money in the long run. Yet this does not alleviate some of the potential uncertainty associated with the LCC results, mainly because LCCAs are typically conducted early in the design process when only estimates of costs and savings are available, rather than specific dollar amounts. This uncertainty in input values means that actual results may differ from estimated outcomes. The LCCA can be applied to any capital investment decision, and is particularly relevant when high initial costs are traded for reduced future cost obligations.

A 2007 study by Davis Langdon updating an earlier study states, "It is clear from the substantial weight of evidence in the marketplace that reasonable levels of sustainable design can be incorporated into most building types at little or no additional cost. In addition, sustainable materials and systems are becoming more affordable, sustainable design elements are becoming widely accepted in the mainstream of project design, and building owners and tenants are beginning to demand and value those features."

Likewise, Ashley Katz, a communications coordinator for the U.S. Green Building Council, says, "Costs associated with building commissioning, energy modeling, and additional professional services typically turn out to be a risk mitigation strategy for owners. While these aspects might add on to the project budget, they will end up saving projects money in the long run, and are also best practices for building design and construction."

<h2 style="text-align:center">8.4 TAX BENEFITS</h2>

State and local governments throughout the country are in the process of drafting new green building regulations to take advantage of incoming stimulus funding. As a follow-up effort to encourage environmentally friendly construction and energy savings, many states have initiated various tax incentive programs for green building. States like New York and Oregon offer state tax credits, while others, such as Nevada, offer property- and sales-tax abatements. The federal government also offers tax credits. The State of Oregon credits vary and are based on building area and LEED™ certification level. At the platinum level for example, a 100,000 square feet building in Oregon can expect to receive a net-present-value tax credit of up to $2 per square foot, which is transferable from public or nonprofit entities to private companies (e.g., contractors or benefactors), making it even more attractive than a credit that applies only to private owners.

A tax credit offered by the State of New York allows builders who meet energy goals and use environmentally preferable materials to apply for up to $3.75 per square foot for interior work and $7.50 per square foot for exterior work against their state tax bill. To qualify for this credit, a building needs to be certified by a licensed architect or engineer in addition to meeting specific requirements for energy use, water use, indoor air quality, waste disposal, and materials selection. This translates to mean that the energy used in new buildings must not exceed 65 percent of that allowed under the New York Energy Code and in rehabilitated buildings energy use cannot exceed 75 percent.

In 2005, the Nevada Legislature passed a poorly considered green building incentive package in an effort to spur private developers in the state. The state offered a property tax abatement of up to 35 percent for up to 10 years to private development projects that achieve LEED™ Silver certification. This means that if the property tax represents 1 percent of value, it could be worth as much as 5 percent of the building cost, which translates to much more than the actual cost of achieving LEED™ Silver on a large project. This has encouraged a large number of Nevada projects to pursue LEED™ certification, including the $7 billion, 17 million square foot Project City Center in Las Vegas, which is one of the world's largest private development projects to date. The hastily written legislation forced the next session of the Nevada Legislature in 2007 to rethink and modify the program because it created an enormous financial crisis for the state. The State of Nevada also provides for sales tax abatement for green materials used in LEED™ Silver certified buildings. South Carolina also introduced a program of tax incentives that meet certain Green Globes or LEED standards for energy efficiency.

In appraising currently existing incentive programs, New York, Oregon, and Maryland preceded Nevada and utilized their state income tax codes as the primary tool to further green buildings in their states. In addition, many jurisdictions have created their own unique programs. Virginia followed the Nevada model by allowing property tax abatements at a local level, New Mexico used the income tax credit approach, and Hawaii tried a new approach by requiring a green building to receive priority processing during governmental reviews for project approvals.

There are also many federal tax incentives available such as the 2005 federal Energy Policy Act, which offers two major tax incentives for differing aspects of green buildings: 1. A tax credit of 30 percent on use of both solar thermal and electric systems, and 2. A tax deduction of up to $1.80 per square foot for projects that reduce energy use for lighting, HVAC, and water-heating systems by at least 50 percent compared with the 2001 baseline standard. These tax deductions may be taken by the design team leader (typically the architect) when applied to government projects.

Following are some of the more prevalent federal tax incentives:

Consumer Incentives

- Homeowners can compile credits for energy improvements to their homes, such as windows, insulation, and envelope and duct sealing.
- Homeowners can acquire credits for installing efficient air conditioners and heat pumps, gas or oil furnaces, and furnace fans. In new or existing homes, credits can be achieved for efficient gas, oil, or electric heat pump water heaters.
- Credits are also available for qualified solar water heating and photovoltaic systems and small wind and geothermal heat pump systems.

Business Incentives

- Businesses can get deductions for new or renovated buildings that save 50 percent or more of projected annual energy costs for heating, cooling, and lighting compared to model national standards, and partial deductions for efficiency improvements to individual lighting, HVAC and water heating, or envelope systems.
- Investment tax credit for combined heat and power systems (CHP).
- Businesses are eligible for tax credits for qualified solar water heating and photovoltaic systems, and for certain solar lighting systems.
- Credits are available to businesses that install qualifying microturbines. These systems, which typically run on natural gas, are small power-producing systems sized to run small to medium size commercial buildings.

Builders and Manufacturers Incentives

- Homebuilders are eligible for credits for homes that exceed national model energy codes by 50 percent, subject to certification. Manufactured home producers are also eligible for a smaller credit for manufactured homes that exceed national model codes by 30 percent or that meet Energy Star standards.
- Credits are available to manufacturers of high-efficiency refrigerators, clothes washers and dishwashers. Due to these manufacturer incentives, special consumer promotions may be available for qualifying products.

It is obvious that a tax credit can provide significant savings. It reduces the amount of income tax you have to pay and unlike a deduction, which reduces the amount of income that is taxable; a tax credit directly reduces the tax itself. In the final analysis, the reader should always check online for the latest tax incentive updates as many new programs are continuously being initiated and older programs expire. For example, the Federal Solar Tax credit has been extended for eight years and on February 17, 2009, President Obama signed into law the American Recovery and Reinvestment Act (ARRA). This act creates new incentives for solar energy, modifies existing incentives, and provides billions of dollars in funding for renewable energy projects. With this act the United States can in the coming years become the largest solar market in the world. For additional details on tax incentives and credits visit: http://energy taxincentives.org; http://www.dsireusa.org/incentives; http://seia.org; http://www.energy.gov/taxbreaks.htm; http://www.aceee.org/energy/index.htm.

8.5 MISCELLANEOUS OTHER GREEN BUILDING COSTS

8.5.1 Energy and Water Costs

Other green building expenses include operational expenses for energy, water, and other utilities. These depend to a large extent on consumption, current rates, and price projections. But since energy (and to a lesser extent water) consumption, building configuration, and building envelope are interdependent elements, energy and water costs are usually assessed for the building as a whole rather than for individual building systems or components. Sometimes the latest, greenest technology just isn't approved yet and may cause both delays and additional costs.

Correct forecasts or predictions of energy costs during the preliminary design phase of a project are rarely simple. Assumptions have to be made regarding use profiles, occupancy rates, and schedules, all of which can have a dramatic impact on energy consumption. There are several suitable computer programs currently on the market like Energy-10 and eQuest that can provide the required information regarding assumptions on the amount of energy consumption for a building. Alternatively, the information and data can come from the engineering analysis. Other software packages, such ENERGY PLUS (DOE), DOE-2.1E, and BLAST are also excellent programs, but require more detailed input not normally available until later in the design process when the design concept is more fully articulated.

It is important to determine prior to program selection whether annual, monthly, or hourly energy consumption estimates are required and whether the program is capable of adequately tracking savings in energy consumption even when design changes take place or when different efficiency levels are simulated. Figure 8.5 provides an example of the typical costs incurred by an HVAC system over 30 years, which represents its useful life.

While estimates vary slightly, the consensus is that green buildings on average use 30 percent less energy than conventional buildings, which is why energy is a substantial and widely recognized cost of building operations that can be reduced through energy efficiency and related measures that are part of green building design. A detailed survey of 60 LEED™ rated buildings demonstrates that green buildings, when compared to conventional buildings, are:

- On average more energy efficient by approximately 25–30 percent
- More likely to generate renewable energy on-site
- Characterized by lower electricity peak consumption
- More likely to purchase grid power generated from renewable energy sources.

Energy savings in sustainable buildings come primarily from reduced electricity purchases and secondarily from reduced peak energy demand. On average, green buildings are estimated to be 28 percent more efficient

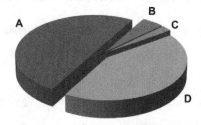

A. Energy Cost 50.0%

B. Maintenance Cost 40.7%

C. Replacement Cost 2.3%

D. HVAC First Cost 43.0%

HVAC System Cost Over 30 Years

FIGURE 8.5   A pie diagram illustrating typical costs (in percentage terms) incurred by an HVAC system over 30 years, which represents its useful life. (Source: Washington State Department of General Administration.)

than conventional buildings and on average generate 2 percent of their power on-site from photovoltaics (PV). The financial benefits that accrue from 30 percent reduced consumption at an electricity price of $0.08/kWh comes to about $0.30 per square foot annually with a 20-year NPV (net present value) of over $5 per square foot, equal to or more than the average additional cost associated with building green.

Jerry Yudelson, author of "The Green Building Revolution," says, "Many green buildings are designed to use 25 to 40 percent less energy than current codes require; some buildings achieve even higher efficiency levels. Translated to an operating cost of $1.60 to $2.50 per square foot for electricity (the most common energy source for buildings), this energy savings could reduce utility operating costs by 40 cents to $1 per square foot per year. Often these savings are achieved for an added investment of just $1 to $3 per square foot. With building costs reaching $150 to $300 per square foot, many developers and building owners are seeing that it is a wise business decision to invest 1 to 2 percent of capital cost to secure long-term savings, particularly with a payback of less than three years. In an 80,000-sq-ft building, the owner's savings translate into $32,000 to $80,000 per year, year after year, at today's prices."

The environmental and health costs associated with air pollution caused by non-renewable electric power generation and on-site fossil fuel use are generally excluded when making investment decisions. Figure 8.6 highlights the reduced energy used in green buildings as compared with conventional buildings.

8.5.2 Operation, Maintenance, and Repair Costs

Numerous studies on sustainability have shown that over the life of the building LEED[TM]-certified buildings typically both cost less and are easier to operate and maintain than conventional buildings. This puts them in a position to command higher lease rates than conventional buildings in their markets. However, maintenance and repair (OM&R) costs and non-fuel

	Certified	Silver	Gold	Average
Energy Efficiency	18%	30%	37%	28%
(above standard code)				
On-Site Renewable Energy	0%	0%	4%	2%
Green Power	10%	0%	7%	6%
Total	28%	30%	48%	36%

FIGURE 8.6   A table showing reduced energy use in green buildings as compared with conventional buildings. (Source: USGBC, Capital E Analysis.)

operating costs are often more difficult to estimate than other building expenditures. Operating schedules and maintenance standards will vary from one building to the next; the variation in these costs is significant even when the buildings are of the same type and age. It is therefore important in estimating these costs to use common sense and good judgment.

Published estimating guides and supplier quotes can sometimes provide relevant information on maintenance and repair costs. Some of the data estimation guides derive their cost data from databases such as Means and BOMA, which typically report average owning and operating costs per square foot, based on the number of square feet in the building, the age of the building, its geographic location, and number of stories.

Green buildings can recoup any added costs within the first one to two years of their life-cycle once they become operational. Studies show that green buildings typically use 30 to 50 percent less energy and 40 percent less water than their conventional counterparts, yielding significant savings in operational costs. The New Buildings Institute (NBI) recently released a new research study indicating that new buildings certified under the U.S. Green Building Council's (USGBC) LEED™ certification program are, on average, performing 25 to 30 percent better than buildings that are not LEED™ certified in terms of energy use. The study also suggests that buildings achieving LEED™ Gold and Platinum categories have average energy savings approaching 50 percent.

8.5.3 Replacement Costs

Lower replacement costs of systems and materials are significant benefits of building green. This is achieved by incorporating durable materials and design that prolong the life of building systems and the building itself. Replacing a building's roof, flooring, HVAC system, or the whole building itself results in the highest cost to the environment and to the owner's bottom line. While many of these features also reduce operating costs, an owner's commitment to proactive maintenance is the key to keeping systems working well into their prime.

To a large extent, the number and timing of capital replacements of building systems are based primarily on the estimated life of the system and the length of the study period. It is expected that the same sources providing the cost estimates for the initial investments will be used to obtain estimates of replacement costs and expected useful lives. Likewise, a good starting point for estimating future replacement costs is to use their initial cost as the base date. The LCCA method is designed to escalate base-year amounts to their future period of occurrence.

The term residual value of a system or component basically represents the value it will have after being depreciated, i.e., its remaining value at the end of the study period, or at the time it is replaced during the study period. According to Sieglinde Fuller, residual values can be based on value in place, resale value, salvage value, or scrap value, net of any selling, conversion, or disposal costs. The residual value of a system with remaining useful life in place can be determined by linearly prorating its initial costs using simple rule-of-thumb calculations.

8.5.4 Other Costs

Finance charges and taxes do not usually apply to federal projects, although finance charges and other payments do apply if a project is financed through an Energy Savings Performance Contract (ESPC) or Utility Energy Services Contract (UESC). These charges are normally included in the contract payments negotiated with the Energy Service Company (ESCO) or the utility.

Non-monetary benefits or costs relate to project-related issues for which there is no meaningful way of assigning a dollar value, and despite efforts to develop quantitative measures of benefits, there are situations that simply do not lend themselves to such an analysis. For example projects may provide certain benefits such as improved quality of the working environment, preservation of cultural and historical resources, or other similar qualitative advantages. By their nature, these benefits are external to the LCCA and are difficult to assess, but if these benefits are considered significant they should be taken into account in the final investment decision and included in a life-cycle cost analysis and also portrayed in the project documentation.

To formalize the inclusion of non-monetary costs or benefits in the decision making process, the analytical hierarchy process (AHP) is used. AHP is one of a set of multi-attribute decision analysis (MADA) methods that can be used when considering qualitative and quantitative non-monetary attributes, in addition to common economic evaluation measures when evaluating project alternatives. The ASTM E 1765 Standard Practice for Applying Analytical Hierarchy Process (AHP) to Multi-attribute Decision Analysis of Investments Related to Buildings and Building Systems that is published by ASTM International presents a general procedure for calculating and interpreting AHP scores of a project's total overall desirability when making building-related capital investment decisions. The WBDG Productive Branch is an excellent source of information for estimating productivity costs.

8.6 DESIGN AND ANALYSIS TOOLS AND METHODS

The nation's largest owner and operator of built facilities is the federal government, which during the energy crisis of the 1970s followed by the

1980's crisis was faced with increasing initial construction costs and ongoing operational and maintenance expenses. As a result, facility planners and designers decided to use economic analysis to evaluate alternative construction materials, assemblies, and building services with the goal of lowering costs. In today's difficult economic climate, building owners wishing to reduce expenses or increase profits are again employing economic analysis to improve their decision making during the course of planning, designing, and constructing a building. Moreover, federal, state, and municipal entities have all enacted legislative mandates requiring the use of building economic analysis to determine the most economically efficient or cost-effective choice among building alternatives. The general steps taken in an economic analysis process are illustrated in Figure 8.7.

8.6.1 Present-Value Analysis

The basic concept of present-value analysis is that the value of a dollar profit today is greater than the value of a dollar profit next year. How much greater is determined by what is called the "discount rate," as in "how much of a discount would you expect if you were buying a dollar's worth of next year's profit." The discount rate used in the NPV calculation is usually the cost of debt, also known as the weighted average cost of debt. Also, net present-value (NPV) allows decision makers to compare

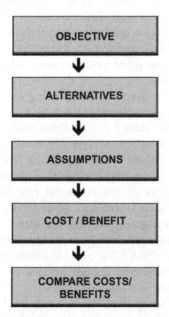

FIGURE 8.7 A diagram illustrating the economic analysis process. (Source: Based on Whole Building Design Guide.)

various alternatives on a similar time scale by converting the various options to current dollar figures. A project is generally considered acceptable if the net present value is positive over the expected lifetime of the project. As an example, take a building that is considering having its lighting changed from traditional incandescent bulbs to fluorescents. The initial investment to change the lights themselves is estimated to be $40,000. After the initial investment, it is estimated to cost $2,000 to operate the lighting system, which will yield $15,000 in savings each year. Thus, this produces an annual cash flow of $13,000 every year after the initial investment. If, for the sake of keeping it simple, a discount rate of 10 percent is assumed and it is calculated that the lighting system will be utilized over a five-year time period, the following net present-value calculations would result:

$$t = 0 \text{ NPV} = (-40,000)/(1 + .10) \, 0 = -40,000.00$$
$$t = 1 \text{ NPV} = (13,000)/(1.10) \, 1 = 11,818.18$$
$$t = 2 \text{ NPV} = (13,000)/(1.10) \, 2 = 10,743.80$$
$$t = 3 \text{ NPV} = (13,000)/(1.10) \, 3 = 9,767.09$$
$$t = 4 \text{ NPV} = (13,000)/(1.10) \, 4 = 8,879.17$$
$$t = 5 \text{ NPV} = (13,000)/(1.10) \, 5 = 8,071.98$$

Based on the above information, the total NPV over the lifetime of the project would come to $9,280.22.

The importance of discounting is that it adjusts costs and benefits to a common point in time. Thus, in order to be able to add and compare cash flows that are incurred at different times during the life cycle of a building, they need to be made time-equivalent. To make cash flows time equivalent, the LCC method converts them to present values by discounting them to a common point in time, which is usually the base date. To some extent, the selection of the discount rate is dependent on its use. The interest rate used for discounting essentially represents the investor's minimum acceptable rate of return.

The Federal Discount Rate FY 2003 Principles and Guidelines states: "Discounting is to be used to convert future monetary values to present values. Calculate present values using the discount rate established annually for the formulation and economic evaluation of plans for water and related land resources." The discount rate for federal energy and water conservation projects is determined annually by the DOE's Federal Energy Management Program (FEMP); for other federal projects, those not primarily concerned with energy or water conservation, the discount rate is determined by the Office of Management and Budget (OMB). These discount rates, however, do not include the general rate of inflation, but rather represent real discount rates.

The length of the study period begins with the base date, which is the date to which all cash flows are discounted. The study period includes any planning, construction, and implementation periods as well as the service or occupancy period. The study period remains unchanged for all of the considered alternatives. The service period, however, essentially begins when the completed building is occupied or when a system is taken into service. This is the period over which operational costs and benefits are evaluated. In FEMP analyses, the service period cannot exceed 25 years. The contract period in ESPC and UESC projects lies within the study period, starting when the project is formally accepted, energy savings begin to accrue, and contract payments begin to be due. With the loan being paid off, the contract period generally ends.

In OMB and FEMP studies, annually recurring cash flows such as operational costs are normally discounted from the end of the year in which they are incurred. In MILCON studies they are typically discounted from the middle of the year. All single amounts, such as replacement costs, and residual values are discounted from their dates of occurrence.

Sieglinde Fuller states that LCCA can be applied to any capital investment decision in which relatively higher initial costs are traded for reduced future cost obligations. Fuller also maintains that, "It is particularly suitable for the evaluation of building design alternatives that satisfy a required level of building performance but may have different initial investment costs, different operating and maintenance and repair costs, and possibly different lives." LCCA is an approach that provides a much better assessment of the long-term cost-effectiveness of a project than alternative economic methods that mainly focus on first costs or on operating related costs in the short run. According to Fuller, LCCA can be performed at various levels of complexity, but its scope could vary from a "back-of-the-envelope" study to a detailed analysis with thoroughly researched input data, supplementary measures of economic evaluation, complex uncertainty assessment, and extensive documentation.

An important attribute of LCCA is that it can be performed in either constant dollars or current dollars. Both methods of calculation produce identical present-value life-cycle costs. However, a constant-dollar analysis does not include the general rate of inflation, which means it has the advantage of not requiring an estimate of the rate of inflation for the years in the study period. A current-dollar analysis on the other hand does include the rate of general inflation in all dollar amounts, discount rates, and price escalation rates. Constant-dollar analysis is generally recommended for federal projects, except for projects financed by the private sector such as through Energy Savings Performance Contracting (ESPC) and the Utility Energy Services Contract (UESC). There are several alternative financing studies available that are usually performed in current dollars if the analyst wants to compare contract payments with actual year-to-year operational or energy cost savings.

8.6.2 Sensitivity Analysis

In general terms, uncertainty and sensitivity analyses investigate the robustness of a study when the study includes some form of mathematical modeling. Sensitivity analysis is a technique recommended by FEMP for energy and water conservation projects. Critical assumptions should be varied and net present value and other outcomes recomputed to determine how sensitive outcomes are to changes in the assumptions. The assumptions that deserve the greatest attention will rely on the dominant benefit and cost elements and the areas of greatest uncertainty of the program being analyzed. In general, a sensitivity analysis is used for estimates of: (a) benefits and costs; (b) the discount rate; (c) the general inflation rate; and (d) distributional assumptions. Models used in the analysis should be well documented and, where possible, available to facilitate independent review.

8.6.3 Break-Even Analysis

Break-even analysis is a useful tool in tracking a business's cash flow. Break-even analysis focuses on the relationship between fixed cost, variable cost, and profit. It is mostly used when decision makers want to know the maximum cost of an input that will allow the project to still break even, or conversely, what minimum benefit a project can produce and still cover the cost of the investment. To perform a break-even analysis, benefits and costs are set equal, all variables are specified, and the break-even variable is solved mathematically. Since we're dealing with cash flow, and depreciation is a non-cash expense, it's subtracted from the operating expenses.

Below are the variables needed to compute a break-even sales analysis for a project:

- Gross profit margin
- Operating expenses (less depreciation)
- Annual debt service (total monthly debt payments for the year).

8.6.4 Computer Estimating Programs

Computer programs can considerably reduce the time and effort spent on formulating the LCCA, performing the computations, and documenting the study. There are a large number of LCCA-related software programs available, all of which can be found on the Internet. Below are some of the more popular and widely used applications:

- ECONPACK (Economic Analysis Package) for Windows is a comprehensive economic analysis computer package incorporating economic analysis calculations, documentation, and reporting capabilities. It is structured to permit it being used by non-economists to prepare complete, properly documented economic analysis (EA) in support of DoD funding requests. The program was developed by the U.S. Army Corps of Engineers. The analytic capabilities of ECONPACK are reportedly generic, providing standardized economic analysis methodologies and calculations to evaluate a broad range of capital investment categories such as hospitals, family housing, information systems, utility plants, maintenance facilities, commercially financed facilities, and equipment.

- Building Life-Cycle Cost (BLCC) Program, version 5.3-09 is a program and economic analysis tool developed by the National Institute of Standards and Technology (NIST) for the U.S. Department of Energy Federal Energy Management Program (FEMP). It is designed to provide computational support for the analysis of capital investments in buildings. BLCC 5.3 conducts economic analyses by evaluating the relative cost effectiveness of alternative buildings and building-related systems or components. Typically, BLCC is used to evaluate alternative designs that have higher initial costs but lower operating-related costs over the project life than the lowest-initial-cost design. It is especially useful for evaluating the costs and benefits of energy and water conservation and renewable energy projects.

- Life-Cycle Cost in Design WinLCCID Program was originally developed for MILCON analyses by the Construction Engineering Research Laboratory of the U.S. Army Corps of Engineers. The program is a life cycle costing tool that is used to evaluate and rank design alternatives for new and existing buildings and carry out "what if" analyses based on variables such as present and future costs and/or maintenance and repair costs.

- ENERGY-10® is a cost estimating program tool that assists architects, builders, and engineers in rapidly (within 20 minutes) identifying the most cost-effective, energy-saving measures to employ in designing a low-energy building. Using climate data that is site-specific, see how different combinations of materials, systems, and orientation yield lesser or greater results based on energy use, comparative costs, and reduced emissions. Using the software at the early phases of a design can reportedly result in energy savings of 40 percent to 70 percent, with little or no increase in construction cost. The software is available through the Sustainable Buildings Industry Council (SBIC).

- Success Estimator Estimating and Cost Management System is a cost estimating tool available from U.S. Cost that gives estimators, project managers, and owners real-time, simultaneous access to their cost data and estimating projects from any Internet-connected computer.

8.6.5 Relevant Codes and Standards

There are many standards that are relevant to green building. These include:

- ASTM E2432—Standard Guide for the General Principles of Sustainability Relative to Building
- Circular No. A-94 Revised—Guidelines and Discount Rates for Benefit-Cost Analysis of Federal Programs
- 10 CFR 436 Subpart A—Federal Energy Management and Planning Programs, Methodology and Procedures for Life-Cycle Cost Analyses
- Energy Policy Act of 2005
- Executive Order 13123—Greening the Government through Efficient Energy Management, DOE Guidance on Life-Cycle Cost Analysis Required by Executive Order 13123
- Executive Order 13423—Strengthening Federal Environmental, Energy, and Transportation Management"
- Facilities Standard for the Public Buildings Service, P100 (GSA)—Chapter 1.8—Life-Cycle Costing
- Standards on Building Economics, 6th ed. ASTM, 2007
- Sustainable Building Technical Manual (DOE/EPA)
- NAVFAC P-442 Economic Analysis Handbook
- Tri-Services Memorandum of Agreement (MOA) on "Criteria/Standards for Economic Analyses/Life-Cycle Costing for MILCON Design" (1991).

Understanding
Specifications

9.1 GENERAL OVERVIEW

Specifications essentially consist of precisely written documentation that describe a project to be constructed, supplementing drawings and forming part of the contract. Specifications describe qualities of materials, their methods of manufacture, and installation into the project, workmanship and mode of construction, in addition to providing other information not shown in the drawings, including description of the final result. Many designers have considerable difficulty preparing a competent set of standard building specifications, partly because it demands a shift of gears using a different medium to express design content—using written documents instead of drawings. It also propels the designer into the technical realm of materials that are not normally dealt with on a daily basis.

Specifications are the written portion of the contract documents that go with the construction drawings and which are used to execute the project. Specifications should complement the drawings, not overlap or duplicate information in the drawings, and normally describe the materials as well as the installation methods. They also prescribe the quality standards of construction expected on the project. Specifications indicate the procedure by means of which it may be determined whether the requirements provided are satisfied. Specifications are considered to be legal documents, and should therefore be comprehensive, accurate, and clear. Specification writing has two principal objectives: 1. Defining the scope of work, and 2. Acting as a set of instructions. Defining the scope of work is at the core of specification writing. The required level of quality of the product and services must be clearly communicated to bidders and the party executing the contract to ensure that the completed project conforms to this specified quality. Projects now generally incorporate the specifications within a project manual that is issued as part of the contract documents package along with the drawings, bidding requirements, and other contract conditions. The specification writer should ensure that the requirements are compatible with the methods that are to be employed and also that the methods selected in one specification are compatible with those selected in another.

A primary function of project specifications is to give detailed information regarding materials and methods of work for a particular construction project. They cover various components relating to the project, including general conditions, scope of work, quality of materials, and standards of workmanship. The drawings, collectively with the project specifications, define the project in detail and clearly delineate exactly how it is to be constructed. The project drawings and specifications are an integral part of the contract documents and are inseparable. The drawings reflect what the project specifications are unlikely to cover; and the project specifications indicate what the drawings are unlikely to portray. Specifications are also sometimes used to further clarify details that are not adequately covered by the drawings and notes on the drawings. Whenever the information on the drawings conflicts with that in the project specifications, the project specifications will take precedence over the drawings.

The Construction Specifications Institute (CSI) has established a widely recognized format of organization for the technical specifications. CSI is a nationwide organization composed of different segments related to the construction industry such as architects, engineers, manufacturers' representatives, contractors, and other interested parties that have closely collaborated to develop this system of identification. These specification standards are noted in the MasterFormat, which in 2004 was expanded from 16 to 50 divisions, as described later in this chapter. It should be noted that the previous 1995 edition of the format is no longer supported by CSI.

Technology and green-related practices have had a tremendous impact on the construction industry and on the general way we conduct our business, including the way in which specifications are written. Examples of this are specification production and reproduction, which in a few short years has witnessed tremendous progress due to these new technologies. Master systems are now commercially available in electronic form, allowing a specifier to simply load the master system into the computer for instant access to drawing checklists and explanation sheets. Upon editing the relevant sections, a printout can be made with an audit trail that informs and records what has been deleted and what decisions remain undetermined.

Green building specifications can be easily incorporated into CSI *MasterFormat*™ in three general ways, including:

- Environmental protection procedures
- Implementing green building materials
- Practical application of environmental specifications.

9.2 THE NEED FOR SPECIFICATIONS

Construction specifications are usually required because drawings alone can rarely define the qualitative issues of a scheme. Well-executed specifications form the written portion of the contract documents that are used to execute the project. Design decisions are continuously made as drawings develop from schematic sketches to detailed design to construction documents. Drawings are intended to depict the general configuration and layout of a design, including its size, shape, and dimensions. It informs the contractor of the quantities of materials needed, their placement, and their general relationship to each other. Technical specifications are a critical component of the contract documents as they reflect the design intent and describe in detail the quality and character of materials, as well as the standards to which the materials and their installation are required to conform, in addition to other issues that are more appropriately represented in written rather than graphic form. The bottom line is that no matter how beautiful a designer's concept is, it is difficult to envisage the project being properly executed without clear, concise, accurate, and easily understood contract documents that include a well-written specification.

Although construction drawings may contain all the information about a structure that can be presented graphically, they nevertheless omit information that the contractor must have, but which is not adaptable to graphic presentation. Information in this category includes quality-related criteria for materials, specified standards of workmanship, prescribed construction methods, etc. In the event of a discrepancy existing between the drawings and the specifications, the specifications must be considered the final authority. For most projects the specification document will consist of a list of 50 divisions that usually starts with a section on general conditions. These are the rules of the job and provide the instructions for what to do in any of the anticipated situations on the project. The general conditions document starts with a general description of the building, including type of foundation, types of windows, character of framing, utilities to be installed, and so on. This is followed by definitions of terms used in the specs and then certain routine declarations of responsibility and other issues.

Douglas D. Harding, a licensed California attorney, says that "Every project manager should be intimately aware of the general conditions to the project as part of his/her project administration effort." Furthermore, without an actual knowledge of the general conditions on each project, contractors and subcontractors are also taking an unacceptable risk that may ultimately cause ruin.

Harding lists the following important general condition clauses that can directly impact the success of a project, if adequate attention is not given to them:

- Progress payments (When are they due? Is there a "condition precedent" clause?)
- Retention (How much and when is it due?)
- Change orders (Overhead and profit; time extensions; inclusions)
- Delay (Notice and time impact analysis)
- Scheduling (Who has scheduling responsibilities? What kind of schedule is required? Is the subcontractor required to complete and maintain a schedule?)
- Order of precedence (What is the order, do specs rule over drawings or do drawings rule over specs?)
- Notice (How many days after a delay do you have to give notice? How is notice to be delivered—verbally, by mail, by registered mail?)

One of the reasons why even well-drawn construction drawings are unable to adequately reveal all the aspects of a construction project is because there are many aspects that cannot be shown graphically. An example of this is trying to describe on a drawing the quality of workmanship required for the installation of electrical equipment or who bears responsibility for supplying the materials. For the majority of projects, the standard procedure then is to supplement construction drawings with written descriptions that define and limit the materials and fabrication according to the intent of the engineer or the designer. The specifications are therefore an important part of the project because they eliminate possible misinterpretation and ensure positive control of the construction.

Time and cost restraints tend to discourage individuals (or small firms) from writing a completely new set of specifications for each project that takes place. Because of this and other issues, specifiers often turn to alternative solutions. In this respect, the superiority and supremacy of using master systems over traditional specification writing is overwhelming. Moreover, because of liability issues, specifiers generally feel more comfortable relying on specifications that have repeatedly proven themselves in the past. Typical advantages of employing master systems include accuracy, the use of correct specification language and format for ease of specification preparation, as well as the many sources, extensive product databases and reference material that is currently available and from which a complete set can be compiled for each new project. The master spec systems are also referred to when modifications are implemented to fit the particular conditions of a given job, or new specifications are incorporated. Master systems contain guide specifications for many materials which are constantly updated; this allows the specifier to edit out unnecessary text rather than generate new information for each project.

Listed below are some of the more popular sources from which specification material can be acquired, much of which can be retrieved from the Internet and public libraries:

- Master specifications (Masterspec®, SPECSystem™, MasterFormat™, SpecText®, BSD Speclink®, ezSPECS On-Line™, 20-20 CAP Studio, and many others)
- Local and national codes and ordinances
- Federal specifications (Specs-In-Tact, G.S.A., N.A.F.V.A.C., N.A.S.A.)
- National standards organizations such as the American National Standards Institute, National Institute of Building Sciences, the National Fire Protection Association, the National Institute of Standards and Technology, and the Association for Contract Textiles
- Manufacturers' industry associations (Fire Equipment Manufacturers' Association, American Plywood Association, The Brick Industry Association, etc.)
- Testing societies (American Society for Testing and Materials, American Society for Nondestructive Testing, Underwriters Laboratories)
- Manufacturers' catalogs (Sweet's Catalog File, Man-U-Spec, Spec-data)

- Industry-related magazines and publications (Construction Specifier, Architecture, Green Magazine On-line, Interior Design, Architectural Lighting, Architectural Record)
- Books on relevant subjects
- Information from files of previously written specifications.

Numerous firms providing online specification writing services have emerged during recent years. These services can easily be found on the Internet.

9.3 SPECIFICATIONS CATEGORIES

In preparing a specification document the specifier has to make determination early in the process on which format or method to use to communicate the desired design intent to the contractor. There are basically two broad categories of specifications, *closed* and *open*, and most products can be specified by either method. Within these two broad categories, there are four generic types of specifying construction products that are industry standards, including:

1. Descriptive
2. Reference standards
3. Performance
4. Proprietary.

The type of specification chosen depends on several factors which are discussed below.

Closed specification

A closed (also called prescriptive or restrictive) specification limits a product to a single manufacturer or a few brand-identified types or models and prohibits substitutions. This type of specification is more often used in the private sector in cases where specifiers feel more comfortable resorting to a specific propriety product with which they are familiar, and which will meet the specific criteria of the project. However, it should be noted that this procedure (particularly when only one product is named) is not competitive, and rarely attracts the most favorable price for the owner. Also, while the closed specification is common in private construction work, it is generally prohibited by the code for public projects and is required by law to be bid under open specifications. An open specification allows products of any manufacturer to be used if the product meets the specified requirements.

The closed proprietary specification method is considered the easiest form to write but the most restrictive in application, because it names a specific manufacturer's product. It generally establishes a narrower definition of acceptable quality than do performance or reference standard methods, and gives the designer complete control over what is installed. The specification can also be transformed into an open proprietary specification in which multiple manufacturers or products are named or alternatives solicited by adding the phrase "or equal." This would increase potential competition and encourage a lower installation price from potential vendors. There are instances where a multiple choice may not be appropriate, as for example in a renovation project where a specific brick is required for repairs to an existing brick facade.

Open specifications

Also called performance or nonrestrictive, this type of specification gives the contractor some choice in how to achieve the desired results and is the type required by the Public Contract Code. Proprietary specifications may also be used as open specifications but with the addition of the "or equal" clause, which allows the contractor to consider other products for bid if they are shown to be equal in performance and specifications. Due to the ambiguity surrounding this clause, and the disagreements it often perpetuates, specifiers generally shy away from incorporating it into the **proprietary specifications**.

Descriptive specifications

This is a method of open specifications that is gaining popularity and is sometimes referred to as prescriptive specifications. As the name implies, this type of specification describes in detail the requirements for the material or product and the workmanship required for its fabrication and installation without providing a trade name. Government agencies sometimes stipulate this type of specifications to allow greater competition among product manufacturers. Descriptive specifications are more difficult to write than proprietary ones because the specifier is required to include all the product's relevant physical characteristics in the specification, bearing in mind the specifier has already decided that the specified product meets functional needs. For an individual product, proprietary, performance, and descriptive specifying techniques may be used.

Reference standard

Reference standards specify standards such as ASTM, ASHRAE, State of California, federal, etc. The various manufacturers must meet these standards. This standard basically describes a material, product, or process referencing a recognized industry standard or test method as the basis for the specification and is often used to specify generic materials such as portland cement or clear glass. Thus, in specifying gypsum wallboard for example, the specification can state that all gypsum wallboard products shall meet the requirements of ASTM C36. It is worth mentioning that a number of construction industry members have voiced the opinion that specifications should not only make references to the applicable standards, but they should also quote the relevant parts of the referenced standards.

With the reference standard specification the product is described in detail so that the specifier is relieved of the necessity to repeat the requirements, but can instead refer to the recognized industry standard. In employing a reference standard, the specifier should not only possess a copy of that standard, but should also know what is required by the standard, including choices that may be contained therein, and which should be enforced by all suppliers. This type of specification is generally short and fairly straightforward and easy to write. In addition, the use of reference standard specifications reduces a firm's liability and the possibility for errors.

Performance specifications

This type of specification has been developed over recent years for many types of construction operations. Rather than specifying the required construction process, performance specifications establish the performance requirements of the finished facility without dictating the methods by which the end results are to be achieved. The precise method by which this performance is obtained is left to the construction contractor. This gives the greatest latitude to contractors because it allows them to use any material or system that meets the required performance criteria, provided the results can be verified by measurement, tests, or other acceptable methods. Performance specifications are difficult to write; the specifier needs to know all the criteria for a product or system, determine an appropriate method for testing compliance, and write a clear and lucid document. This requires sufficient data to be provided to ensure that the product can be adequately demonstrated. Performance specifications are primarily used in cases where a specifier wants to inspire new ways of achieving a particular result in specifying complex systems.

Proprietary (product) specifications

This type of construction specification for a product often uses a combination of methods to convey the designer's intent. It is normally written by referencing specific products by manufacturer and brand or model name and applies to materials and equipment. For example, a specification for a terra cotta tile would use a proprietary specification to name the product or products selected by the specifier, a descriptive specification to specify the size and design, and a reference standard to specify the ASTM standard, grade, and type required. It is distinguished from prescriptive specifications in that the physical characteristics are inferred rather than explicitly stated. For an individual product, proprietary, performance, and descriptive specifying techniques may be used. Proprietary specifications can be made "open" by adding the phrase "or equal."

9.4 THE PROJECT MANUAL

In 1963 the CSI developed a format for organizing the written documents, and in 1964 the American Institute of Architects (AIA) first developed the concept of the "project manual," primarily to meet the pressing need for a consistent arrangement of building construction specifications. The project manual consists of an assemblage of documents related to the construction work on a project, which typically include bidding requirements (contract forms, bonds, certificates, etc.), sample documents, conditions of the contract, and the technical specifications, which together with the drawings, constitute the contract documents. The project manual has gained general acceptance in the industry and is greatly preferred to the previous method, which was based on individual preference by the design firm and resulted in a wide variety of methods around the country, causing much confusion. As design firms and contractors became increasingly nationwide in their operations, the project manual continued to develop, and while it may differ depending on the size and type of project, a typical project manual may include, but not be limited to, the following:

- General project information:
    1. Title page to include names and addresses of all parties responsible for the development of the project (owners, architects, civil engineers, mechanical engineers, electrical engineers, and structural engineers) in addition to a statement of compliance by the architect or engineer of record.
    2. Table of contents.
    3. Schedule of drawings.
- Bidding requirements. This applies where contracts are awarded through the bidding process. These would include:
    4. Invitation to bid and advertisement for bids.
    5. Instructions to bidders, including prequalification forms, bid forms, information available to bidders, date and time of bid opening, and notice of pre-bid conference,
- Contract forms, may include:
    6. Sample Forms. *Include public entity crime form, owner/contractor agreement, performance and payment bond, change order, bid form,*

which may require the general contractor's license number, may include a subcontractors list and license numbers, and other project forms.

7. Bonding requirements. Labor and materials payment bonds are required on projects costing above a certain amount.

   a. *Bid security to be submitted in form of a certified check, cashier's check, treasurer's check, or bank draft of any national or state bank.*

   b. *Performance bond and materials and payment bond. Each bond shall equal one hundred (100%) percent of the contract amount.*

8. Insurance requirements:

   a. *Workers' compensation and employers' liability.*

   b. *Public liability to include personal injury, bodily injury, and property damage.*

   c. *Products and completed operations liability.*

   d. *Owner's protective liability.*

   e. *Business automobile liability, including owned, non-owned, and hired automobiles.*

   f. *Property all-risks coverage to one hundred (100%) percent of the value at risk, subject to acceptable deductibles.*

- Contract conditions: General conditions of the contract such as AIA Form 201 or similar preprinted forms. Supplementary conditions include anything that is not covered in the general conditions, such as addenda (changes made before contract signing), and change orders (changes made after contract signing). In addition, contract conditions to include:

9. General conditions and supplementary conditions including, but not limited to, the following:

   a. *Deductive alternates must be used if bidding is to take place on a project where funds are in jeopardy of reversion and a rebid process would not be possible within remaining time available, and when the client wants to preserve the option to negotiate with the apparent low bidder.*

   b. *Notice of time limit and method of payment to the contractor, including final payment.*

   c. *Time limit in which the construction is to be completed.*

   d. *The penalty to be paid by the contractor for failure to comply with the time limits of the contract.*

   e. *Federal wage rates and hourly scales shall be used where applicable. Federal wage rates are not required for construction projects financed totally from local or state funds.*

   f. *A provision setting forth who should pay for standard tests of concrete, plumbing, electrical, steel, and others as required by industry standards.*

   g. *The client may include an incentive in the contract for early completion of the project.*

- Technical specifications: These provide written technical requirements concerning building materials, components, systems and equipment shown on the drawings with regard to standards, workmanship quality, performance of related services, and stipulated results to be achieved by application of construction methods (Figure 9.1).

As legal documents, the specification language must be written in a clear, precise, and unambiguous manner in order to communicate the intended concept. In this respect, a convention has developed over the years as to what specific information should be shown on the drawings and what should more appropriately be included in the specifications. Drawings should depict information that can be most aptly and effectively expressed graphically by means of drawings and diagrams. This would include relevant information such as dimensions, sizes, proportions, gauges, arrangements, locations, and interrelationships. Additionally, drawings are used to express quantity, whereas specifications normally describe quality. Also, drawings would denote type (e.g., wood), whereas specifications would clarify the species (e.g., oak). Well-written specifications, on the other hand, are essentially based on a number of broad general principles, including:

- Specifications should only transmit information that lends itself to the written word, such as standards, descriptions, procedures, guarantees, and names.
- Specifications should be clear, concise, and technically correct.
- Specifications should avoid the use of ambiguous words that could lead to misinterpretation.
- Specifications should be written using simple words in short, easy-to-understand sentences.
- Specifications should use technically correct terms, and avoid slang or "field" words.
- Specifications should avoid fielding conflicting requirements.
- Specifications should avoid repeating requirements stated elsewhere in the contract.

Sometimes exceptions to these understandings may create confusion. For example, building departments of the majority of municipalities will only accept drawings with applications for building permits, and refuse to accept a project manual with specifications. Additionally, all data demonstrating building code compliance must be indicated on the drawings.

MASTERSPEC® SMALL PROJECT™ 2005

COMBINED TABLE OF CONTENTS - MASTERFORMAT 2004 (Section Text Only)

© 2005 The American Institute of Architects

Issue Date	Sect. No.	SECTION TITLE	SECTION DESCRIPTION
DIVISION 01 - GENERAL REQUIREMENTS			
2003	011000	SUMMARY	Summary of the Work, Owner-furnished products, use of premises, and work restrictions.
2003	012000	PRICE AND PAYMENT PROCEDURES	Allowances, alternates, unit prices, contract modification procedures, and payment procedures.
2005	013000	ADMINISTRATIVE REQUIREMENTS	Project management and coordination, submittal procedures, delegated design, and Contractor's construction schedule.
2003	014000	QUALITY REQUIREMENTS	Testing and inspecting procedures.
2005	014200	REFERENCES	Abbreviations, acronyms, and trade names referenced in
2003	015000	TEMPORARY FACILITIES AND CONTROL	Temporary utilities and facilities for support, security, and protection.
2003	016000	PRODUCT REQUIREMENTS	Product selection and handling and product substitutions.
2003	017000	EXECUTION AND CLOSEOUT REQUIREMENTS	Examination and preparation, cutting and patching, installation, and closeout requirements.
DIVISION 02 - EXISTING CONDITIONS			
2003	024119	SELECTIVE STRUCTURE DEMOLITION	Demolition and removal of selected portions of buildings and site elements.
DIVISION 03 - CONCRETE			
2003	033000	CAST-IN-PLACE CONCRETE	General building and structural applications.
2005	033713	SHOTCRETE	Pneumatically projected mortar and concrete.
2005	034100	PRECAST STRUCTURAL CONCRETE	Conventional precast units.
2005	034500	PRECAST ARCHITECTURAL CONCRETE	Exposed surface units.
2003	034713	TILT-UP CONCRETE	Wall panels.
DIVISION 04 - MASONRY			
2003	042000	UNIT MASONRY	General applications, walls, partitions.
2005	042300	GLASS UNIT MASONRY	Glass block.
2005	044300	STONE MASONRY	Stone veneer laid in mortar.
2003	047200	CAST STONE MASONRY	Architectural units set in mortar.
DIVISION 05 - METALS			
2005	051200	STRUCTURAL STEEL FRAMING	Framing systems
2005	052100	STEEL JOIST FRAMING	Standard SJI units.
2005	053100	STEEL DECKING	Roof, floor, composite types.
2005	054000	COLD-FORMED METAL FRAMING	Load-bearing and curtain-wall studs; floor, ceiling, and roof joists.
2003	055000	METAL FABRICATIONS	Iron, steel, stainless steel, and aluminum items (not sheet metal).
2003	055100	METAL STAIRS	Steel; with pan, abrasive-coated, and floor plate treads and tube railings.
2003	055200	METAL RAILINGS	Metal railings, including glass panels and wood rails.
DIVISION 06 - WOOD, PLASTICS, AND COMPOSITES			
2005	061000	ROUGH CARPENTRY	Framing, sheathing, subflooring, etc.
2005	061053	MISCELLANEOUS ROUGH CARPENTRY	Rough carpentry for minor applications.
2005	061600	SHEATHING	Wall and roof sheathing, subflooring, underlayment, and related products.
2005	061753	SHOP-FABRICATED WOOD TRUSSES	Metal-plate-connected members.
2003	061800	GLUED-LAMINATED CONSTRUCTION	Glued-laminated beams, arches, and columns.
2005	062000	FINISH CARPENTRY	Exterior and interior trim, siding, paneling, shelving, and stairs.
2005	064013	EXTERIOR ARCHITECTURAL WOODWORK	Trim, door frames, shutters, and ornamental items.
2005	064023	INTERIOR ARCHITECTURAL WOODWORK	Trim, custom cabinets, counter tops, flush paneling, and stairwork and rails.
2003	066113	CULTURED MARBLE FABRICATIONS	Vanity tops, shower walls, and tub surrounds.
DIVISION 07 - THERMAL AND MOISTURE PROTECTION			

FIGURE 9.1   The MASTERSPEC® SMALL PROJECT™ 2005 combined table of contents for a small project—MasterFormat™ 2004—section text only. (Source: American Institute of Architects.)

MASTERSPEC® SMALL PROJECT™ 2005
COMBINED TABLE OF CONTENTS - MASTERFORMAT 2004 (Section Text Only)
© 2005 The American Institute of Architects

Issue Date	Sect. No.	SECTION TITLE	SECTION DESCRIPTION
2005	071113	BITUMINOUS DAMPPROOFING	Hot- and cold-applied dampproofing.
2005	071326	SELF-ADHERING SHEET WATERPROOFING	Self-adhering sheet types.
2005	072100	THERMAL INSULATION	Common building types, excluding roof.
2005	072419	WATER-DRAINAGE EXTERIOR INSULATION AND FINISH SYSTEM (EIFS)	Thincoat stucco over foam insulation.
2005	073113	ASPHALT SHINGLES	Both organic and fiberglass based.
2005	073129	WOOD SHINGLES AND SHAKES	Shingles and shakes for roofing and siding.
2005	073200	ROOF TILES	Clay and concrete units.
2005	074113	METAL ROOF PANELS	Factory-finished units, steel and aluminum, insulated and noninsulated.
2005	074213	METAL WALL PANELS	Factory-finished units, steel and aluminum, insulated and noninsulated, and liner panels.
2003	074600	SIDING	Aluminum, fiber-cement, and vinyl siding and soffit.
2005	075113	BUILT-UP ASPHALT ROOFING	Hot asphalt systems; insulation.
2005	075216	STYRENE-BUTADIENE-STYRENE (SBS) MODIFIED BITUMINOUS MEMBRANE ROOFING	Styrene-butadiene-styrene roofing systems; insulation.
2005	075323	ETHYLENE-PROPYLENE-DIENE-MONOMER (EPDM) ROOFING	Adhered, mechanically-fastened, and loosely-laid EPDM systems; insulation.
2005	076100	SHEET METAL ROOFING	Flat-seam, standing-seam, and batten-seam sheet metal roofing.
2005	076200	SHEET METAL FLASHING AND TRIM	Mostly for roofing systems.
2003	077100	ROOF SPECIALTIES	Copings, fasciae, gravel stops, gutters, and downspouts.
2003	077200	ROOF ACCESSORIES	Roof curbs, equipment supports, roof hatches, gravity ventilators, and ridge vents.
2005	078100	APPLIED FIREPROOFING	Concealed cementitious and sprayed fiber.
2003	078413	PENETRATION FIRESTOPPING	Through-penetration firestop systems.
2005	079200	JOINT SEALANTS	Elastomeric, latex, and acoustic sealants.

DIVISION 08 - OPENINGS

Issue Date	Sect. No.	SECTION TITLE	SECTION DESCRIPTION
2005	081113	HOLLOW METAL DOORS AND FRAMES	Standard steel doors and frames.
2005	081416	FLUSH WOOD DOORS	Wood-veneer, plastic-laminate, and hardboard-faced units.
2003	081423	CLAD WOOD DOORS	Solid-core and hollow-core units with stile-and-rail appearance.
2003	081433	STILE AND RAIL WOOD DOORS	Stock stile and rail units.
2005	083113	ACCESS DOORS AND FRAMES	Wall and ceiling units.
2005	083213	SLIDING ALUMINUM-FRAMED GLASS DOORS	Residential and commercial grades.
2005	083219	SLIDING WOOD-FRAMED GLASS DOORS	Residential and light-commercial grades.
2003	083323	OVERHEAD COILING DOORS	Steel, aluminum, and stainless steel curtains.
2005	083513	FOLDING DOORS	Accordion folding, panel folding, metal bifold, and bifold mirror doors.
2003	083613	SECTIONAL DOORS	Steel, aluminum, and wood and hardboard types.
2003	084113	ALUMINUM-FRAMED ENTRANCES AND STOREFRONTS	Standard systems including hardware.
2005	084433	SLOPED GLAZING ASSEMBLIES	Standard systems, mechanically or structural-sealant glazed.
2005	085113	ALUMINUM WINDOWS	Most standard types.
2005	085200	WOOD WINDOWS	Most standard types.
2005	085313	VINYL WINDOWS	Most standard types.
2003	086100	ROOF WINDOWS	Wood, aluminum, and vinyl flat-glass units.
2005	086200	UNIT SKYLIGHTS	Single- and double-dome acrylic and polycarbonate units.
2005	087100	DOOR HARDWARE	Hinges, locksets, latchsets, closers, stops, accessories, etc.
2003	087113	AUTOMATIC DOOR OPERATORS	Swinging and sliding types with controls.
2005	088000	GLAZING	General applications including mirror glass.
2005	089000	LOUVERS AND VENTS	Fixed, extruded-aluminum and formed-metal louvers; wall vents.

DIVISION 09 - FINISHES

Issue Date	Sect. No.	SECTION TITLE	SECTION DESCRIPTION
2005	092216	NON-STRUCTURAL METAL FRAMING	Non-structural metal furring and framing.
2005	092300	GYPSUM PLASTERING	Includes gypsum lath, metal lath, and gypsum plaster.

FIGURE 9.1 *(Continued)*

MASTERSPEC® SMALL PROJECT™ 2005
COMBINED TABLE OF CONTENTS - MASTERFORMAT 2004 (Section Text Only)
© 2005 The American Institute of Architects

Issue Date	Sect. No.	SECTION TITLE	SECTION DESCRIPTION
2005	092400	PORTLAND CEMENT PLASTERING	Includes metal lath and portland cement plaster.
2005	092613	GYPSUM VENEERPLASTERING	Includes gypsum base, cementitious backer units, and gypsum plaster.
2005	092713	GLASS-FIBERRE INFORCED PLASTER (GFRP) FABRICATIONS	Fabricated units for interior use.
2005	092900	GYPSUM BOARD	Gypsum board, glass mat gypsum backing board, and cementitious backer units.
2005	093000	TILING	Typical installations; includes stone thresholds and cemetitious backer units.
2003	093033	STONE TILING	Thin, modular, cut stone units; includes stone thresholds and cemetitious backer units.
2005	095113	ACOUSTICAL PANEL CEILINGS	Mineral-base and glass-fiber-base panels with exposed suspension systems.
2005	095123	ACOUSTICAL TILE CEILINGS	Mineral-base tile with concealed suspension systems.
2005	096340	STONE FLOORING	Exterior and interior stone traffic surfaces.
2005	096400	WOOD FLOORING	Solid-and engineered-wood flooring.
2005	096513	RESILIENT BASE AND ACCESSORIES	Vinyl and rubber wall base, treads, nosings, and edgins.
2005	096516	RESILIENT SHEET FLOORING	Unbacked and backed sheet vinyl products, and linoleum sheet.
2005	096519	RESILIENT TILE FLOORING	Solid vinyl, rubber, vinyl composition, and linoleum floor tiles.
2005	096813	TILE CARPETING	Modular tile for commercial applications.
2005	096816	SHEET CARPETING	Direct glue-down and installations including cushion.
2005	097200	WALLCOVERINGS	Vinyl wall coverings and wallpaper.
2005	097500	STONE FACING	Dimension stone wall facings, trim, and countertops.
2005	097723	FABRIC-WRAPPED PANELS	Decorative, tackable, and acoustic units.
2005	099100	PAINTING	Exterior and interior substrates; includes stained and transparent finished wood.

DIVISION 10 - SPECIALTIES

Issue Date	Sect. No.	SECTION TITLE	SECTION DESCRIPTION
2005	101400	SIGNAGE	Exterior and interior signs, letters, and plaques.
2005	101700	TELEPHONE SPECIALTIES	Telephone enclosures and directory storage units.
2005	102113	TOILET COMPARTMENTS	Color coated steel, stainless steel, plastic laminate, and solid-plastic types.
2005	102226	OPERABLE PARTITIONS	Acoustically rated, manually and electrically operated, flat-panel partitions.
2005	102600	WALL AND DOOR PROTECTION	Wall guards, hand rails, bed locators, corner guards, and wall covering.
2005	102800	TOILET, BATH, AND LAUNDRY ACCESSORIES	Standard commercial and residential units.
2003	103100	MANUFACTURED FIREPLACES	Fabricated metal units, wood-burning and gas, and accessories.
2005	104413	FIRE EXTINGUISHER CABINETS	Fire extinguisher cabinets including hose and valve cabinets.
2005	104416	FIRE EXTINGUISHERS	Fire extinguishers and mounting brackets.
2003	105143	WIRE MESH STORAGE LOCKERS	Fabricated storage units.
2003	105723	CLOSET AND UTILITY SHELVING	Coated wire units, fixed and adjustable.

DIVISION 11 - EQUIPMENT

Issue Date	Sect. No.	SECTION TITLE	SECTION DESCRIPTION
2005	112600	UNIT KITCHENS	Standard manufactured units.
2003	113100	RESIDENTIAL APPLIANCES	Kitchen and laundry appliances.
2003	115213	PROJECTION SCREENS	Front projection screens, manual and electrically operated units.

DIVISION 12 - FURNISHINGS

Issue Date	Sect. No.	SECTION TITLE	SECTION DESCRIPTION
2005	122113	HORIZONTAL LOUVER BLINDS	Manually operated blinds.
2005	122116	VERTICAL LOUVER BLINDS	Manually operated blinds.
2005	122200	CURTAINS AND DRAPES	Curtains and drapes including manual and motorized tracks.
2003	123530	RESIDENTIAL CASEWORK	Manufactured cabinets; plastic-laminate and solid-surface-material countertops.
2005	123640	STONE COUNTERTOPS	Granite, marble, serpentine, and slate countertops.

MASTERSPEC SMALL PROJECT SPECIFICATIONS TABLE OF CONTENTS - COMBINED - Page 3 of 6

FIGURE 9.1   *(Continued)*

MASTERSPEC® SMALL PROJECT™ 2005
COMBINED TABLE OF CONTENTS - MASTERFORMAT 2004 (Section Text Only)
© 2005 The American Institute of Architects

FIGURE 9.1 *(Continued)*

MASTERSPEC® SMALL PROJECT™ 2005

COMBINED TABLE OF CONTENTS - MASTERFORMAT 2004 (Section Text Only)

© 2005 The American Institute of Architects

Issue Date	Sect. No.	SECTION TITLE	SECTION DESCRIPTION
2005	235216	CONDENSING BOILERS	Pulse-combustion hot-water boilers.
2005	235223	CAST-IRON BOILERS	Gas- and oil-fired hot-water boilers.
2005	235400	FURNACES	Gas-, electric-, and oil-fired furnaces.
2003	236200	PACKAGED COMPRESSOR AND CONDENSER UNITS	Air-cooled units.
2005	236419	RECIPROCATING WATER CHILLERS	Packaged air-cooled units with factory mounted controls.
2005	236423	SCROLL WATER CHILLERS	Packaged air-cooled units with factory mounted controls.
2005	236500	COOLING TOWERS	Packaged, closed-circuit cooling towers.
2003	237333	INDOOR, INDIRECT-FUEL-FIRED HEATING AND VENTILATING UNITS	Packaged units with gas- and oil-fired furnace.
2003	237339	INDOOR, DIRECT GAS-FIRED HEATING AND VENTILATING UNITS	Packaged units; natural gas and propane fired.
2005	237413	PACKAGED, OUTDOOR, CENTRAL-STATION AIR-HANDLING UNITS	Packaged units with compressors, condensers, evaporator coils, fans, controls, filters, and dampers.
2003	238113	PACKAGED TERMINAL AIR-CONDITIONERS	Self-contained, through-the-wall terminal units with controls.
2005	238119	SELF-CONTAINED AIR-CONDITIONERS	Packaged cooling and heating units, with filters and controls; suitable for exposed installations.
2003	238146	WATER-SOURCE UNITARY HEAT PUMPS	Concealed and exposed console and rooftop heat-pump units.
2005	238219	FAN COIL UNITS	Heating and cooling, vertical and horizontal units.
2005	238229	RADIATORS	Hot water and electric baseboard units and convectors.
2005	238239	UNIT HEATERS	Hot water and electric, cabinet and propeller unit heaters.
2005	238316	RADIANT-HEATING HYDRONIC PIPING	Embedded pipe, fittings, specialties, and controls.

DIVISION 26 - ELECTRICAL

Issue Date	Sect. No.	SECTION TITLE	SECTION DESCRIPTION
2005	260500	COMMON WORK RESULTS FOR ELECTRICAL	Raceways, identification, supporting devices, and seismic controls.
2005	260923	LIGHTING CONTROL DEVICES	Modular dimming controls, time switches, photoelectric switches, occupancy sensors, and conductors and cables.
2005	260936	MODULAR DIMMING CONTROLS	Modular dimming controls.
2005	262200	LOW-VOLTAGE TRANSFORMERS	General purpose and specialty dry type, 600 V and less.
2003	262413	SWITCHBOARDS	Front-connected, front-accessible switchboards, 600V and less, and TVSS and overcurrent protective devices.
2003	262416	PANELBOARDS	Lighting and appliance and distribution panel boards.
2003	262713	ELECTRICITY METERING	Energy and demand metering by utility.
2005	262726	WIRING DEVICES	Receptacles, switches, service fittings, and multioutlet assemblies.
2003	262813	FUSES	Cartridge type (600 V and less) and spare fuse cabinets.
2003	262816	ENCLOSED SWITCHES AND CIRCUIT BREAKERS	Fusible and nonfusible switches, molded-case circuit breakers, and enclosures.
2003	262913	ENCLOSED CONTROLLERS	Manual and magnetic controllers and enclosures.
2003	263533	POWER FACTOR CORRECTION EQUIPMENT	Capacitors for 600 V and less.
2005	264113	LIGHTNING PROTECTION FOR STRUCTURES	Roof- and stack-mounting air terminals and ground rods.
2003	264313	TRANSIENT-VOLTAGE SUPPRESSION FOR LOW-VOLTAGE ELECTRICAL POWER CIRCUITS	Service entrance and panelboard suppressors and enclosures for low-voltage power, control, and communication equipment.
2005	265000	LIGHTING	Exterior and interior luminaires, ballasts, exit signs, emergency lighting units, and lamps.

DIVISION 27 - COMMUNICATIONS

Issue Date	Sect. No.	SECTION TITLE	SECTION DESCRIPTION
2005	270500	COMMON WORK RESULTS FOR COMMUNICATIONS	Raceways, identification, supporting devices, and seismic controls.
2005	271013	RESIDENTIAL STRUCTURED CABLING	Communication system distribution devices, terminals, panels, raceways, and cables.
2005	271100	COMMUNICATIONS EQUIPMENT ROOM FITTINGS	Connections, backboards, hardware, and grounding.
2005	271500	COMMUNICATIONS HORIZONTAL CABLING	Unshielded twisted pair cable, outlets, and accessories.

DIVISION 28 - ELECTRONIC SAFETY AND SECURITY

Issue Date	Sect. No.	SECTION TITLE	SECTION DESCRIPTION
2005	280500	COMMON WORK RESULTS FOR ELECTRONIC SAFETY AND SECURITY	Raceways, identification, supporting devices, and seismic controls.

MASTERSPEC SMALL PROJECT SPECIFICATIONS TABLE OF CONTENTS - COMBINED - Page 5 of 6

FIGURE 9.1   *(Continued)*

MASTERSPEC® SMALL PROJECT™ 2005

COMBINED TABLE OF CONTENTS - MASTERFORMAT 2004 (Section Text Only)

© 2005 The American Institute of Architects

Issue Date	Sect No.	SECTION TITLE	SECTION DESCRIPTION
2005	280513	CONDUCTORS AND CABLES FOR ELECTRONIC SAFETY AND SECURITY	Cabling, control wire and cable, optical fiber, and conductors.
2005	283100	FIRE DETECTION AND ALARM	Noncoded, zoned and addressable multitplex systems.

DIVISION 31 - EARTHWORK

2003	311000	SITE CLEARING	Clearing and grubbing, and topsoil removal.
2003	312000	EARTH MOVING	Excavating, filling, and grading.
2003	313116	TERMITE CONTROL	Soil treatment, borate wood treatment, bait stations, and metal mesh barriers.

DIVISION 32 - EXTERIOR IMPROVEMENTS

2005	321216	ASPHALT PAVING	Asphalt paving, pavement-marking paint, and concrete wheel stops.
2003	321313	CONCRETE PAVING	Concrete paving and pavement-marking paint.
2005	321400	UNIT PAVING	Brick, concrete, and rough-stone pavers in aggregate setting beds.
2003	323113	CHAIN-LINK FENCES AND GATES	Residential and commercial types.
2005	323223	SEGMENTAL RETAINING WALLS	Dry-laid concrete masonry unit walls.
2003	328400	PLANTING IRRIGATION	Piping, sprinklers, and controls.
2005	329200	TURF AND GRASSES	Seeded and sodded lawns.
2005	329300	PLANTS	Trees, shrubs, ground cover, and plants.

DIVISION 33 - UTILITIES

2005	332100	WATER SUPPLY WELLS	Domestic water wells.
2003	334100	STORM UTILITY DRAINAGE PIPING	Piping and utility structures for gravity stormwater systems.
2005	334600	SUBDRAINAGE	Drainage piping, panels, and geotextiles.

FIGURE 9.1 *(Continued)*

However, the repetition of identical data on both the specifications and the drawings exposes the documents to potential errors and inconsistency. To achieve better communication, the specifier should:

- Avoid specifying standards that cannot be measured or phrases that are subject to wide interpretation.
- Acquire a thorough understanding of the most current standards and test methods referred to and the sections that are applicable to the project. Use accepted standards to specify quality of materials or workmanship required, such as Portland Cement: Conform to ASTM C150, Type I or Type II, low alkali. Maximum total alkali shall not exceed 0.6 percent.
- Avoid specifications that are impossible for the contractor to execute.
- Use clear, simple, direct statements, concise use of terms, and attention to grammar and punctuation. Avoid use of words or phrases that are ambiguous and imply a choice that may not be intended.
- Be impartial in designating responsibility. Avoid exculpatory clauses such as, "the General Contractor shall be totally responsible for all…," which try to shift responsibility.
- Describe only one important idea per paragraph to make reading easier while facilitating comprehension, editing, and modifying at a later date. Specifications to be kept as short and concise as possible, omitting words like *all, the, an,* and *a.*
- Capitalize the following: 1. The Contract Documents, such as specifications, Working Drawings, Contract, Clause, Section, Supplementary Conditions; 2. Major parties to the contract, such as Contractor, Client, Owner, Architect; 3. Specific rooms within the building, such as Living Room, Kitchen, Office; 4. Grade of materials, such as No.1 Douglas Fir, FAS White Oak; and, of course, 5. All proper names.
- Avoid underlining anything in a specification, as this implies that the remaining material can be ignored.
- Ensure that the terms shall and will are used correctly. "Shall" designates a command: "The Contractor shall…." whereas "will" implies choice: "The owner or architect will…..".

It is imperative that the specifications and construction drawings are fully coordinated as they complement each other. Moreover, they should not contain conflicting requirements, errors, omissions, or duplications. Below is a summary of project manual requirements for a new construction project.

Summary—General Project Manual Requirements (to be edited as required)

1. List of contacts
2. Location map/site plan/building plans/elevations (reduced scale)
3. Borrowers loan agreement (BLA)

 4. A/E Agreement (design Services)
 5. CM agreement
 6. Construction agreement
 7. Consultant services agreement
 8. Additional service billings
 9. Project analysis report (PAR)
10. Project status report template
11. Borrower's draw requests
12. Construction schedule
13. GC/CM applications for payment (current and log)
14. Change order/pending change order log
15. Change orders
16. RFI Log
17. Submittal log
18. Buy out/subcontractor log
19. Vendor log
20. Allowances.

9.5 SPECIFICATION FORMATION AND ORGANIZATION

The Construction Specifications Institute and Construction Specifications Canada originally created the 16-division MasterFormat™ in 1963. Today, this format is widely used both in the United States and Canada for preparing construction specifications concerning non-residential building projects. MasterFormat™ has become the standard for titling and arranging construction project manuals containing bidding requirements, contracting requirements, and specifications. Since its inception the Construction Specification Institute (CSI) struggled to try to standardize the specification numbering system and the format of the sections, producing a modified MasterFormat™ in 1995. During recent years the CSI actively sought to further improve MasterFormat by adding new divisions to the system. In a concerted effort to address rapidly evolving and growing computer and communications technology, a modified MasterFormat™ was introduced in 2004. In the new MasterFormat™ edition division numbers are increased from 16 to 50, of which 13 divisions are left blank to provide room for future revisions and to allow construction products and technology to evolve (Figure 9.2).

MasterFormat™ is a well-structured system employed by specifiers for organizing information into project manuals, for organizing cost data, for filing product information and other technical data, as well as for identifying drawing objects and presenting construction market data. The CSI describes it as a master list of numbers and titles for organizing information relating to construction requirements, products, and activities into a standard sequence.

PROCUREMENT AND CONTRACTING REQUIREMENTS GROUP

- Division 00 — Procurement and Contracting Requirements

SPECIFICATIONS GROUP

General Requirements Subgroup

- Division 01 — General Requirements

Facility Construction Subgroup

- Division 02 — Existing Conditions

- Division 03 — Concrete

- Division 04 — Masonry

- Division 05 — Metals

- Division 06 — Wood, Plastics, and Composites

- Division 07 — Thermal and Moisture Protection

- Division 08 — Openings

- Division 09 — Finishes

- Division 10 — Specialties

- Division 11 — Equipment

- Division 12 — Furnishings

- Division 13 — Special Construction

- Division 14 — Conveying Equipment

- Division 15 — RESERVED FOR FUTURE EXPANSION

- Division 16 — RESERVED FOR FUTURE EXPANSION

FIGURE 9.2   The newly revised 2004 MasterFormat™ edition, which has replaced the 1995 edition. (Source: Construction Specification Institute, Inc.)

- Division 17 — RESERVED FOR FUTURE EXPANSION

- Division 18 — RESERVED FOR FUTURE EXPANSION

- Division 19 — RESERVED FOR FUTURE EXPANSION

Facility Services Subgroup:

- Division 20 — RESERVED FOR FUTURE EXPANSION

- Division 21 — Fire Suppression

- Division 22 — Plumbing

- Division 23 — Heating, Ventilating, and Air Conditioning

- Division 24 — RESERVED FOR FUTURE EXPANSION

- Division 25 — Integrated Automation

- Division 26 — Electrical

- Division 27 — Communications

- Division 28 — Electronic Safety and Security

- Division 29 — RESERVED FOR FUTURE EXPANSION

Site and Infrastructure Subgroup:

- Division 30 — RESERVED FOR FUTURE EXPANSION

- Division 31 — Earthwork

- Division 32 — Exterior Improvements

- Division 33 — Utilities

- Division 34 — Transportation

- Division 35 — Waterways and Marine Construction

- Division 36 — RESERVED FOR FUTURE EXPANSION

FIGURE 9.2 *(Continued)*

- Division 37 — RESERVED FOR FUTURE EXPANSION

- Division 38 — RESERVED FOR FUTURE EXPANSION

- Division 39 — RESERVED FOR FUTURE EXPANSION

Process Equipment Subgroup:

- Division 40 — Process Integration

- Division 41 — Material Processing and Handling Equipment

- Division 42 — Process Heating, Cooling, and Drying Equipment

- Division 43 — Process Gas and Liquid Handling, Purification, and Storage Equipment

- Division 44 — Pollution Control Equipment

- Division 45 — Industry-Specific Manufacturing Equipment

- Division 46 — RESERVED FOR FUTURE EXPANSION

- Division 47 — RESERVED FOR FUTURE EXPANSION

- Division 48 — Electrical Power Generation

- Division 49 — RESERVED FOR FUTURE EXPANSION

FIGURE 9.2   *(Continued)*

Although construction projects use a number of different delivery methods, products, and installation, effective communication among the people involved on a project is crucial to its successful completion. MasterFormat™ facilitates standard filing and retrieval schemes throughout the construction industry, since without a standard filing system familiar to each user, information retrieval would be almost impossible. This continuous effort to modify and enhance the MasterFormat™ is driven in part by the radical changes in the construction marketplace, particularly with the development of new technologies.

The new MasterFormat™ standard provides a master list of divisions, and section numbers and titles within each division, to follow in organizing information about a facility's construction requirements and associated activities. Standardizing the presentation of information improves communication among all parties involved in construction projects. A full explanation of the titles used in MasterFormat™ is provided, giving a general description of the coverage for each title. A keyword index of requirements, products, and activities is also provided to help users find appropriate numbers and titles for construction subjects. The current MasterFormat™

groups, subgroups, and divisions consist essentially of dividing the specs into 50 divisions. The MasterFormat™ 2004 Edition divisions are:

Specification section format

This provides a uniform standard for arranging specification text in a project manual's sections using a three-part format, and reduces the probability of omissions or duplications in a specification section. According to the CSI, "Rather than grouping administrative, product requirements, and execution requirements under each product separately, *SectionFormat* provides a uniform approach to organizing specification text within each section. *SectionFormat* is based upon the principle that a section should be organized by grouping the administrative requirements, product requirements, and execution requirements for each product together." Thus, each specification section covers a particular trade or sub-trade (e.g., drywall, carpet, ceiling tiles). Furthermore, each section is divided into three basic parts, each of which contains the specifications about a particular aspect of each trade or sub trade (Figure 9.3). The updates are intended to reflect changes in the industry related to advancements in information technology and electronic publishing.

CSI notes that "PageFormat offers a recommended arrangement of text on a specification page within a project manual by providing a framework for consistently formatting and designating articles, paragraphs, and subparagraphs. It also includes guidance for page numbers and margins." CSI states that recent updates to PageFormat focused on making documents easier to read without limiting specific technology or processing methods. The updated version offers greater freedom to use sophisticated publishing and electronic media techniques on a wider variety of display devices.

Part 1: General

This part of the specification outlines the general requirements for the section and describes the administrative, procedural, and temporary requirements unique to the section. Part 1 is an extension of subjects covered in Division 1 and amplifies information unique to the section. In general, it outlines quality control requirements, delivery and job conditions requirements, trades with which this section needs to be coordinated, and specifies what submittals are required for review prior to ordering, fabricating, or installing material for that section. It generally consists of the following:

1. *Description and scope*: This article should include administrative and procedural requirements specific to the section. It should also specify the scope of the work and the interrelationships between work in this section and the other sections. It should also include

Product Specification

Franchise: Hilton   Project: Evolution '05 Prototype
Brand: Hilton Garden Inn   Location: N/A

Item Code:	CA-11 Guestrooms
Item:	Guestroom

Issue Date:	07/22/2005
Print Date:	08/25/2006

Description:	Enhanced loop - tip sheer		
Pattern Name:	Custom		
Color:	Custom		
Pat Design No:	Custom	**Aprvd Strike-Off No:**	M45104
Dr No:	N/A	**Reference No:**	N/A

Dimensions

Carpet Width:	12 ft 0 in
Ptrn Repeat:	Width: N/A Length: N/A

Specs

Contents:	100% solution dyed nylon
Dye Method:	Solution dyed

Guage: 1/10   **Pile Height:** .250"   **Oz Wt-SY:** 32   **Tot Oz Wt-SY:** 74.10

Prm Backing:	Woven Polypropylene
Sec Backing:	ActionBack
Carpet Pad:	Pad to be Hartex Super pad. Fiber: Polyester; Weight: 40 oz; Density: 8.8; Thickness: 0.25-0.375; Breaking Strength: MD 32 lbs, CMD 31 lbs, ASTM 2262; Sound Absorption: NAC .61 ASTM C423-77; Thermal conduct: 2.02 ASTM C518-76.
Installation:	Stretch-in
Locations:	Guestroom
Fire Safety:	Material must meet NFPA-253 or ASTM E-648 as required by local and state fire codes for use in hotels and other public areas.

Notes:	1. Carpet installation shall comply with all carpet manufacturer standards.
	2. Certification of compliance required verifying that carpet meets or exceeds all local and state fire codes.
	3. Manufacturer to provide maintenance instructions to owner.
Special Notes:	1. Templeton approved strike-off number: 45104.
Quantity:	Quantity to be verified by installer before purchasing.
Vendors:	

Templeton Carpet, 1900 Willowdale Road NW, Dalton, GA, 30720
Ph: 706-275-8665 Fax: 706-275-0687
Contact: Kelly Barker, Templeton Carpet Ph: 281-807-7927 Fax: 281-807-7158

Changes to this specification must be approved by the Designer.   CA-11

FIGURE 9.3   An example of a product specification, in this case a carpet for a Hilton Garden Inn hotel chosen for the guest bedrooms. The vendor is Templeton Carpet. (Source: Paradigm Design Group.)

a list of important generic types of products, work, and specified requirements. In addition, it should include the following:

  a. Products supplied but not installed: List products that are only supplied under this section but whose installation is specified in other sections.

  b. Products installed but not supplied: List products that are only installed under this section but furnished under other sections.

  c. Allowances: List products and work included in the section that are covered by quantity allowances or cash allowances. Do not include cash amounts. Descriptions of items should be included in Part 2 or Part 3.

  d. Unit prices: Include statements relating to the products and work covered by unit prices and the method to be used for measuring the quantities.

  e. Measurement procedures: State the method to be used for measurement of quantities. Complete technical information for products and types of work should be specified in the appropriate articles of Part 2 and Part 3.

  f. Payment procedures: Describe the payment procedures to be used for measurement of quantities used in unit price work. Complete technical information for products and types of work should be incorporated in the appropriate articles of Part 2 and Part 3.

  g. References: List standards referenced elsewhere in the section, complete with designations and titles. Industry standards and associations may be identified here. This article does not require compliance with standards but merely a listing of those used.

  h. Definitions: Define unusual terms not explained in the contracting requirements and that are utilized in unique ways not included in standard references.

  i. Alternates: Check whether the acceptability of alternatives is detailed in the general requirements.

2. *Quality assurances*: To include prerequisites, standards, limitations, and criteria that establishes an acceptable level of quality for products and workmanship. To achieve this the following are required to be considered:

  a. Qualifications: List statements of qualifications of consultants, contractors, subcontractors, manufacturers, fabricators, installers, and applicators of products and completed work.

  b. Regulatory requirements: Describes obligations for compliance with specific code requirements for contractor designed items and for public authorities and regulatory agencies including product environmental requirements.

    c. Certifications: Includes statements to certify compliance with specific requirements.

    d. Field samples: Includes statements to establish standards used to evaluate the work with the assistance of field samples. These are physical examples representing finishes, coatings, or a material finish such as wood, brick, or concrete.

    e. Mock-ups: Includes statements to establish standards by which the work will be evaluated by the use of full mock-ups. These are full size assemblies erected for construction review, testing, operation, coordination of specified work, and training of the trades.

    f. Pre-installation meetings: Determine requirements for meetings to coordinate products and techniques, and to sequence related work for sensitive and complex items.

3. *Submittals*: Includes, but is not limited to, requests for certain types of documentary data and affirmations of the manufacturer or contractor to be furnished as per the contract. Includes requests for specific types of product data and shop drawings for review as well as submittal of product samples and other relevant information, including warranties, test and field reports, environmental certifications, maintenance information, installation instructions, and specifics for closeout submittals.

4. *Product handling, delivery, and storage*: To furnish instructions for various activities to include:

    a. Packing, shipping, handling, and unloading: Specify requirements for packing, shipping, handling, and unloading that pertain to products, materials, equipment, and components specified in the section.

    b. Acceptance at site: Describe the conditions of acceptance of items at the project site. This normally applies to owner provided products.

    c. Storage and protection: Outline special measures including temperature control that are needed to prevent damage to specific products prior to application or installation.

    d. Waste management and disposal: Affirm any special measures required to minimize waste and dispose of waste for specific products.

5. *Project and site conditions*: Determine physical or environmental conditions or criteria that are to be in place prior to installation, including temperature control, humidity, ventilation, and illumination required to achieve proper installation or application. Statements that reference documents where information may be found pertaining to such items as existing structures or geophysical reports. For example, all wall tiling should be completed prior to cabinet installation.

6. *Sequencing and scheduling*: This is required where timing is critical and where tasks and/or scheduling need to be coordinated and follow a specific sequence.
7. *Maintenance*: List items to be supplied by the contractor to the owner for future maintenance and repair. Delineate provisions for maintenance services applicable to critical systems, equipment and landscaping.
8. *Warranties*: Terms and conditions of special or warranty bonds covering the conformance and performance of the work should be spelled out and the owner should be provided with copies.
9. *System startup, owner's instructions, commissioning*: List applicable requirements to the startup of the various systems. Include requirements for the instruction of the owner's personnel in the operation of equipment and systems. State requirements for commissioning of applicable systems to ensure installation and operation are in full compliance with design criteria.
10. *System description*: Describe performance or design requirements and functional requirements of a complete system. Limit descriptions to composite and operational properties to the extent necessary to link multiple components of a system together, and to interface with other systems.

Part 2: Products

This section describes the materials, products, equipment, fabrications, components, mixes systems, and assemblies being specified and that are required for incorporation into the project. This section also details the standards to which the materials or products must conform so as to fulfill the specifications, and similar concerns. Materials and products are included with the quality level required. Included in the itemized sub-sections are:

1. *Manufacturers*: This section is used when writing a proprietary specification, and will include a list of approved manufacturers. The section should be coordinated with the product options and substitutions section. Names of manufacturers may be supplemented by the addition of brand names, model numbers, or other product designations.
2. *Materials, furnishings, and equipment*: A list should be provided of materials to be used. If writing descriptive or performance specifications, detail the performance criteria for materials, furnishings, and equipment. Describe the function, operation, and other specific requirements of equipment. This article may be omitted and the materials included with the description of a particular manufactured unit, equipment, component, or accessory. Environmental concerns such as toxicity, recycled content, and recyclability can be addressed here.

3. *Manufactured units*: Fully describe the complete manufactured unit, such as standard catalog items.

4. *Components*: Describe the specific components of a system, manufactured unit, or type of equipment.

5. *Accessories*: Describe requirements for secondary items that aid and assist specified primary products or are necessary for preparation or installation of those items. This article should not include basic options available for manufactured units and equipment.

6. *Mixes*: This section specifies the procedures and proportions of materials to be used when site mixing a particular product. This article relates mainly to materials such as mortar and plaster.

7. *Fabrication*: Manufacturing, shop fabrication, shop assembly of equipment and components, and construction details should be given. Specify allowable variations from specified requirements.

8. *Finishes*: Describe shop or factory finishing here.

9. *Source quality control*: Indicate requirements for quality control at off-site fabrication plants.

   a. Tests, inspection: Describe tests and inspections of products that are required at the source, i.e., plant, factory, mill, or shop.

   b. Verification of performance: State requirements for procedures and methods for verification of performance or compliance with specified criteria before items leave the shop or plant.

10. *Existing products*: Create a list of characteristics of assemblies, components, products, or materials that must match existing work, including matching material, finish, style, or dimensions. Specify compatibility between new and existing in-place products.

Part 3: Execution

The section describes the quality of work—the standards and requirements specified in the installation of the products and materials. Site-built assemblies and site-manufactured products and systems are included. This part of the specification specifies basic on-site work and includes provisions for incorporating products in the project. It also describes the conditions under which the products are to be installed, the protection required, and the closeout and post-installation cleaning and protection procedures. The sub-headings in this section include:

1. *Inspection*: Define what the contractor is required to do, for example to the subsurface, prior to installation. Sample wording may include, "The moisture content of the concrete should meet manufacturer's specifications prior to installation of the flooring material."

2. *Preparation*: Specify actions required to prepare the surface, area, or site to incorporate the primary products of the section. Also stipulate the

improvements to be made prior to installation, application, or erection of primary products. Describe protection methods for existing work.

3. *Installation, construction, and performance*: The specific requirements for each finish should be specified, as well as the quality of work to be achieved and includes:

   a. Special techniques: Describe special procedures for incorporating products which may include spacings, patterns, or unique treatments.

   b. Interface with other work: Include descriptions specific to compatibility and transition to other materials. This may include incorporating accessories, anchorage, as well any special separation or bonding.

   c. Sequences of operations: Describe the required sequence of operations for each system or piece of equipment.

   d. Site tolerances: State allowable variations in application thicknesses or from indicated locations.

4. *Field quality control*: State quality control requirements for on-site activities and installed materials, manufactured units, equipment, components, and accessories. Specify the tests and inspection procedures to be used to determine the quality of the finished work.

5. *Protection*: Where special protection is necessary for a particular installation, such as marble flooring, this section must be included. Provisions for protecting the work after installation but prior to acceptance by the owner should be cited.

6. *Adjust*: Describe final requirements to prepare installed products to perform properly.

7. *Clean*: Describe in detail the cleaning requirements for the installed products.

8. *Schedules*: To be used only if deemed necessary. When schedules are included indicate item/element/product/equipment and their location.

9. *Demonstration*: State installer or manufacturer requirements to demonstrate or to train the owner's personnel in the operation and maintenance of equipment.

The CSI also gives the following sample outline to illustrate its three-part SectionFormat™ with respect to environmental specifications of products:

Part 1. General

1. Environmental Requirements

   1. List applicable environmental standards, regulations, and requirements.

   2. Include VOC requirements.

   3. List recycled content requirements.

4. Identify reuse, recycling, and salvaging methods.

5. Reference Division 1 Environmental Procedures for Construction.

i) VOCs or chemicals to avoid.

ii) General environmental procedures.

iii) Reuse, recycling, or salvaging requirements.

iv) Healthful building maintenance.

Part 2. Products

2. Specific Environmental Product Attributes

1. Product contains no xxxx chemicals (list and identify).

2. Product contains xx percent recycled content:

i) Identify postindustrial recycled content.

ii) Identify postconsumer recycled content.

3. Product is recyclable after useful life.

4. Product is certified by an independent third party.

i) Recycled content.

ii) Sustainably harvested.

5. Product is durable (list warranty).

6. Product is moisture resistant (if applicable).

7. Include any other environmental attributes.

Part 3. Execution

3. Environmental Procedures

1. Address environmental installation of materials.

2. Include protection of materials.

3. Identify environmental methods of cleanup.

4. Include recycling of scrap during construction.

5. Reference Division 1 Environmental Procedures.

The primary purpose of sample specifications is basically to supplement rather than replace the standard specifications. Model green specs are designed to be edited, adapted, and incorporated into the standard specifications of building projects and generally augment standard specifications by providing additional environmental information, such as sustainable building criteria, definitions, and performance requirements.

9.6 COMPUTERIZED SPECIFICATION WRITING SYSTEMS

The era of manually prepared building specifications is rapidly entering the annals of history. Indeed, over the past decade, a number of firms have emerged that have developed various versions of automated specification writing systems and many now offer these services on-line to architects, interior designers, engineers, and others. And while CAD has revolutionized the drawing process, many architects and designers no longer possess the knowledge, expertise, or interest to write specifications for their projects. Computer software has reduced the need for some of the traditional skills shaped by years of experience in writing construction specifications. Computer resources offer practitioners greater efficiency and the ability to deal with continually expanding compilations of information and provide almost immediate access to information from thousands of applicable electronic databases. This is coupled with an increasingly complex construction industry, changing methods of procurement, and the tremendous pressure on architects and designers to prepare contract documentation, including specifications of the highest quality and in less time.

This is further complicated by the fact that the complexity of new commercial construction requires continually updated knowledge. The changing nature of the industry with new energy-efficient products, laws, environmental regulations, techniques for assembling products, and building industry practices cannot be adequately presented in educational courses for engineers and architects. Specialists are therefore urgently needed to manage and update information gathered from the many new products and innovations in the construction industry.

Of the more popular automated CAD specification packages on the market is Building Systems Design's (BSD) SpecLink®. This is an automated master guide and specification management system for production of specifications, with built-in intelligence designed to help you significantly speed up editing tasks and reduce specification production time while minimizing errors and omissions. Combined with the industry's most comprehensive and up-to-date master database, SpecLink enables you to accelerate your specification development with tremendous accuracy and integrity.

The software system includes specification sections designed for use in construction documents, short form specifications, and design criteria documents. SpecLink uses master guide specifications in CSI three-part format, and contains a database of over 780 master specification sections and over 120,000 data links that automatically include related requirements and exclude incompatible options as you select specification text. BSD also developed the PerSpective® early design performance specifications organized by CSI UniFormat, which is the industry's first commercially available database of performance-based specifications (Figure 9.4).

InterSpec, LLC is another firm that uses a proprietary technology that provides construction document management solutions and services built on its patented e-SPECS® specification management technology

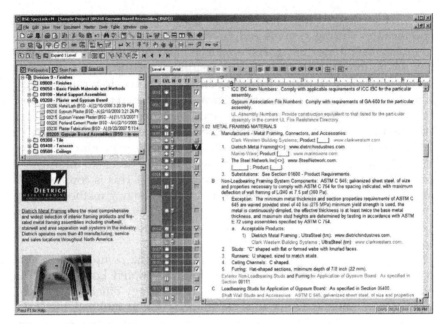

FIGURE 9.4 BSD's SpecLink summary catalog listing and computer screen printout. SpecLink is one of the many electronic specification services that have emerged in recent years. (Source: Building Systems Design, Inc.)

(Figure 9.5). e-SPECS software automates the specification process by extracting the product and material requirements directly from the project drawings; it also connects a large database of building specifications to an electronic architectural drawing of the project. Furthermore, e-SPECS integrates directly with Autodesk's AutoCAD®, AutoCAD® Architecture, AutoCAD® MEP and all Revit®-based applications in addition to supporting all libraries of MasterSpec®. Interspec also has a do-it-yourself program designed for architects and designers with small projects.

For architects and engineers who spend many hours on every project preparing construction specifications, e-SPECS software is a proprietary browser-based specification management system that saves time and money while ensuring that the specifications agree with the construction documents. With this system the designer can also access the specs through the Internet and make alterations as the specs are being written. Like other automated systems, e-SPECS will enable design professionals to increase their productivity while simultaneously reduce their costs. Also, by linking the architect's CAD drawings to the master guide specifications, there is no longer the need to mail or deliver large blueprint drawings to the spec writer. Moreover, with these automated systems, the designer can input all required information at the earliest phases of the project, before any

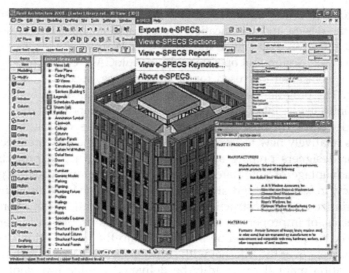

FIGURE 9.5 e-SPECS is an automated computer software system that integrates specifications with CAD and BIM applications. It can extract all the project requirements from the Revit model and allows direct access to the specs directly within Revit, facilitating redlining and commenting in collaboration with the team members. (Source: InterSpec, Inc.)

drawings are even available, in addition to being able to almost instantly obtain an outline or preliminary specification of the project. InterSpec, Inc. has recently released version 4.2 of e-SPECS, with new functionality and Building Product Manufacturer (BPM) BIM content.

SpecsIntact® System (Specifications-Kept-Intact) is another automated software system available for preparing standardized facility construction specifications used in facility construction projects worldwide. SpecsIntact was initially developed by NASA to help architects, engineers, specification writers, project managers, construction managers, and other professionals doing business with the three government agencies using it—the National Aeronautics and Space Administration (NASA), the U.S. Naval Facilities Engineering Command (NAVFAC), and the U.S. Army Corps of Engineers (USACE).

According to its authors, the system provides quality assurance reports and automated functions that reduce the time required to complete project specifications. The Unified Facilities Guide Specifications (UFGS) Master that is employed by SpecsIntact is divided into functional divisions according to the Construction Specification Institute (CSI) format, with each division containing related specification sections. The principal elements within the sections are annotated using SpecsIntact's application of the Extensible Markup Language (XML). This tagging scheme provides the intelligence that SpecsIntact uses in automatically processing these sections.

20-20 CAP Studio is an integrated package of applications that automates the design and specification process, and contains two base applications—20-20 CAP Designer and 20-20 CAP Worksheet. CAP Designer is a CAD-based design tool operated within AutoCAD. CAP Worksheet is a power specification tool used for product pricing, specification, and estimating. The program has the ability to import complete large-scale space plans and layouts into CAP Worksheet for full specification, discounting, and order entry. The program specializes in furniture specification.

This relatively new technology is transforming the way architects and interior designers prepare specifications for construction projects. The main advantage of automated systems is that they can provide greater accuracy, in less time, and at a lower cost. These systems also eliminate or minimize costly construction modifications caused by omissions, discrepancies, or improper quality controls. A firm's proprietary interactive online editing systems can be integrated into the specification development process over the Internet with secure password access. A completed specification manual can readily be delivered on-line for client downloading, and can easily be printed and bound, or presented on CD-ROM. For smaller design firms that lack resources, outsourcing may be worth considering as an effective way to proceed.

9.7 LIABILITY CONCERNS

All professionals, including architects and engineers, are expected to exercise reasonable care and skill in the implementation and execution of their work. The level of performance by professionals should be consistent with that normally provided by other qualified practitioners under similar circumstances. However, this does not imply that projects will be executed to 100 percent perfection at all times.

Law relating to professional responsibility and liability has in recent years become very active and has assumed unprecedented urgency. The parameters of risk and exposure have expanded dramatically in professional practice, so that under current law, if a professional designer enters into a contractual agreement and specifies a subsystem of a commercial or institutional space, he/she assumes responsibility for the satisfactory performance of that system. Another area that is emerging and that is causing significant concern is exposure relating to the liability of the professional designer to third parties who have no connection with the contract for claims of negligence or design errors that allegedly lead to injury of persons using the building. The legal bases for the majority of liability suits often overlap, but generally include professional negligence, breach of contract, implied warranty or misrepresentation, joint and several liability, and liability without fault for design defects. Designers may now be considered professionally negligent and in breach of contract if they fail to reject defective work by a contractor or supplier.

Product liability, i.e., building product performance, is another area of exposure in which the architect is held responsible for damages caused by faulty materials and components and sometimes for the cost of their replacement. This additional burden places a heavy emphasis on the appropriate selection and specification of building products with long records of satisfactory performance. This discourages the introduction of new materials (e.g., green products), and methods.

Product liability is primarily concerned with negligence and is discussed in Chapter 11. It especially affects manufacturers, retailers, wholesalers, and distributors. Furthermore, with the upsurge of green delivery systems, designers and specifiers are increasingly finding themselves involved in product liability suits. The best way to minimize these product liability actions is by specifying products that are manufactured for the intended use and that have been adequately tested.

Finally, design professionals can protect themselves from potential liability suits by working within their area of expertise, using concise contracts and specifications, complying with codes and regulations, using reputable and licensed contractors, maintaining accurate records, and maintaining legal counsel and ensuring that adequate and appropriate liability insurance is in place.

10

Litigation and Liability Issues

10.1 GENERAL OVERVIEW

In recent decades American society has witnessed a "litigation explosion." This increasing litigation has forced the construction industry to change in a manner that actually has the potential to improve the quality of the constructed project. The dramatic rise in building litigation has

Doi: 10.1016/B978-1-85617-676-7.00010-5

also produced an increased urgency for finding specialist lawyers, forensic experts, and consultants. Lawsuits have become a grave concern, although the vast majority of construction disputes continue to be settled before they go to court, partly due to the skyrocketing costs of traditional litigation. This has encouraged the formation of risk management programs that focus increasingly on partnering, quality assurance, and quality control.

Despite more than a decade of research and litigation and billions of dollars lost to insurance claims and lost productivity, issues like SBS, BRI, and IAQ are still prevalent in the commercial buildings sector. Most of these efforts have been reactive, i.e., a problem is reported, its cause is identified, and then it is corrected. And while that may be the less expensive, apparently practical solution, it is failing to stem the rising tide of these costly, debilitating, and otherwise avertable painful solutions.

Mold-related claims have increased dramatically and have become another primary concern for the insurance industry, as have the number and size of water-related property claims. Many insurance companies are struggling to find ways to address mold claims, and in most cases are now entirely excluding mold from their coverage. The insurance industry is further responding by changing policy language, claims-handling procedures, and loss reserving, while continuing to try to manage the regulators. Some architects and engineers have also reacted to this problem by eliminating the term "supervision" from their contractual responsibility during construction and replacing it in their contracts with the terms "observation" or "inspection" (which more correctly describes their services) in an effort to reduce their professional liability.

Building projects are generally required to adhere to zoning and building code requirements. Projects that fail to do so can adversely impact the consultant and expose him/her to multi-million dollar litigation. The construction industry in recent years has had to deal with higher premiums for all types of insurance, but following 9/11, those costs have skyrocketed beyond all expectations. Legal claims for all types of building envelope failures continue to rise and are typically made against developers, contractors, property management corporations, architects, engineers, building trades, and government authorities, among others.

Although the green building movement is having a very significant and positive impact on the construction industry as a whole, there are aspects such as the risk of liability and "greenwashing" (misleading environmental benefits of a product or service) that are causing considerable disquiet and which need to be addressed. Thus, although green buildings are generally more efficient users of energy and materials, resulting in reduced safety issues for the different systems, they nevertheless sometimes use non-standard materials and systems, which may result in an increased risk of failure for the materials or systems incorporated in green buildings. To minimize these risks, qualified designers should be employed to ensure that the design process is correctly implemented. In this respect, Ward Hubbell, president, Green Building, says, "One of our most pressing issues is the fact that some buildings designed to be green fail to live up to expectations. And in business, as we all know, where there are failed expectations there are lawsuits. All practices and/ or products that could possibly result in a firm's exposure and liability should be clearly identified. The good news is that this period of increased legal action, or the threat thereof, will in fact motivate the kind of clarity and measurement that both reduces liability risks and results in better buildings."

In addition to contractors and subcontractors, design professionals are also highly vulnerable to claims from clients, owners, and users. This vulnerability is partly due to the failure by some architects and engineers to understand the challenges that professionalism impose. It goes without saying that design professionals need to be above reproach in every aspect of dealings with others and in the management of the firm. Moreover, design professionals may need to concentrate more than most professionals on maintaining good relationships with colleagues and coworkers to meet the many potential challenges they will face. Not only will this minimize claims, but it will also help to attract clients.

Communication failures are often at the center of lawsuits. Litigation in its many forms is usually the result of a breakdown in understanding between the parties involved, either in the interpretation of the contract documents or in the practical working communications between the various parties and others on the construction scene. When the conflicting parties are unable to reach agreement, the courts are the final resort to resolve the situation and called upon to decide what was communicated based upon case law. In the construction industry, problems and failures are most often the result of defects due to design, negligence, or execution or else due to deterioration, which is a natural process unless excessive or rapid deterioration takes place.

According to Carl de Stefanis, president of Inspection & Valuation International, Inc. (IVI), a prominent construction consulting and due diligence firm, claims against firms providing due diligence services is an increasing concern. De Stefanis states that roughly 80 percent of the claims are building envelope related, including roofs, EIFS, windows, masonry, etc. Not far behind are claims relating to building code issues, followed by mold-related claims. The number and size of both mold litigation and water-related property claims have witnessed a dramatic rise over the last two decades, forcing many insurance companies to grapple with the challenges of how to address these issues. Several carriers have decided that the most expedient way may be to entirely exclude mold from their coverage (Figure 10.1).

FIGURE 10.1 Illustrations showing heavy mold infestation caused by water penetration left unattended. Mold litigation has emerged as a major issue in recent years. (Source: Servpro Industries, Inc.)

The frequency with which acronyms and phrases like IAQ (indoor air quality), IEQ (indoor environmental quality), Sick Building Syndrome (SBS), and Building Related Illness (BRI) are tossed around have encouraged building owners and managers to just shrug them off. This is surprising since all indications are that the incidence of commercial buildings with poor IAQ and the increase in litigation over the consequences of poor IAQ are quite significant. These increases will have an obvious impact on insurance carriers, which pay for many of the costs of health care and general commercial liability.

In cases where an action is brought against an architect or engineer, an expert in the same discipline will likely be required to give an opinion as to whether negligence was a factor in the design, execution, or performance of duty. In the majority of cases however, the investigation will involve much more than expert opinion; for example, laboratory and other tests may be recommended to help determine the cause of failure. The role that experts are required to play will therefore vary depending on the case.

Among the more commonly requested legal services involving architectural and engineering experts are issues such as building envelope investigations, structural failure assessments, exposure reconstruction, assessments involving mold growth in buildings, and construction defect evaluations. Sometimes consultants are asked to attempt reconstructing events that took place years ago. The expert consultant's conclusions will normally be used by the client to evaluate a claim's strength, as well as an evidentiary tool in ensuing dispute resolution proceedings. But in addition to consulting and expert testimony services for both defendants and plaintiffs, the expert may be required to perform case evaluations, assist with settlements, and provide advice on alternative dispute resolution (ADR) and litigation avoidance.

It is indeed fortunate that in today's litigious environment, it is estimated that more than 90 percent of disputes in the construction industry find resolution before they ever see the inside of a courtroom. Most construction-related disputes are resolved through some form of settlement or alternative dispute resolution (ADR). In fact in some jurisdictions, courts have now made it mandatory for some parties to resort to ADR of some type, usually mediation, before allowing the parties' cases to be tried in court. ADR's rising popularity is due to some degree to the tremendous overload of cases in the courts, and the knowledge that ADR is normally less expensive than traditional litigation. Added to this is the fact that parties sometimes feel a need to have greater control over the selection of those who will decide their dispute, in addition to the preference for confidentiality.

10.2 LIABILITY ISSUES

Due to the extreme complexity of liability issues, it is not possible in this section to address the many concerns and legal matters that may arise with regard to liability topics, and builders, manufacturers, and designers are strongly advised to consult their attorneys and professional liability insurance carriers and agents for advice on these matters. While building owners and managers are rarely expected to guarantee the safety or well-being of their tenants, visitors, and guests, they are required to exercise reasonable care to protect them from foreseeable events.

Some of this increased concern toward liability issues may be due to the enormous upsurge in interest in green buildings, which has resulted in many misconceptions and exaggerations put forth by owners, designers, manufactures, and distributors. "Greenwash" is a general term often used within the industry for this form of misconception. It is a term that can be applied to building materials, systems, buildings, or companies, among other things. Greenwash has the potential to ultimately discredit the entire green building industry, in addition to being the source of numerous lawsuits, when the ultimate declared goals of green buildings are not achieved. These claims can be categorized into two basic groups: 1. Materials, and 2. Performance related.

1. Material Claims

Because of the lack of preciseness in what constitutes a green building, material, or system, providers will frequently find a material property with limited green characteristics, and often market this property, and the material, as being "green." As an example, a material that is composed of recycled content might also contain unacceptable amounts of urea-formaldehyde in its production; and even though the material's overall impact on the environment is negative, this material is erroneously marketed as green. Such false claims have also occurred when material or system providers base their claims on unreliable and inaccurate information. But as green products and the green building industry becomes better understood, and as processes such as life cycle analyses become more mainstream, these risks should begin to decline. The employment of reliable material rating systems should also reduce the plethora of false claims that currently plague the market.

2. Performance Related Claims

A frequently cited phenomenon within the green building industry is the misrepresentation of a person or building stakeholder company's knowledge and expertise regarding green building. When building owners and other building stakeholders rely on this professed expertise, the result can be a dismal failure of the green building to achieve its stated goals.

In considering a building's operational performance, Ward Hubbell, president of the Green Building Initiative (GBI) notes, "There is an expectation that green buildings will, in addition to reducing environmental impacts, offer lower energy and water costs, less maintenance, and other long-term benefits to the building owner. However, while the design may incorporate a wide range of green features, there are, of course, a tremendous number of variables between a building's design and occupancy that can impact operational performance. These potential areas of misunderstanding can be mitigated by following good business practices that facilitate clear communication and common expectations between building owner, designer, and rating organizations."

When building owners, designers, and builders disagree in their interpretation of what constitutes a successful green building, problems tend
to emerge, particularly when building owners fail to explicitly delineate
and communicate their thoughts at the commencement of the project.
These issues and problems tend to compound when the parties lack an
understanding of the concepts of the green building process. However, a
building's failure to achieve a promised level of green building certification can be problematic in that it could impact the building owner's ability to qualify for a grant or tax incentive, on which the owner may have
relied to assist in offsetting the project's initial costs. As for public buildings, certification may indeed be required by law.

Herbert Leon Macdonell, author of "The Evidence Never Lies," rightly
says, "You can lead a jury to the truth, but you can't make them believe
it," which is why good field notes and photographs are imperative as they
form the basis of solid documentation. In addition to being accurate and
articulate, field notes should be written in a manner that is clear, neat, leg-
ible, and self-explanatory as they provide the first hand recorded observations and are irreplaceable. Likewise, photographs are usually required to
provide a visual record and are cardinal to forensic investigations, particularly with issues such as mold, failures, etc. Photographs should be of
the highest quality (preferably high quality digital) and taken from different positions so as to get a comprehensive overview of the scene in question. Photographs should also be well annotated and filed appropriately.
Digital photographs may be stored on a computer or on CD. Whether it's
a failure-related or performance-related issue, it may be advisable for the
forensic/consultant expert to supplement documentation of the project or
scene with video photography. This will depend largely on the circumstances prevailing at the time and the documentation already in hand.

Experience has shown that assigning culpability for green disputes
most often boils down to a matter of negligence, ignorance, or incompetence. American courts often require that qualified experts testify to the
standard of care that is applicable to the case in dispute, and to the professional's performance as measured by that applicable standard of care.
However, the principle of standard of care may not apply to building
contractors, since they are not legally considered to be "professionals"
with respect to making independent evaluations and judgments based on
learning and skill. Builders are nevertheless held to a "duty to perform,"
meaning that the provisions of their contracts require them to strictly follow the plans, specifications, and contract documents.

Dispute Resolution

In past AIA agreements the standard dispute resolution provisions
called for non-binding mediation as a condition precedent to binding
arbitration. Under the new design-build documents, this requirement of

mediation as a condition precedent to other forms of dispute resolution remains, although the new documents now offer the parties three methods to choose from for dispute resolution. The parties have a choice of binding arbitration, litigation, or a third method to be decided upon by the parties themselves. If the parties fail to choose a binding method of dispute resolution, litigation becomes the selection by default.

It should be noted that while the terms and conditions in the 1996 family of design-build documents incorporated the AIA A201 General Conditions of the Contract for Construction, the AIA documents A141 Agreement between Owner and Design-Builder no longer incorporates the A201 form. It now contains its own general conditions as Exhibit A to the agreement. Likewise, the AIA document A142 Agreement between Design-Builder and Contractor also contains its own general conditions.

The architect traditionally serves as the owner's representative and handles much of the project administration, including the certification of substantial completion of the project. On a design-build project however, the administrative functions of the architect who is part of the design-build team are significantly different. Recognizing this, the AIA has prepared a new form entitled "G704/db Acknowledgment of Substantial Completion of a Design-Build Project." which requires the owner to inspect the project and acknowledge the date when substantial completion occurs. The form is a variation of the G704-2000 Certificate of Substantial Completion.

10.3 RESPONSIBILITY FOR FAILURE, NEGLIGENCE, AND STANDARD OF CARE

Most failures are the result of interrelated multiple causes. Even when there is agreement among the parties that the failure was due to a single technical cause, the responsibility may still not be clear-cut given the complexity of project delivery systems. And as new high technology methods of project delivery are introduced we will witness a corresponding evolution in legal liability interpretation.

Liability issues are further complicated by the fact that architects, engineers, and builders all have obligations to third parties beyond those included in the project contracts. It is the third parties that usually are the ones to submit injury and property damage claims in the event of physical, product, or performance failures. The legal responsibility of design professionals clearly extends beyond the party with whom they have contracted to any other party who may be injured by an alleged act of negligence.

The principal factors that can impact the legal issues and interests associated with construction failures are:

1. Type of failure.
2. Cause of failure.
3. Affected parties and their interests in the failure.

Although each issue is important, the strategy to be employed to effectively represent the party's legal interests should be based mainly on the facts relevant to the failure. These facts can usually be determined through documentation, interviews, and public records, and represent the baseline from which any legal analysis begins. This is one of the reasons why it is so important for the legal and technical teams to be interwoven from the outset of any investigation. It is also important from the outset to determine the ultimate goal of conducting a legal analysis regarding the failure. For example, it could be to determine the cause of the failure, legal responsibility, and whether any steps need be taken to avoid causing further harm or damage.

In the majority of these types of cases, a methodical team approach of both legal and forensic expertise is adopted to help achieve the best possible outcome. The overall goals are interdependent and their individual importance may vary at different points in the investigation progress. Moreover, the legal concerns surrounding a construction failure will vary with the roles and responsibilities of the different parties. The main interest of project owners is in preserving their rights against potentially responsible parties, and in having the project returned to its pre-failure condition, whereas design professionals may be concerned about their legal exposure to injured parties. And although the various parties have different interests, there are a number of actions that the interested parties need to observe and act upon when a failure occurs. The initial steps each party should typically take include the formation of an investigative team; the team will develop a plan of action and deal with public agencies and the media as well as protect confidentiality.

The elements of the investigative team will depend on several factors, such as the type and size of the failure in question. Moreover, the leader of the team should be a senior level member of the client organization who will have the respect and support of the organization. All persons involved in the investigation should take an objective view during the investigation of the cause of failure. The investigative team should include independently retained qualified consultants to assist in the investigation and in the analysis of the various components of the project. Depending on the magnitude and type of failure, additions to the team may include an attorney as lead counsel to assess the viability of legal claims and defenses and one or more forensic experts to study the technical causes of the failure. Likewise, the team should include persons who have personal knowledge of the project, particularly concerning work relating to the failure in question.

While not all disputes and litigation are related to construction failures, a good percentage of the major cases are. Thus when a structural failure does occur, the contractor needs to be prepared. The first steps following a failure are critical to conducting the investigation as well as to preventing further potential damage or loss of life. The key factor to avoiding or mitigating the impact of conflicts is having the appropriate mechanisms in place to manage related issues, and acting proactively before such issues arise. Disputes in the construction industry are common, but they are not inevitable.

The actions taken immediately following a failure can have a tremendous impact on the outcome of any subsequent technical investigation. This is because much of the evidence associated with a collapse or structural failure is often of a perishable nature and needs to be preserved and protected. An investigation cannot succeed in achieving its objectives to determine the cause of failure and correctly attribute responsibility for the failure if the evidence scene is tampered with because the existing failure scene, condition, and other circumstances on the site become critical evidence. This evidence will play a crucial role in determining the most important factors that may have caused the failure, and thus directly impact an investigation's outcome. Moreover, it provides a major element in assisting in the development of hypotheses and theories to the cause of failure, which is why the site should be immediately protected, completely documented, and appropriately recorded. This process can be greatly facilitated if the owner and contractor cooperate with the investigation.

A typical failure is demonstrated by the partial roof collapse in July 2007 of a luxury high-rise building in Greenwood Village, Colorado being constructed by Beck Development, a construction company from Dallas, Texas (Figure 10.2). The permits show the building will have 261 units. The plans also call for restaurants, shops, a spa, and a theater. The cause is under investigation.

Document Collection

One of the earliest and top priority actions to be taken when a failure occurs is to compile project documents. The primary sources for these documents are obviously the design and engineering consultants, the contractor(s), and the owner. In addition to the contract documents (construction drawings: architectural, structural, electrical, etc. and specifications), the principal documents required to conduct a preliminary evaluation include:

- Shop drawings, assembly drawings, and other contractor submittals
- Test reports

FIGURE 10.2 Fourteen people were injured when a section of the roof of the "The Landmark," a $140 million dollar high-rise building in Greenwood Village, Colorado collapsed onto the 13th and 12th floors as concrete was being poured for the building's roof which was under construction. (Source: CBS Broadcasting Inc.)

- Boring logs
- Engineer of record calculations
- Change orders, warranties, etc.
- Construction monitoring photographs and reports
- Relevant correspondence.

Interviews

Once an investigation team has been commissioned to investigate a defect, system failure or collapse, the person in charge should immediately make efforts to interview eyewitnesses and other persons such as project personnel who can provide useful information regarding the investigation and which may assist in determining the probable causes of the failure. It is important to conduct these interviews as soon after a reported incident as possible, but especially prior to its "contamination" and while the recollection is still fresh. The specific information that is sought from interviewees depends largely on the type of failure and the circumstances of the failure.

Cooperation between Experts

Opportunities often avail themselves during an investigation for experts to cooperate with other experts with whom there may be common interests. It allows the pooling of resources and avoids unnecessary duplication of effort. Other areas of potential collaboration include possible destructive testing, identification of debris and relevant components, as well as the possible sharing of interview information. But while sharing basic information and fostering cooperation is often desirable and may greatly facilitate an investigation, nevertheless, in situations where there is a possibility of litigation, the client or the client's attorney should be consulted to ensure that any sharing of information does not inadvertently compromise the interests of the client.

Preliminary Assessment

A preliminary evaluation should emerge from the initial investigation of possible failure scenarios and possible contributing factors after the input of the various consultants, eyewitnesses, and staff members are taken into consideration and recorded. This should be followed up by initiatory structural analyses and tests to try and determine the viability of initial hypotheses as well as identify other possibilities that may have triggered the failure such as excessive occupancy loading or environmental factors (e.g., excessive snow load on roof), strong winds—greater than the building envelope was designed to withstand, etc.

It is prudent for the expert to fully understand the client's and attorney's strategy and objectives before proceeding to advise the client and attorney regarding photographs, testing, or measurements, which must be taken as soon as possible to preserve the evidence. The expert must also confer with the client regarding analytical procedures and obtain approval to implement them, and also to identify important documents that need to be obtained. But whatever the case, forensic investigations should be conducted on the assumption that the consultant will be required to offer sworn testimony on the investigation and conclusions reached in a court proceeding. As court procedure rules vary from one jurisdiction to another, the client's attorney should provide specific guidance in this regard.

Negligence

This occurs when a person fails to exercise the standard of care that a reasonable, prudent person would have exercised in a similar circumstance. Thus, the bottom line in assigning culpability for failure issues or construction disputes is often narrowed down to a result of negligence, ignorance, or incompetence. Gross negligence is when a person or party shows unrestrained disregard of consequences; where ordinary care is not taken in circumstances where, as a result, injury or grave damage is likely. Negligence is said to occur when something is omitted that ought to be done. With respect to design professionals, evidence of negligence is often detected in the preparation of contract documents, for example in the lack of coordination between construction drawings and site conditions or in discrepancies between building plans or specifications and shop drawings. Should a design professional be shown to be negligent, his/her license to practice may be temporarily or permanently revoked. However, while design professionals are required to possess and apply the same degree of skill, knowledge, and ability of other members of their profession and are required to exercise a standard of care and their best judgment in executing the assignment, they cannot be expected to guarantee a perfect set of plans or contract documents or provide guarantees that the results will always be perfect.

It is important for design professionals not to undertake projects that are clearly beyond their technical abilities or those of the personnel available to work on the assignment. Only experienced, competent, and qualified staff should be assigned to a task. Junior and inexperienced personnel must be carefully supervised by fully qualified professionals. Outside consultants sometimes need to be retained to supplement the firm's own capabilities.

Standard of Care and Duty to Perform

The concept of the standard of care has been described as the line between negligent and non-negligent error. The doctrine of reasonable standard of care basically implies that one who undertakes to render services in the practice of a profession is required to exercise the skill and knowledge that members of that profession (whether, architect, engineer, contractor, etc.) normally possess. This is a promise, which is really all a client should expect from a professional; be it an architect, engineer, or other. Since the designer by providing professional services is essentially exercising judgment gained from experience and learning, and typically provides those services in situations in which certain unknown or uncontrollable factors are common, a certain degree of error in those services is allowed and should be expected. Thus, with respect to engineers for example, key points to remember relating to standard of care include:

- Possess learning and skill ordinarily possessed by reputable engineers practicing in the same or similar locality and under similar circumstances
- Exercise care and skill ordinarily possessed by reputable engineers practicing in the same or similar locality and under similar circumstances
- Use reasonable diligence and best judgment in the execution of the project
- The objective is to accomplish the purpose for which the engineer was employed.

Regarding contractors, California courts have consistently held that:

- The standard of care applicable to negligence claims against a contractor is that standard of a licensed contractor under the circumstances; and,
- Therefore, expert testimony is required as to the standard of care itself, as well as to a defendant's compliance with it.

Therefore, any reference to what a contractor defendant "should" have done under a reasonable person standard would not only be irrelevant, but it would also be improper and prejudicial.

Raymond T. Mellon, Esq., senior partner with the law firm Zetlin & De Chiara, LLP, says, "It is imperative to note that the standard of care is kinetic and continually evolving. Both events and technology can and do affect and change the standard of care. Recent technological changes in the last 20 years have mandated revisions to building construction and safety. For example, many building codes now require various types of computerized fire safety devices, smoke detectors, strobe lights, and other safety features not available 20 years ago. While complying with statutory requirements for safety in building design is straightforward, the important issue raised by 9/11 is what design changes and technological advances must be incorporated into a building in the absence of statutory mandates. The fact that a building code in a particular municipality has not yet been amended to include new safety features or technological advances, does not, by itself, provide a safe haven from liability for damages incurred by a terrorist attack."

The investigation and testimony of an expert consultant/witness can have a devastating impact on reputations and people's livelihoods even though the service is required to determine the cause, assign responsibility, and prevent the repeat occurrence of a failure. For this, and other reasons, it is most important that forensic experts understand the accepted standards that designers and contractors are expected to meet. For example, is the standard of care bar raised when the designer or contractor is a LEED™ AP? The fact that a consultant makes a mistake, and that mistake causes injury or damage, is normally insufficient to lead to professional liability on the part of the consultant. In order for there to be professional liability, it must be proven that the services offered were professionally negligent, meaning that they fell beneath the expected standard of care of the profession. While it may be obvious that design professionals today have a duty to practice sustainable design, that professional standard of care may or may not rise to the level of promising the achievement of, e.g., a certain LEED certification level.

R.T. Ratay, a forensic engineer expert, defines the standard of care simply as, "That level and quality of service ordinarily provided by other normally competent practitioners of good standing in that field, when providing similar services with reasonable diligence and best judgment in the same locality at the same time and under similar circumstances."

Qualified experts are required to testify in American courts to the standard of care that was applicable to the case on trial, and these qualified experts testify to the professional's performance as measured by that applicable standard of care. The principle standard of care, however, does not apply to building contractors, since they are not deemed to be "professionals" in the sense of making independent evaluations and judgments based on learning and skill. Nevertheless, they are held to a "duty to perform," which basically means that they must strictly adhere to the plans, specifications, and provisions of their contracts and the contract documents.

10.4 TRADITIONAL LITIGATION VERSUS ALTERNATIVE DISPUTE RESOLUTION

10.4.1 Traditional Litigation

The traditional litigation process can be complex and drag on for years; it is extremely costly and continues on an upward trend. Traditional litigation necessitates the observance of certain protocols regarding rules of evidence and procedures for such things as reports, pretrial discovery techniques, interrogatories, depositions, and direct/redirect and cross/re-cross examination. It is helpful if the consultant, building owner, and product manufacturer have a basic understanding of the different stages and procedural details of civil lawsuits. Litigation procedures are typically governed by statute in each jurisdiction. Federal and civil trials are normally governed by the *Federal Rules of Civil Procedure* (FRCP). Although each state has its own rules of civil procedure, in many states these are similar to the FRCP. Moreover, many courts have their own local rules of procedures that supplement federal or state rules.

Traditional litigation lawsuits can be divided into two basic stages: 1. Pretrial, and 2. Trial.

In American law, discovery is the pretrial phase in a lawsuit in which the parties are identified and the issues in dispute are clarified. In addition, each party is given an opportunity to learn about the other party's witnesses and potential evidence. Each party can request documents and other evidence from the other parties or can use a subpoena or other discovery devices to compel the production of evidence and depositions.

If a lawsuit has been filed against you or your firm, the next thing the plaintiff must do is "serve" you with the "summons" and "complaint." In most jurisdictions, traditional litigation starts with the service of a summons. The complaint consists of the plaintiff's factual and legal allegations against you or your firm. The summons is basically a document that notifies you that you are required to appear in the lawsuit and file a response to the complaint. This would usually take place sometime after a failure or deficiency has actually occurred. In federal court, the suit starts when the summons and complaint are filed with the court prior to service on the defendant. This gives the attorneys for the defendants an opportunity to reply to the complaint and possibly decide to file a counterclaim. If a counterclaim is filed, the plaintiff's attorney is expected to answer the counterclaim. Either way, the defendant must respond to the plaintiff's complaint by denying or admitting the allegations. Failure by the defendant to respond to the plaintiff's allegations is usually an admission of guilt.

Occasionally the defendant decides to utilize third-party practice to bring an outside party into the suit who may be liable to the defendant if the defendant is liable to the plaintiff. Thus, a defendant owner may serve a third-party complaint on a contractor or consultant designer claiming that should the owner be found liable to the plaintiff, then the contractor or consultant designer would be liable to the owner to the extent that the liability was caused by defective construction or design.

Attorneys often use discovery, inspection, and disclosure techniques to reveal details of the adverse party's claim. This also allows a party to inspect an adversary's files, and can be initiated by the attorney of the defendant, plaintiff, or third party, serving a document demand on the adversary to view certain documents that the party wishes to inspect such as minutes of job meetings, correspondence, photographs, and tests.

Interrogatories consist of written questions formulated by one party and served upon the other, which are required to be answered in writing, and which may only be served on a party to the lawsuit. Interrogatories are intended to cover issues that help prove or disprove the presence of material fact. To facilitate this, experts may be brought in to assist where the issues are technically complex. The party receiving the interrogatories must answer them in writing and under oath within a specified time period (30 days under FRCP 33).

Depositions testimonies are oral questions, given under oath and which may be taken of witnesses as well as parties. They normally take place prior to trial commencement, and consist of in-depth questions of a party or witness by the attorneys of the various parties involved in the case. They consist mainly of a cross-examination given under oath and recorded verbatim, but in the absence of judge and jury. Depositions often assume cardinal importance at a trial and can be used for any relevant purpose including discrediting the trial testimony, and impeaching or throwing doubt on the credibility of a witness.

Prior to trial commencement, the attorneys for both sides often meet with the trial judge in a pretrial conference at which the attorneys and judge decide whether a settlement is possible or not. If settlement is not possible, the attorneys and judge decide how the trial will be conducted and what types of evidence are to be admissible. This is followed by jury selection (called *voir dire*), which depends largely on the jurisdiction of the trial. It can either be accomplished by the attorneys themselves or by the judge, depending on the constrain rules of the court where the case is being heard. Once jury selection is completed, the trial can begin with opening statements by the attorneys for the plaintiff and defendant. These statements typically outline the strategy that will be used by the respective attorneys to prove their case.

Following the opening statements, the plaintiff's and defendant's witnesses take the stand. The witnesses are examined and cross-examined by the adversary's attorney. Closing arguments are then made by the defendant's attorney and the plaintiff's attorney. After the judge advises the jury on the laws that are applicable to the facts of the case, the jury is allowed to deliberate to make a judgment. This gives the jury an opportunity to inspect all the relevant documents entered into evidence prior to a final verdict being reached. When the jury has reached a decision, it issues a verdict in favor of one party or another.

10.4.2 Alternative Dispute Resolution

The construction industry has undergone a significant transformation over recent years and the dissatisfaction by many in the industry with the current state of traditional civil litigation has provided an important impetus to the development of various alternative dispute resolution (ADR) techniques. Most persons try and avoid being involved in lawsuits. Litigation can entail lengthy delays, high costs, unwanted publicity, and ill will. Appeals may also be filed, causing additional delay, after a decision has been rendered. ADR techniques such as arbitration, on the other hand, are usually faster and less expensive, as well as conclusive.

ADR techniques such as arbitration, mediation, negotiation, and other out-of-court settlement procedures are now widely employed in the vast majority of construction disputes. They are particularly useful with minor defects or failures (Figure 10.3). ADR techniques are essentially based on the premise that disputes can best be resolved through negotiation or mediation immediately after a conflict comes to light rather than through the tedious, costly, and time consuming route of traditional civil litigation. On occasion there are cases when one of the parties to a dispute insists on litigation; in these circumstances it is usually because there are legal precedents have been shown to be favorable to that party. Settlement discussions are often the best, least bruising, most private, and least expensive ways of resolving disputes. In the United States there are several methods of resolving construction disputes as shown in Figure 10.4.

Both arbitration and mediation have proven to be viable, cost-effective alternatives to litigation. The American Arbitration Association (AAA) states that, "Arbitration is the submission of a dispute to one or more

FIGURE 10.3 Examples of common failures that are not excessive in their magnitude and that can perhaps be resolved more readily through ADR procedures.

RESOLUTION PROCESS	ADVANTAGES	DISADVANTAGES
NEGOTIATION/ASSISTED NEGOTIATION	• Parties have control • Confidential	• No structure • Entrenched bargaining positions likely
MEDIATION	• Structured • Skilled mediator helps avoid entrenched positions • Control and resolution lies with parties • Helps maintain future commercial relationship for parties • Costs less than litigation • Quick result • Confidential	• No decision if parties do not agree • A resolution may not be reached
ARBITRATION	• Structured • Can be quick, timetable controlled by parties • Costs may be less than litigation • Confidential	• Parties do not have control • Imposed decision • May jeopardize future relationship of parties
LITIGATION (COURT ACTION)	• Structured	• Timetable controlled by court • Costs may be significant • Parties do not have control • Imposed decision • May jeopardize future relationship of parties • Long waiting times • Goes on public record (no confidentiality)

FIGURE 10.4 A table showing the principal advantages and disadvantages of different forms of conflict resolution.

impartial persons for a final and binding decision, known as an award. Awards are made in writing and are generally final and binding on the parties in the case. Mediation, on the other hand, is a process in which an impartial third party facilitates communication and negotiation and promotes voluntary decision making by the parties to the dispute. This process can be effective for resolving disputes prior to arbitration or litigation."

It is now mandatory in several jurisdictions to use ADR methods prior to accepting a case, and then only if ADR methods have failed to resolve the dispute in question. Most attorneys however, advise caution in choosing ADR methods over traditional litigation (especially when it is in their interest), and when ADR is decided upon, preference is usually given to methods that are voluntary and non-binding.

Arbitration

This is one of the oldest forms of ADR of which there are two basic types—binding and non-binding arbitration. This form of ADR includes participation of the parties on a typically voluntary basis and is only legally enforceable and binding if agreed beforehand. To become binding by the arbitrator's decision, a separate written agreement must be in place certifying that the arbitration agreement has been read and understood, and that both parties agree to be bound by it. In arbitration a third party is agreed upon to act as a private judge and is authorized to render a decision. The feuding parties control the range of issues to be resolved by arbitration, the scope of the relief to be awarded, and many of the procedural aspects of the process. The contract between the owner and builder or designer may very well include an arbitration clause, although even if there isn't one, a disputant party can still choose to take the case to arbitration. The main attraction of arbitration is that it costs less than a trial and has the added advantage of often having a speedy outcome, as well as relative privacy of the parties. Also, since the parties control the process, they enjoy tremendous flexibility. Still, it is always advisable to obtain legal advice before making a determination to proceed with arbitration proceedings.

Once the decision is made to go to arbitration, all the facts have to be sorted out and the real issue in dispute established. A decision must also be made about potential witnesses for your side. A determination is needed on who the arbitrator should be and the venue for the arbitration hearing.

Expert consultants do not normally attend settlement discussions, although the parties may rely on their respective expertise in formulating their respective positions. Expert consultants, however, do usually participate in mediations and are often the primary presenters of their clients' technical positions, and are relied upon to provide continuous input throughout the ongoing arguments. The arbitration process is rarely able to proceed without the use of expert witnesses, partly because arbitrators often probe deeply into technical matters of the case, and also the witnesses provide the parties with credible and authoritative presentations of their technical positions.

On the other hand, expert testimony at a trial involving architecture, engineering, and construction is typically required by the court for the purpose of elucidating to the judge or jury the technical aspects of the case; and both of the feuding parties want the judge or jury to hear their own technical positions as expressed by their own experts, as well as challenges to opposing experts' opinions and conclusions.

Mediation

This is another ADR nonbinding, facilitated negotiation process. The main objective of mediation is to assist the parties in voluntarily reaching an acceptable resolution of issues in dispute with the aid of a neutral third party—the mediator. The mediator's role is advisory. The mediator may offer suggestions but does not impose a resolution on the parties. Mediation proceedings are confidential and private. Mediation has several advantages and fewer disadvantages than the other options and is becoming the ADR method of choice for resolving disputes. The advantages of mediation include lower costs and potentially speedier dispute resolution.

A mediator, like a facilitator, makes primarily procedural suggestions regarding how parties can reach agreement, because in the end it is all about reaching a compromise. Often, clients may have to forgo some of what they consider their legal rights in the matter. Some mediators try to set the stage for bargaining by making minimal procedural suggestions, and intervening in the negotiations only to avoid or overcome a deadlock. Other mediators get much more personally involved in forging the details of a resolution. In either case, experienced and knowledgeable mediators are often necessary to assist the disputants in brainstorming alternatives. Mediation is particularly useful in highly-polarized disputes where the parties have either been unable to initiate a productive dialogue, or where the parties have been talking and have reached a seemingly insurmountable impasse. If it is determined that the parties cannot reach an amicable agreement, the case will then have to go to court, which will mean significant costs, as well as considerable time and energy. Even if mediation fails, the parties may still try to settle just to avoid the exorbitant court process, with neither party admitting fault. In some countries like the United Kingdom, ADR is synonymous with what is generally referred to as mediation in other countries.

Negotiation

This process is voluntary and there is no third party to facilitate the resolution process or impose a resolution. This informal process is one of the most fundamental methods of dispute resolution, offering parties maximum control over the process. But as with any endeavor, negotiation can be effective or unproductive. Negotiation is a time-proven approach to resolving disputes between feuding parties through discussions and mutual agreements. The most appropriate and successful approaches in negotiations are those in which the negotiators conduct discussions that focus on the common interests of the parties and not the traditional approach of focusing on the parties' relative power or positions. A construction industry negotiation variant is the "step negotiation" procedure which is a multitiered process sometimes used when the information that the parties have in place is incomplete, or as a mechanism to break a deadlock.

Normally there are time limitations to a civil action relating to building work in that it cannot be instigated against anyone after 10 years from the date when the work that caused the problem was executed. This law is stipulated in section 393 of the Building Act 2004.

10.4.3 Professional Ethics and Confidentiality

Many specialty fields have their own rules of professional conduct which the memberships' leaders design to police their own professions. Thus, contractors and professional architects or engineers remain governed by ethical obligations of their own, as the vast majority of professional associations and organizations have their own code of ethics. For example, the American Institute of Architects (AIA) and the American Society of Civil Engineers (ASCE) each has its own code of ethics.

Experts and professionals in civil litigation, who are or were retained by an attorney on a client's behalf, should always observe client confidences and confidentiality. There is an obligation to treat all information obtained from the client directly or through the attorney in complete confidence. This obligation arises from the fact that the attorney hired the expert/professional to act for the client and as the attorney is sworn by his code of conduct to maintain and preserve the client's confidences, so must the expert/professional. It should be noted that litigants do not normally file discovery documents in court in civil litigation, unless a matter is intended to go to trial. Such documents and recordings should not be released to third parties without the attorney's permission, although a release may waive the privilege.

The law pertaining to conflict of interest is extremely complicated, but a potential conflict obviously exists when an attorney attempts to oppose a party he/she previously represented. A potential conflict of interest exists whenever a prior representation may in any way influence the attorney's loyalty to a current client. In fact, attorneys, experts, and professionals may be privy to customer lists and other proprietary business information, hold government secret classification status, and be committed to keep the information confidential. All such prior confidential commitments could come into conflict when an expert or professional agrees to act as a consultant and possibly testify. Attorneys, as well as experts and professionals, must therefore be vigilant and avoid even the appearance of impropriety.

Whenever a potential conflict does exist, a professional's obligation is to disclose the nature of the conflict to the attorney to allow the attorney to determine whether the professional should continue to function on the client's behalf. As a technical specialist, a consulting expert has a professional obligation to uncover all relevant facts in connection with the issue being investigated, whether beneficial or harmful.

Should a construction failure occur, typically each of the involved parties (e.g., owner, contractor, design consultant, etc.) will retain an expert consultant to investigate the causes of and responsibilities for the failure. If it is later deemed necessary, the expert may be required to provide technical support in the litigation of potential claims. Experts and consultants can play critically important roles in litigation. In many cases, a technical expert's participation in a construction-related lawsuit can strongly impact, and in many cases determine, its success or failure. The service and role played by a particular expert will vary from case to case.

Experts are called just that because they have special knowledge, by virtue of their skill, experience, training, and/or education, well beyond the normal experience of the general public, to an extent that others may officially (and legally) rely on their opinion on such matters. Additionally, the expert witness is formally allowed to offer an opinion as testimony in court without having been a witness to any occurrence relating to the lawsuit. The expert witness should also be able to present highly technical matters relating to architecture and engineering in language that can easily be comprehended by non-experts, and be able to justify it under cross examination. This makes the expert a valuable and necessary professional in the resolution of construction related claims.

10.5 INSURANCE NEEDS

Liability can be extremely complex and this section cannot possibly address the endless concerns and legal matters that may arise with regard to liability issues, which is why builders, manufacturers, and designers are strongly advised to consult their attorneys and professional liability insurance carriers for advice on such matters. While building owners and managers are not expected to guarantee the safety or well-being of their tenants, visitors, and guests, they are required to exercise reasonable care to protect them from foreseeable events. The number of liability lawsuits filed against American companies has increased dramatically over the last decade and is a major concern to all involved in the construction industry.

Attitudes about insurance vary. Few understand it. Some have a strong belief in the protection of their business, which insurance provides, while others believe it to be a "necessary evil." An insurance policy consists of a legal contract expressed in relatively complex legal terms. It is a promise to provide compensation for something which may never happen, although in the construction industry these things can and do happen. Indeed, contractors without adequate insurance risk the success of their business. Independent and expert advice is imperative. Insurance should be considered as crucial to the continued operation of the business, as few companies and proprietors in the building trade have sufficient capital from their own resources to address adversity. While there are many inherent risks in building construction projects, properly arranged insurance can remove or mitigate many of the risks for a small cost compared to the potential liabilities.

There are several types of insurance policies commonly used in the construction industry to cover potential losses. They include:

- Commercial general liability policies
- Builder's risk policies
- Errors and omissions policies
- Workmans' compensation insurance
- Professional liability insurance.

10.5.1 Commercial General Liability (CGL)

CGL policies are intended to protect the policyholder and others insured through the policy from claims instituted by third parties. For construction companies, liability insurance is one of the most important types of insurance to carry. It protects the contractor and the company from damages or injuries sustained by others on-site. In most cases, the construction contract will specify that the subcontractors name the general contractor as an additional insured on their respective CGL policies. Both general contractors and subcontractors typically hold CGL policies that cover personal injury and property damage on either an occurrence or claims-made basis. Also, CGL policies contain a general aggregate limit that states the maximum limit that the insurer will pay out during the policy period for damages resulting from bodily injury, property damage, and personal and advertising injury.

In recent years many contractors have experienced increasing difficulty in obtaining general liability insurance due to the construction defect litigation crisis. To address this crisis, the insurance industry has started restricting coverage in their general liability policies by placing numerous exclusions in the "fine print" that may result in no coverage for many types of lawsuits. These exclusions include but are not limited to EIFS, mold, lead, products-completed operations, prior completed work, damage to work performed by subcontractors on your behalf, subsidence, contractual liability limitation, independent contractor exclusion, roofing operations exclusions, etc. It is imperative to fully understand the coverage implications of these exclusions if they are found on your general liability policy and decide on the action to be taken.

Seth D. Lamden, an attorney and partner with Howrey LLP, says, "After nearly three decades of litigation and scores of decisions from nearly every jurisdiction, courts still do not agree on whether a commercial general liability (CGL) policy protects a general contractor against claims for property damage occurring after the construction project is complete when the damage is caused by the defective work of a subcontractor." Lamden goes on to say, "The battle still rages on as to whether a general contractor should be entitled to coverage for liability resulting from damage to the completed project caused by the defective work of a subcontractor. A close reading of the standard form CGL policy reveals that the policy should provide such coverage. Until courts agree on this issue, however, general contractors should be aware of the insurance law in their jurisdiction regarding how CGL policies are interpreted in the construction context to make sure that they truly are protected from the main risk they face on construction projects."

10.5.2 Builder's Risk

These policies focus on losses incurred by the policyholder as a result of damage to the project during construction and differ from CGL policies, which focus on losses incurred by third parties. Builder's risk insurance is generally purchased on a project-specific basis and usually indemnifies against losses due to fire, vandalism, wind, hail, lightning, explosion, and similar forces. Notably, faulty workmanship and construction defects are excluded from coverage, as are earthquakes, flood, acts of war, or intentional acts of the owner. These insurance policies are designed to insure construction projects, and cover buildings and other structures while being built, including building materials and equipment designed to become part of the building or structure, and will usually last from the start of construction to acceptance of the completed project (Figure 10.5). This policy is usually purchased by the project owner but the general contractor constructing the building may buy it if it is required as a condition of the contract.

Builder's risk policies typically include less exclusions and many cover flood, earthquake, testing, and additionally provide broader transit and off-premises coverage. However, a potential problem that may arise with a separate builder's risk policy is securing permanent coverage when the builder's risk policy expires. Builder's risk policies typically contain

FIGURE 10.5 There were more than 1500 people in this luxury department store in Seoul, South Korea, when a section of the five-story building collapsed in June 1995, killing more than 500 people. The collapse of the building, which was constructed using steel-reinforced concrete pillars, was blamed on faulty construction.

a provision stating when coverage will come to an end, but this provision will vary according to the policy. This may create potential problems if the permanent commercial property insurance fails to be placed in a timely manner. The owner also has the option to delegate the responsibility for obtaining the required builders risk insurance to the GC. This is not uncommon with larger contractors because the latter are often more familiar with the market and coverage needed, which is why they prefer to have a degree of control over who will end up insuring the project.

Obtaining sufficient builders risk insurance should be carefully considered and insurance consultants need to be familiar with the construction contract, particularly the responsibility for procuring adequate and appropriate insurance. It is incumbent on all stakeholders to know not only what coverage is required by the construction contract, but also what coverage is available in the marketplace, so that the proposed insurance policy provides appropriate protection to meet the many potential exposures that may be faced.

10.5.3 Errors and Omissions (E&O)

E&O policies are basically malpractice insurance and cover the liability of design and other professionals for malpractice in the execution of their work. Owners often expect the architect to produce a perfect set of design and contract documents. Design professionals know this is virtually impossible, yet hesitate to discuss the potential of errors and omissions with the client. This discussion should take place, for the design professional's protection.

E&O coverage extends to the payment of defense costs, court costs, and any resulting judgments up to the policy limit. Liability for a design professional's negligence claims are not addressed in CGL policies, leaving the E&O policy as the primary protection tool for these professionals. The policy covers architects, engineers, accountants, and attorneys, etc. for alleged professional errors and omissions which amount to negligence. Errors and omissions insurance protects a company from claims when a client holds the company responsible for errors, or the failure of executed work to perform as promised in the contract.

A generally acceptable method for dealing with errors and omissions should be worked out between the owner and the design professional in advance. The client should also be advised of the increased potential for errors, omissions, and concealed conditions, which are inherent to certain types of construction, such as renovation and fast-track. No design drawing can be perfect, and thus the potential for errors and omissions is always present in design work. The prudent administrator or consultant therefore tries to provide a "safety net" to ensure that mistakes are caught and corrected before they cause major problems. Checklists are very valuable, as are construction reviews.

However, perhaps the most important thing the administrator can do to control professional liability in this area is to reach a clear understanding with his client that errors will occur, and that by working together such errors can be corrected in a timely manner.

10.5.4 Workmans' Compensation Insurance

This is not required by every employer. In some states small businesses with fewer than three to five employees may not be required to carry workers' compensation insurance. However, in today's litigious environment it is strongly recommended to take out this insurance if possible. The alternative may be a massive legal bill and unlimited personal and or corporate financial liability, including punitive damages that could run in the millions.

Workmans' compensation insurance provides compensation medical care for employees who are injured in the course of employment, in exchange for mandatory relinquishment of the employees' right to sue their employers for the tort of negligence. General damages for pain and suffering and punitive damages for employer negligence are generally not included in worker compensation plans.

10.5.5 Professional Liability Insurance

In this environment of increasing litigation, the private practice of a design professional or contractor can be particularly vulnerable, because of the damage that even an unfounded lawsuit can do to a person's reputation and financial stability. Business insurance is imperative because it protects you and your business with coverage for claims related to allegations of negligent activities or failure to use reasonable care. It is important to manage professional liability to mitigate potential risks.

What steps can be taken to minimize these risks to litigation? We will begin by looking at the five major areas of the interaction between the consultant/administrator and the other members of the project team. These are the areas where the consultant/administrator can do the most to protect himself/herself from liability:

- Professionalism
- Interpersonal relationships
- Business procedures
- Technical procedures
- Professional liability insurance.

10.5.6 Standard Documents and Related Issues

Construction projects in the United States often involve standard documents published by organizations such as the American Institute of Architects (AIA) or the Associated General Contractors (AGC), to name only two. However, project owners sometimes prefer to prepare their own documents, using a combination of standard and other forms. Project owners are required in both the AIA and AGC documents to provide builder's risk insurance covering the interests of all those involved in the project. Generally, the policy should provide: 1. "All-risk" or comparable coverage; 2. Coverage for property at the job site, material stored off site or in transit; 3. Coverage for all parties/stakeholders to the contract: owner, lenders, contractors, subcontractors, architect/engineers, etc.; 4. Permission for waivers of subrogation among the parties; 5. Coverage for the duration of the project. The construction contract itself, although not a party, remains a crucial element of the construction project. Coverage for construction equipment, such as forklifts, bulldozers, mobile tools, etc., would be provided by contractors' equipment insurance.

The project owner can also add the construction project to its regular commercial property policy, ensuring compliance with the above criteria, or alternatively purchase separate builder's risk coverage. The builder's risk approach has the advantage of having significantly broader coverage than that provided by standard commercial property insurance.

It is important to note that in some contract documents such as the AIA document, for example, if the owner does not intend to purchase the builder's risk insurance, he/she must inform the GC in writing prior to commencement of the project. The contractor then has the option of obtaining the necessary insurance, to protect the interests of all parties, and charge the cost to the owner.

In the UK the Joint Contracts Tribunal's (JCT) Standard Form of Building Contract is one of the most common standard contracts used to procure building work, and is updated regularly to incorporate changes in legislation and industry practice and relevant court decisions from litigation. The new JCT 2005 contract looks very different from the JCT 1998, however the insurance provisions have not significantly changed, and today most commercial building work in the UK is carried out under a standard form of contract, with or without amendments.

There are many factors to consider when taking out insurance for a project, such as:

1. It is clear that there are several parties that have insurable interests in the overall construction project.
2. Materials and equipment can normally be located on or off the job site, and at different times may belong to the owner, general contractor (GC), or subcontractors. Some materials and equipment may be owned by suppliers, but these individuals or entities are not considered to be subcontractors. Because their interests are rarely covered by the course of construction policies, they must be specifically added.
3. To make it easier to purchase insurance and to avoid potential gaps in coverage, one of the parties to the contract is usually required to assume responsibility for insuring the project on behalf of all parties.
4. Responsibilities of the various stakeholders, including responsibility for obtaining insurance, are generally delineated in the construction contract.

In building construction, losses can be caused by many factors such as:

- The building construction process—i.e., materials, workmanship, security, heat work, health and safety procedures, etc.
- Environmental factors including weather, ground conditions, proximity to hazards
- Negligence, lack of skill and care and failings in design, specification, workmanship or materials
- The inability to complete a contract on time, resulting in financial losses (whether due to insured perils or insolvency of contractor)
- Close on-site proximity of various contractors, subcontractors, and professionals, leading to a significant public liability risk in addition

to employer's liability exposures, the result of carrying, lifting,
working at height, confined spaces, collapse, dropping, toppling, etc.

Should the GC sustain a loss due to the owner's failure to obtain or
maintain coverage, without notifying the GC, the owner will then have to
assume responsibility for all reasonable damages sustained by the GC. To
meet the contract requirements regarding insurance, the insurance consultant may be required to negotiate modifications in coverage to comply
with the contract. It may not always be possible to obtain specific coverage, such as flood or earthquake, and modifications to the construction
contract may be necessary to delete such a requirement.

10.5.7 Influence of Green Features on Insurance Policies

As green building moves deeper into the mainstream, many insurance
carriers have started to take notice. For example, Lexington Insurance
Company, a Chartis company, which is a world leading property-casualty
and general insurance organization serving more than 40 million clients in over 160 countries and jurisdictions, introduced in August 2009
Upgrade to Green®—Builder's Risk, which provides coverage that supports green building construction and renovation projects registered with
the LEED or the GREEN GLOBES® rating systems. Upgrade to Green is
available as an endorsement to Lexington's Completed Value Builder's
Risk Policy, and extends coverage to address the risks green buildings
face during three key areas of construction: project management and
administration, site ecological impacts, and consumption of resources.

This is emphasized by Liz Carmody, senior vice president, Lexington,
who says, "We are committed to supporting sustainable development through innovative green products and services," and "Upgrade
to Green, which is part of our Ecosurance® portfolio of green products, addresses the unique risks that green construction and renovation
presents to property owners." In the event of a covered loss to a green
building, Upgrade to Green provides coverage for the fees of qualified
professionals related to the building's design and restoration; it also covers the costs of re-commissioning building systems and replacing vegeta-
tive roofs. Furthermore, the product is designed to respond to changes to
the relevant rating system criteria or loss of anticipated rating points as a
result of a covered loss. Within the insurance industry this appears to be
a bandwagon that will rapidly gather speed.

But while the green building movement is having a very significant and positive impact on the construction industry, there are serious concerns relating to the risk of liability that need to be addressed.
For example, green buildings are generally more efficient users of
energy and materials, resulting in reduced safety factors for the different

systems. However, green buildings are also sometimes prone to the use of non-standard materials and systems that may result in a heightened risk of failure of these materials or systems when incorporated into green buildings. To minimize these risks qualified and experienced designers should be employed to ensure that the design process is correctly implemented. In this respect, Ward Hubbell, president, Green Building Initiative, says, "One of our most pressing issues is the fact that some buildings designed to be green fail to live up to expectations. And in business, as we all know, where there are failed expectations there are lawsuits. All practices and/or products that could possibly result in a firm's exposure and liability should be clearly identified. The good news is that this period of increased legal action, or the threat thereof, will in fact motivate the kind of clarity and measurement that both reduces liability risks and results in better buildings."

Unfortunately, acronyms and phrases such as IAQ (indoor air quality), IEQ (indoor environmental quality), Sick Building Syndrome (SBS), and Building Related Illness (BRI) are tossed around to the degree that building owners and managers just ignore them. This is unfortunate since recent studies indicate that the incidence of commercial buildings with poor IAQ and the frequency of litigation over the effects of poor IAQ are increasing substantially. These increases will have obvious ramifications for insurance carriers, who pay for many of the costs of health care and general commercial liability.

A not uncommon phenomenon within the green building industry is the misrepresentation of a person or building stakeholder company's knowledge and expertise regarding green building capabilities. When building owners and other building stakeholders rely on this expertise, the result can be a dismal failure of the green building to achieve its intended goals. Misconceptions of this type have been found to permeate many building stakeholder groups.

Many of these problems arise when building owners, designers, and builders differ in their interpretation of what constitutes a successful green building, and particularly when building owners fail to explicitly communicate their thoughts at the commencement of the project. Issues of this kind are compounded when the parties are relatively new to the concepts of the green building process. Two of the main areas that typically need to be addressed are: 1. A building's failure to achieve a promised level of green building certification, and 2. A building's operational performance.

The first area may be the most problematic in that it could impact the building owner's ability to qualify for a grant or tax incentive, on which the owner may (and possibly the Lender) have relied on to assist in offsetting the project's initial costs and viability of the project. As for public buildings, certification may indeed be required by law.

In considering a building's operational performance, there is some expectation that green buildings will, in addition to reducing environmental impacts, also reduce energy and water costs, require less maintenance and other long-term benefits to the building owner. The point to note however is that while a design may incorporate a wide range of green features, there are numerous considerations between a building's design and occupancy that can invariably impact the building's operational performance. These potential problem areas can be minimized by basically following good business practices that facilitate clear communication and common expectations between building owners, designers, and rating organizations (such as LEED™ or Green Globes).

Legal actions may be brought against the building owner, the builder, the architect or engineer, or the product manufacturer. In such cases, an expert(s) will likely be required to give an opinion as to whether there has been negligence in the design, execution, or performance of duty. However, often the investigation may involve much more than expert opinion; for example, laboratory and other tests may be recommended to help determine the cause of failed performance. The role that experts may be required to play will vary depending very much on the case in question. Sometimes an expert will serve solely as a consultant to the lawyer, and remain in the background. At other times an expert may be used in the pre-trial stages, possibly to give an affidavit supporting one or more issues of the case. The expert may also serve solely as an expert witness at trial or the expert may play a combination of these roles.

Assigning culpability for green disputes often boils down to a matter of negligence, ignorance, or incompetence. American courts often require bringing in qualified experts to testify to the standard of care that is applicable to the case in dispute, and that the qualified experts testify to the professional's performance as measured by that applicable standard of care. It should be noted that the principle of standard of care may not apply to building contractors, because they are not considered to be "professionals" in the sense of making independent evaluations and judgments based on learning and skill. However, builders are held to a "duty to perform," which means they must strictly follow the plans, specifications, and provisions of their contracts.

The element of utmost good faith is an important principle by which insurance contracts are governed. The main impact of this is that there is a duty on any entity taking out insurance to disclose all material facts and to expeditiously notify the insurer of any events that may lead to a claim. Insurers typically rely on any breaches of this duty to avoid liability. An insured must have an "insurable interest" in the subject matter of the insurance—if not the insurance policy is considered to be null and void. The insurable interest may pertain to an interest in property, or pertain to a liability or potential liability; for example damages caused by negligence or breach of contract. But whatever the case, the insured can usually recover only what has actually been lost and based on the indemnity principle is not permitted to legally make a profit out of the insurance policy. Linked to this indemnity principle is the concept of subrogation which basically allows the insurer to take over any claims that the insured might have in place against third parties and to receive any payments or compensation made to the insured by third parties.

Finally, the most common problems relating to insurance policies is the failure to make full disclosure of all material facts when taking out a policy and failure to promptly report possible claims. Parties taking out insurance should always carefully consider the policy wording to ensure that it adequately serves its purpose. Policies are frequently vague or inconsistently worded and may contain exclusions which limit their usefulness. It is strongly advised to consult an attorney before finalizing an insurance policy.

11

Green Project Commissioning

11.1 GENERAL OVERVIEW

Building commissioning has grown significantly over recent years and is being embraced today by public and private organizations alike because of its benefits in improved project delivery results and because building commissioning undoubtedly enhances a building's value to the owner. Recent studies clearly demonstrate that commissioning is considered the single-most cost-effective strategy for reducing energy, costs, and greenhouse gas emissions in buildings today. For this reason building owners are demanding higher performance in their buildings from their architects, engineers, and contractors. The impetus for increased momentum for commissioning is also coming from energy and environmental policymakers as well as the private sector, and is increasingly resonating with building owners' interest in greening their properties and those seeking LEED™ or Green Globes™ certification. ASHRAE Guideline 0, The Commissioning Process defines commissioning as "a quality-oriented process for achieving, verifying, and documenting that the performance of facilities, systems, and assemblies meets defined objectives and criteria."

Building commissioning (Cx) is an all inclusive process encompassing all the planning, delivery, verification, and managing of risks to critical functions performed in, or by, facilities. It is a systematic process that ensures all building systems are installed and perform interactively according to the design intent and meet the owner's operational needs. Cx also promotes higher energy efficiency, improved environmental health and occupant safety, and improved indoor air quality by making sure the building components are working correctly and that the plans are implemented with the greatest efficiency. It basically confirms that the systems are efficient and cost effective and the installation is adequately documented according to requirements written into the project contract documents and that the operators are adequately trained. Ideally it starts at a project's inception, i.e., the beginning of the design process, and proceeds through design, construction, startup, inspection, testing, balancing, acceptance, training, and an agreed warranty period.

Commissioning is a quality assurance-based process that delivers preventive and predictive maintenance plans, tailored operating manuals, and training procedures for users to follow. The main function of the commissioning process therefore is to ensure that the HVAC&R systems and associated controls, domestic hot water systems, lighting controls, renewable energy systems (PV, wind, solar, etc.) and other energy-using building systems meet the owner's performance requirements, and perform and operate as intended and at maximum efficiency (Figure 11.1). Successful project commissioning can help dramatically reduce operating and maintenance costs, provide healthier occupant conditions, and facilitate upgrades, as well as fulfill LEED certification requirements.

What systems need commissioning

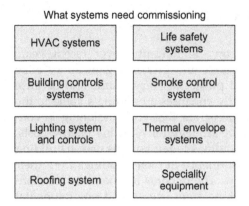

FIGURE 11.1 A diagram depicting the general scope for building commissioning and the major systems that typically need commissioning.

Many modern buildings today contain highly sophisticated conservation and environmental control technologies which, to function properly, require careful supervision of installations, testing and calibration, and adequate training of building operators. Modern sustainable (and conventional) buildings may possess high technology electrical or air-conditioning systems, or employ certain sustainable features that may require specialized attention to ensure they operate as designed. Cx can also reduce operating and maintenance costs, and extend the useful life of equipment. Cx provides better planning, coordination, and communication between the various stakeholders, resulting in fewer change orders, shorter punch lists, and fewer callbacks. In addition, on new construction projects, Cx will help reduce construction delays, ensure the correct equipment is properly installed, and increase productivity and reduce employee absenteeism. Once the project is completed, it is important that complete as-built information and operating and maintenance information be passed on to owners and operating staff. Case studies show that returns for these commissioning services often pay for themselves in energy savings within a year of completing a project.

Modern building commissioning is a fairly recent concept that includes what was historically referred to as "testing, adjusting, and balancing." But commissioning today goes much further; it acquires particular importance when complex mechanical and electrical systems are involved, where there is a need to ensure that these systems operate as intended, and to achieve energy savings and an improved building environment that justifies the incorporation and installation of more complex systems. Cx is also crucial to achieving optimum performance when special building features are installed to generate renewable energy, recycle waste, or reduce other environmental impacts. More importantly, commissioning practices should be specially tailored to address the size and complexity of the building, its systems and components in order to verify their performance and to confirm that all requirements are met as per the construction documents and specifications. In addition to verification of the installation and performance of systems, commissioning will culminate in the production of a commissioning report for the owner.

Few things are more frustrating than learning that essential systems in a new building fail to operate according to specifications. This sort of outcome can be prevented by incorporating a total building commissioning approach as part of the design, construction, and operation process. The level of commissioning applied should be appropriate to the complexity of the project and its systems, and the owner's need for assurances, as well as the budget and time available. It is prudent to request a number of quotations. For example, HVAC commissioning costs will vary, but are usually in the range of 1 to 4 percent of the value of the mechanical contract.

An important study entitled "Building Commissioning: A Golden Opportunity for Reducing Energy Costs and Greenhouse-gas Emissions" by Evan Mills (July 21, 2009) responds to an apparent anxiety that end-users lack confidence in the nature and level of energy savings that can be achieved through the commissioning process. The report addresses this issue by assembling diverse case studies and previously unpublished data, and incorporating performance benchmarks using standardized assumptions. Depicted below are some of the key findings outlined in the report:

- Median commissioning costs: $0.30 and $1.16 per square foot for existing buildings and new construction, respectively (and 0.4 percent of total construction costs for new buildings)
- Median whole-building energy savings: 16 percent and 13 percent
- Median payback times: 1.1 and 4.2 years
- Median benefit-cost ratios: 4.5 and 1.1
- Cash-on-cash returns: 91 percent and 23 percent
- Very considerable reductions in greenhouse-gas emissions were achieved, at a negative cost of -$110 and -$25/tonne CO_2-equivalent
- High-tech buildings are particularly cost-effective, and saved large amounts of energy and emissions due to their energy-intensiveness
- Projects employing a comprehensive approach to commissioning attained nearly twice the overall median level of savings, and five times the savings of projects with a constrained approach
- Non-energy benefits are extensive and often offset part or all of the commissioning cost
- Limited multi-year post-commissioning data indicate that savings often continue for a period of at least five years
- Uniformly applying our median whole-building energy-savings value to the stock of U.S. non-residential buildings yields an energy savings potential of $30 billion by the year 2030, and annual greenhouse

gas emissions reductions of about 340 megatons of CO_2 each year. An industry equipped to deliver these benefits would have a sales volume of $4 billion per year and support approximately 24,000 jobs.

Moreover, while commissioning of building systems will vary from one project to another, most projects will generally entail equipment startup, HVAC systems, electrical, plumbing, communications, and security and fire management systems and their controls and calibration. Large or complex projects may require other systems and components to be included. Commissioning usually begins with checking the documentation and design intent for reference. Performance testing of components is conducted upon first arrival on the job site and again after installation is complete. Providing maintenance training and manuals is typically the final step of commissioning.

There are currently no building code requirements at a national level that mandate building commissioning. However, studies repeatedly show that proper commissioning is cost effective and benefits all new or renovation building programs. Recent case studies conducted in private sector facilities have concluded that the building commissioning process can improve building energy performance by 8 to 30 percent. Formal building commissioning processes are even more imperative for complex building types with highly integrated building systems, as they can provide dramatic compounding benefits. Moreover, some governmental agencies, such as the GSA, NAVFAC, and USACE have adopted formal requirements, standards, or criteria for commissioning of their capital construction projects. However, the level of commissioning utilized will depend on several factors, including available project funds.

11.2 COMMISSIONING OBJECTIVES AND GOALS

11.2.1 Why Commission?

Empirical studies clearly show that the vast majority of building energy systems fail to function to their full potential. Poor communication of design intent, inadequate equipment capacity, inferior equipment installation, insufficient maintenance, and improper system operation all adversely impact energy cost savings.

The nature of deficiencies frequently found in non-commissioned energy projects are quite diverse and include:

- Significant air flow problems
- Inadequate documentation of project installation/operational requirements

- Underutilized energy management systems for optimum comfort and efficiency
- Inappropriate cooling and heating sequences
- Short cycling of HVAC equipment
- Erroneous lighting and equipment schedules
- Improperly installed or missing equipment
- Erroneous calibration of controls and sensors
- Malfunctioning economizers
- Lack of building operator training.

11.2.2 Benefits of Commissioning

As all modern building systems are integrated, a deficiency in one or more components can adversely impact the operation and performance of the other components. Rectifying these deficiencies therefore can result in a variety of benefits, including:

- Energy savings generally means lower utility bills (Figure 11.2)
- Increased occupant comfort and productivity
- Improved functioning of systems and equipment
- Faster and smoother equipment start-up due to systematic equipment and control testing procedures
- Increased owner satisfaction
- Increased occupant safety
- Significant life-cycle extension of equipment/systems
- Enhanced environmental/health conditions
- Improved building operation and maintenance
- Improved building documentation
- Shortened occupancy transition period.

Commissioning generally facilitates the delivery of a project that provides a safe and healthy facility; optimizes energy use; reduces operating costs; ensures adequate O&M staff orientation and training; and improves installed building systems documentation. In addition, commissioning benefits owners through improved energy efficiency, improved workplace performance due to higher IAQ, reduced threat risks, and prevention of business losses. Some industry sources estimate that on average the operating costs of a commissioned building are between 8 and 20 percent below that of a non-commissioned building. Meanwhile, the cost of not commissioning is equal to the costs of correcting deficiencies plus the costs of inefficient operations. Commissioning is even more crucial for mission-critical facilities, as the cost of not commissioning can be measured by the cost of wasted downtime and lack of required facility use.

Commissioning improves coordination between building systems, thereby improving general building performance. It also provides the

Illustrative relationships between commissioning and
energy efficiency measures

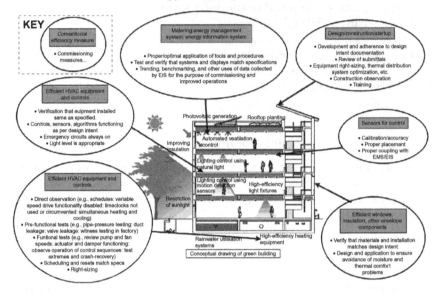

FIGURE 11.2 An illustration from a July 2009 report prepared for the California Energy Commission Public Interest Energy Research (PIER) showing the relationships between commissioning and energy efficiency measures. (Source: Evan Mills, Lawrence Berkeley National Laboratory.)

owner with additional in-house knowledge for optimizing equipment, system, and control efficiencies. Perhaps the most valuable benefit from commissioning comes from better building control which extends equipment life in addition to improving operation efficiency through frequent equipment cycling avoidance, and the ensuing improvements to thermal comfort and indoor air quality. This helps minimize occupant complaints and employee absenteeism, and increase staff retention. While difficult to quantify, it is estimated that the health and productivity benefits of a building with good IAQ is likely to be worth more than five times the energy and operating cost savings.

Commissioning provides better up-front performance accountability since problem prevention is known to be less expensive than problem correction. Providing front-end performance accountability and quality control allows frequent comparison of consistent project construction with project design, thus providing rapid feedback to design professionals on the dynamic performance of their design. Proper commissioning also minimizes the risk of liability from environmental hazards and/or equipment failure.

Most governmental projects employ commissioning because mission critical facilities support essential public infrastructures. Corporations most often use commissioning on projects to ensure peak performance to positively impact the bottom line and business continuity. Manufacturers find commissioning essential because of the high levels of environmental controls required in their processes and to ensure occupational safety. While it is evident that projects with special performance needs require commissioning, in fact all projects require some level of commissioning if they are to perform to their full potential.

There are several factors driving the demand for commissioning of modern facilities, such as performance needs and the desire to obtain certification through the LEED and Green Globes programs. These rating systems have been developed to improve energy efficiency and environmental performance in buildings. Commissioning is a prerequisite for LEED certification (although enhanced commissioning is a credit) and a requirement in Green Globes. A building certified to these rating systems is likely to include highly efficient power and lighting systems, photovoltaic and active/passive solar technologies, and from an owner's perspective, such sophisticated building technologies should be accompanied by strict construction quality assurance and performance verification measurement, such as provided by the commissioning process.

An October 2003 study prepared for a group of more than 40 California government agencies concluded that investing in green construction will pay for itself 10 times over. The study, conducted by the Capital E Group at Lawrence Berkeley National Laboratory, with input from a number of state agencies, reflects the most definitive cost-benefit analysis of green building to date.

11.2.3 Commissioning Goals

Commissioning is frequently misinterpreted to mean focusing solely on testing upon completion of the construction phase. But, in fact, commissioning is principally a collaborative process for planning, delivering, and operating buildings that work as intended. ASHRAE (The American Society of Heating, Refrigeration, and Air-Conditioning Engineers) defines commissioning as "...the process of ensuring that systems are designed, installed, functionally tested, and capable of being operated and maintained to perform in conformity with the design intent... Commissioning begins with planning and includes design, construction, startup, acceptance, and training, and can be applied throughout the life of the building." This definition accurately depicts commissioning as a holistic process that spans from pre-design planning to post-construction operation. It consists basically of a checks-and-balances system.

Accordingly, the main objectives of commissioning are to:

1. Define and document requirements at the commencement of each phase and provide appropriate updates throughout the process.

2. Establish and document commissioning process tasks and responsibilities for subsequent phase delivery team members.
3. Verify and document compliance as each phase is completed.
4. Deliver construction projects that meet the owner's needs, at the time of completion.
5. Verify that operation and maintenance personnel and occupants are properly trained.
6. Maintain facility performance across its life-cycle.

All new construction project commissioning goes through pre-design and design phases to establish an owner's needs, goals, scope, and design solutions for a proposed project. The evaluation of proposed designs and constructed work can only be made by comparison with objective criteria and measures that can be found in well-documented project requirements. Project development is a continuous learning process where building performance decisions undergo continuous refinement over the course of a project's life-cycle. Key commissioning activities supporting this principle include:

- Understanding the needs of special building types
- Determining key program goals and objectives
- Evaluating key threats, risks, and consequences
- Critical analysis of systems to facilitate achieving goals
- Conducting important commissioning programming activities.

11.2.4 Factors Affecting Cost of Commissioning

There are many factors that can impact the overall cost of commissioning, such as:

- Complexity of systems to be commissioned
- Number of systems to be commissioned and the sample rate of like systems and equipment
- Building size
- At what phase commissioning commences (e.g. during design, construction, or post-construction)
- Degree to which the CxA actively performs testing (as opposed to passively observing testing)
- Required deliverables (design intent document, commissioning plan, commissioning specification, O&M manuals, training plans, final report, etc.)
- Commissioning process protocol (does it include documenting and witnessing all equipment pre-startup and startup activities, pre-functional test procedures, functional test procedures, spot check tests, etc.)

- Cost allocations (e.g., does it include commissioning consultant's fees, increased contractor bids, increased designer fees, O&M personnel time, etc.)
- What tools are available such as installed sensors, meters, trend logs, etc.
- To what extent will operators assist in testing (including the future building operators in testing can help reduce the time required by the CxA).

11.2.5 Long Term Cost Implications

Commissioning has tremendous potential for long-term cost savings which, theoretically, could induce owners to perform system commissioning with payback being a major consideration. However, studies show that commissioning costs per square foot tend to be higher for more complex buildings such as hospitals, and as a result of their relatively high energy intensity, commissioning payback has also been found to be lowest in these building types. In regard to existing buildings, the median whole-building energy cost savings associated with commissioning was found to be about 15 percent. Following are some of the potential long-range cost benefits of conducting an effective commissioning process.

- A properly documented building will be easier and less time-consuming to maintain which translates into significantly lower operating and maintenance costs.
- A commissioned building is generally more energy efficient and is therefore likely to consume less energy than if the same building had not been commissioned.
- If IAQ controls have been commissioned and are operating properly, tenants and employees will be more productive, have less absenteeism, and be less likely to develop "Sick Building Syndrome" and "Building Related Illness" symptoms.
- For industrial, research, or other specialized facilities, the value of their processes, experiments, and/or collections can be far greater than the cost of commissioning or a potential product loss caused by improper control or malfunction of those systems.

11.3 PLANNING THE COMMISSIONING PROCESS

11.3.1 The Commissioning Plan Process

The commissioning plan provides guidance in the execution of the commissioning process and preferably commences early in the design process. It also contains a process for identifying planning delivery team member roles and responsibilities and tasks for the various project phases and activities. These include development and approval of commissioning plans, overview of review and acceptance procedures, documentation compliance, checking commissioning schedules, and testing and inspection plans. The process also includes identification of special testing needs for unique or innovative assemblies and measures that will ensure appropriate O&M training. It forms part of the bid and contract documents and is binding on the contractor; it also outlines many of the contractor's responsibilities, procedures, and tasks throughout the Cx process. The specifications will take precedence over the commissioning plan. The commissioning plan also describes the functional performance testing (FPT) that will be performed during the acceptance phase, and culminating with staff training and warranty monitoring.

The commissioning process generally culminates with a final complete commissioning report that is prepared and submitted to the owners along with drawings and relevant equipment manuals. This report should contain all the documentation pertaining to the commissioning process, procedures and testing results, in addition to any deficiencies and records of accepted corrections of these deficiencies. System commissioning requires specialized knowledge, which is why it is usually conducted by a mechanical consultant with appropriate experience and training. This person preferably is hired by and responsible directly to the project's owner and is independent of the mechanical consultant firm and general contractor. Where very large or complex projects are involved, it may be necessary to designate a special commissioning coordinator to be responsible for conducting the commissioning process. The architect or designer of record (DOR) is normally designated with the responsibility of overseeing completion of the commissioning process.

A suggested structure of the commissioning plan is shown below, bearing in mind that all information contained in the commissioning plan must be project specific.

Introduction	Purpose and general summary of the plan.
General Project Information	Overview of the project, emphasizing key project information and delivery method characteristics.
Commissioning Scope	The commissioning scope including which building assemblies, systems, subsystems, and equipment will be commissioned on this project.
Team Contacts	Project specific commissioning team members and contact information.
Communication Plan & Protocols	Documentation of the communication channels to be used throughout the project.

Commissioning Process	Detailed description of the project specific tasks to be accomplished during the planning, design, construction and tenant occupancy stages with associated roles and responsibilities.
Commissioning Documentation	List of commissioning documents required to identify expectations, track conditions and decisions, and validate/certify performance.
Commissioning Schedule	Specific sequences of events and relative timeframes, dates, and durations.

Source: U.S. General Services Administration—Building Commissioning Guide.

11.3.2 Documentation-Compliance and Acceptance

Commissioning serves as a general record of the owner's expectations for project performance during the project delivery process. Commissioning records and confirms the establishment of standards of performance for building systems, and verifies that designed and installed systems meet those standards.

A.   The contractor shall deliver to the commissioning authority one copy of the following as specified in the Cx Plan and other sections of the specification:

   1. Shop drawings and product data relating to systems or equipment to be commissioned. The CxA shall review and incorporate any comments via the designated design engineer.

   2. Startup checklists along with the manufacturers startup procedures for installed equipment. CxA will review, assist, and recommend approval if appropriate.

   3. Provide all system test reports. CxA will review and compile prior to FPT.

   4. Completed equipment startup certification forms in addition to the manufacturer's field or factory performance and startup test documentation. CxA will review prior to FPT.

   5. Completed test and balance reports. CxA will review prior to FPT.

   6. Equipment and other warrantees.

   7. Proposed training plans.

   8. O&M information per the requirements of the Cx plan, division 1 requirements.

   9. Record drawings.

B.   Record drawings: The contractor is to maintain at the site an updated set of record or "as-built" documents reflecting actual conditions of installed systems.

The following checklist is a guide to commissioning activities and documentation provided by the U.S. Department of Energy—Energy Efficiency and Renewable Energy:

Owner's requirements—List and describe the owner's requirements and basis of design intent with performance criteria.

Commissioning plan—Create the commissioning plan as early in the design phase as possible, including the management strategy and list of all features and systems to be commissioned.

Bid documents—Integrate commissioning requirements in the construction bid and contract documents. Designate the Construction Specifications Institute (CSI) Construction Specification Section 01810 in Division 1 for general commissioning requirements. Use the unassigned Sections 01811 through 01819 to address requirements specific to individual systems. Notify the mechanical and electrical subcontractors of Division 15 and 16 commissioning requirements in Sections 15995 and 16995.

Functional performance test procedures and checklists—Develop functional performance test procedures or performance criteria verification checklists for each of the elements identified in the commissioning plan.

Commissioning report—Complete a final commissioning report and submit it to the owner. The commissioning report should summarize all the tasks, findings, and documentation of the commissioning process and address the actual performance of the building systems in reference to the design documents. The report should identify each component, equipment, system, or feature, including the results of installation observation, startup and checkout, operation sampling, functional performance testing, and performance criteria verification. All test reports by various subcontractors, manufacturers, and controlling authorities will be incorporated into the final report.

Training—Assemble written verification that training was conducted for appropriate personnel on all commissioned features and systems.

Operation and maintenance manuals—Review operation and maintenance manuals for completeness including instructions for installation, maintenance, replacement, and startup; replacement sources; parts list; special tools; performance data; and warranty details.

Recommissioning management manual—Develop an indexed recommissioning management manual with components such as guidelines for establishing and tracking benchmarks for whole building energy use and equipment efficiencies; recommendations for recalibration frequency of sensors; list of all user adjustable set-points and reset schedules; and list of diagnostic tools.

Acceptance phase: For this phase, the facility and its systems and equipment are inspected, tested, and verified, etc. It is during this phase that most of the formal training occurs, which is generally after the construction phase is complete. A/E and contractor now finalize the as-built or record documentation. The end of this phase is marked by an approved functional completion.

This section of the Whole Building Design Guide is based primarily on the commissioning process recommended in ASHRAE Guideline 0-2005. It is highly recommended that project teams that employ the building commissioning process should follow the process outlined in ASHRAE Guideline 0. Guideline 0 has been adopted by both ASHRAE and the National Institute of Building Sciences (NIBS) and does not focus upon specific systems or assemblies, but adheres to a standard process that can be used to commission any building system critical to the function of a project. The NIBS Total Building Commissioning Program is currently working with industry organizations to develop commissioning guidelines for various systems and assemblies.

11.4 COMMISSIONING AUTHORITY

Many stakeholders have crucial roles to play in the Cx process, but the key role is played by the commissioning authority (CxA). This consists of a team of senior specialists that direct and oversee the process. The CxA should be retained early on in the programming phase of the project. The CxA has many roles to play in the process, such as providing technical and procedural oversight during the different phases of the project, and conducting functional performance testing during the construction, acceptance, and warranty phases of the project. The CxA is also required to aid and provide support in the training, documentation, and handover of the facility to the owner. Upon completion of the project, the warranty phase comes into play and the CxA is expected to periodically monitor the facility during this phase in order to optimize it with the actual occupancy.

11.4.1 Commissioning Authority Responsibilities

A. Construction Phase
   1. Conduct Cx meetings.
   2. Review relevant project documentation such as shop drawings, TAB reports, product data, record drawings, O&M data, etc. for compliance and to ensure system functionality.
   3. Installation to be monitored and periodically inspected.
   4. Review and approve startup checklist forms.

5. Attend progress meetings as required to observe progress and assist other parties, facilitate the Cx process, and help expedite completion. The CxA however does not direct work nor approve/accept materials, systems, or equipment.

6. Compile O&M information and systems overview and format the O&M manuals.

7. Witness selected test startups and equipment training.

B. Acceptance Phase

1. Verify, test, adjust, and balance by spot check all TAB reports, control component. Calibration and equipment performance certifications (test 100 percent of key systems or test sample percentage).

2. Analyze all trend logs.

3. Test systems and equipment to verify correct system operation and that they are functioning as per specifications, including failure and safety modes.

4. Review training plan and coordinate training activities between the O & M staff and the contractors/vendors to assure the training is appropriate for the staff.

5. Record commissioning procedures and provide Cx report with testing documentation.

6. Verify that contractors/vendors provide proper O&M material (fan curves, pump curves, operating parameters, etc.), and not just equipment mounting information.

7. Follow through to ensure that commissioning issues are resolved.

C. Warranty Phase

1. Talk with building users to identify any problem areas that have developed after building acceptance.

2. Assist owner in resolving outstanding issues with contractor and design firms.

3. Provide followup training to O&M staff, especially if new staff that was assigned to the building did not previously receive vendor training.

4. Check building performance and conduct seasonal testing on systems.

The Cx process is designed to ensure and validate that the design intent meets the owner's needs and the installation meets the design intent, and that the operation and use of the facility is in accordance with the design intent. It should be noted that the construction contractor is not responsible to deliver the design intent nor is the A/E responsible for the installation.

11.4.2 Commissioning Team

The commissioning team consists of individuals (owners, users, occupants, design professionals, contractors, operations and maintenance staff) who through coordinated actions are responsible for implementing the commissioning process and are led by the CxA. All traditional parties to the design and construction process are vital to the Cx process and have a role to play as part of the CxA Team (CT). The various roles will generally provide an extra focus to their efforts and in some cases delineate required assignments and rules that are normally included in the traditional process but are often ignored or poorly executed. It is important for all commissioning team members to be involved as early as possible in the project to allow the valuable input of their knowledge and experience in the design and to allow them to become active participants in the initial check out and acceptance of the facility.

The first step in the commissioning process is for the commissioning team leader (the CxA) to develop a commissioning plan (preferably at the project inception phase) and then to identify and layout the composition of the commissioning team. It is necessary that the commissioning teamwork as a cohesive unit so that all steps in the commissioning process are completed and the facility objectives are met. In general, roles and responsibilities within the commissioning team do not change. The facility owner/user is normally responsible for clearly communicating the facility needs and for understanding the design and functional intent. The A/E is responsible for designing a facility that accomplishes what the user requested and is in compliance with all regulations and accepted practices. Construction contractors, subcontractors, and vendors are responsible for supplying and installing the facility per the contract documents.

The size and number of members that comprise the commissioning team will fluctuate depending on the project's type, complexity, and size. However, in many cases the team will include:

- Owner
- Commissioning agent (CxA)
- Project manager
- Users
- Operating personnel
- Architect/engineer (A/E)
- Technical experts such as structural, mechanical, electrical, LEED/sustainability, elevator, fire protection, seismic, and other specialists
- Construction manager agent (CMa)
- Construction contractor and subcontractors.

General descriptions of the commissioning roles and responsibilities according to the USACE LEED Commissioning Plan Template (based on PECI model commissioning plans) are as follows:

CxA: Coordinates the Cx process, develops and updates the Cx Plan, assists, reviews, and approves incorporation of commissioning requirements in construction documents. Writes or approves tests, oversees and documents performance tests. Develops commissioning report.

PE: Facilitates the Cx process. Coordinates between GC and CxA. Approves test plans and signs-off on performance. Performs construction observation, approves O&M manuals (design-bid-build contracts).

GC: Facilitates the Cx process, ensures that subs perform their responsibilities, and integrates Cx into the construction process and schedule.

Subs: Demonstrate proper system performance.

DOR: Develops and updates basis of design, incorporates commissioning requirements in construction documents. Performs construction observation, approves O&M manuals (design-build contracts), and assists in resolving problems

PM: Facilitates and supports the Cx process

Mfr.   The equipment manufacturers and vendors provide documentation to facilitate the commissioning work and perform contracted startup.

11.4.3 Commissioning Authority Qualifications and Certification

While the industry is not unanimous on which party should be the CxA, it is strongly recommended that an independent party be the CxA, i.e., neither the contractor, A/E, nor the CM. The CxA should be motivated solely by the needs of the owner and the facility user, and should not normally be competitors of the A/E or contractors. Individuals chosen should be highly specialized in the types of facilities and systems to be commissioned (LEED for example, has specific requirements in this respect). Because of the level of technical oversight that is expected, individuals should be a certified commissioning professional (CCP), licensed professional engineer, or have applicable experience in specialized systems/facilities being installed, in addition to extensive experience in the design, optimization, remediation, and acceptance testing of applicable systems, as well as training and building manual preparation.

Building projects are increasingly requiring performance certifications such as LEED, Green Globes, Energy Star, OSHA, and others. To achieve this certification, requirements have to be determined during the planning and design phases so that a commissioning for certifications can be included in the OPR and commissioning plans. Several organizations, including The American Institute of Architects, are formulating new programs, training, and contract documents to assist their members in providing building commissioning as additional services to their clients. The Building Commissioning Association (BCxA) has also created the certified commissioning professional (CCP) program to raise professional standards and provide a way for certification in the building commissioning industry. To earn a CCP certification, individuals are required to complete an application form that is reviewed by the Building Commissioning Certification Board in addition to passing a two-hour written examination.

Finally, commissioning has in recent years greatly increased in importance as a quality assurance measure due to the elevated complexity in today's building designs, equipment, and fast-paced construction timeline. The cost ramifications for delayed occupancy and the early detection of design and installation faults on their own provide more than adequate economic justification for the majority of today's commissioning projects. The commissioning process can be employed using various methods that focus on building systems and assemblies and can be readily customized to suit specific project needs. However, whatever the commissioning approach and system focus decided upon, a clear articulation of performance expectations, rigorous planning and execution, and comprehensive project testing, operational training, and documentation is crucial to a successful result.

Commissioning or retro-commissioning a building should be seen as more than just a tool to save energy or reduce the payback period of investments; amongst other things, it also helps the environment, produces healthier buildings, improves the economic performance of a building, and increases productivity. Moreover, if a building is seeking a LEED certification for new construction or an existing building, commissioning is most likely to be required. Owners should look at the latest version of LEED (currently version 3) released by the USGBC to learn what changes have been made to commissioning requirements.

11.5 THE COMMISSIONING PROCESS

Formal commissioning has now become a prerequisite as most modern buildings today incorporate complex and digitally-controlled HVAC systems, or natural ventilation systems integrated with HVAC systems; others, especially if they are "green," incorporate renewable energy, on-site water treatment systems, occupancy sensor lighting controls, or other high technology systems. Commissioning is not usually requested for projects with minimal mechanical or electrical complexity, such as typical residential projects.

11.5.1 Commissioning Process Requirements

The following is from SECTION 01 91 00—General Commissioning Requirements. It is guidance for designers and specifiers, with suggested language to be modified and incorporated into project specifications, and provides a brief overview of the typical commissioning tasks during construction and the general order in which they occur (Source: BuildingGreen, Inc. 2007):

1. Commissioning during construction begins with an initial commissioning meeting conducted by the CxA where the commissioning process is reviewed with the commissioning team members.
2. Additional meetings will be required throughout construction, scheduled by the CxA with necessary parties attending, to plan, coordinate, schedule future activities, and resolve problems.
3. Equipment documentation is distributed by the A/E to the CxA during the normal submittal process, including detailed startup procedures.
4. The CxA works with the contractor in each discipline in developing startup plans and startup documentation formats, including providing the contractor with construction checklists to be completed during the installation and startup process.
5. In general, the checkout and performance verification proceeds from simple to complex; from component level to equipment to systems and intersystem levels, with construction checklists being completed before functional testing occurs.
6. The contractors, under their own direction, will execute and document the completion of construction checklists and perform startup and initial checkout. The CxA documents that the checklists and startup were completed according to the approved plans. This may include the CxA witnessing startup of selected equipment.
7. The CxA develops specific equipment and system functional performance test procedures.
8. The functional test procedures are reviewed with the A/E, CxA, and contractors.
9. The functional testing and procedures are executed by the contractors under the direction of, and documented by, the CxA.
10. During initial functional tests and for critical equipment, the engineer will witness the testing.
11. Items of non-compliance in material, installation, or setup are corrected at the contractor's expense, and the system is retested.
12. The CxA reviews the O&M documentation for completeness.
13. The project will not be considered substantially complete until the conclusion of commissioning functional testing procedures as defined in the commissioning plan.

14. The CxA reviews and coordinates the training provided by the contractors and verifies that it was completed.
15. Deferred testing is conducted as specified or required.

For U.S. Green Building Council (USGBC) LEED™ certification, commissioning is an integral and prerequisite component. For new construction, commercial interiors, schools and core and shell categories, LEED™ has two commissioning components:

1. Fundamental commissioning of building systems, which is a prerequisite, (i.e., obligatory), and
2. Enhanced commissioning, which receives a credit but is not a prerequisite.

Commissioning should ideally take place through all phases of a building project. A commissioning agent should be designated as early as possible in the project time line, ideally during the pre-design phase. While it is beneficial to employ a third party commissioning authority to provide a more comprehensive design and construction review, it is nevertheless acceptable for a project to use a qualified member of the design team as the commissioning agent (CxA), providing there is no conflict of interest. The commissioning authority (CxA) is required to serve as an objective advocate of the owner, direct the commissioning process, and present the owner with final recommendations regarding the performance of commissioned building systems. The CxA is expected to lead the commissioning process and to introduce standards and strategies early in the design process. Additionally, the CxA should ensure the implementation of selected measures by clearly stating all requirements in the construction documents. Upon completion of construction, the CxA verifies that the minimum requirements have been met. The CxA should also provide guidance on how to operate the building at maximum efficiency.

11.5.2 Fundamental Commissioning

The LEED intent of fundamental commissioning is to ensure that installation, calibration, and performance of energy systems meet the owner's project requirements, basis of design, and construction documents. Commissioning is a prerequisite for LEED certification (unlike enhanced commissioning which is a credit and not a prerequisite) and is required for both new construction and major retrofits, as well as for medium or large energy management control systems that incorporate in excess of 50 control points. Commissioning is also essential when large or very complex mechanical or electrical systems are in place or where on-site renewable energy generation systems, such as solar hot water heaters or photovoltaic arrays, are in place. Commissioning should also be considered when innovative water-conservation strategies, such as gray water irrigation systems or composting toilets, are installed.

If a building is targeting LEED certification, then the following commissioning process activities need to be implemented by the commissioning team:

1. The owner or project team must designate an individual as the commissioning authority (CxA) to lead, review, and oversee the commissioning process activities until completion. This individual should be independent of the project's design or construction management unless the project is less than 50,000 sq ft.
2. The designated CxA should have documented commissioning authority experience in at least two building projects. Additionally, he/she should ideally meet the minimum qualifications of having an appropriate level of experience in energy systems design, installation and operation, as well as commissioning planning and process management. LEED also recommends that a designated CxA have hands-on field experience with energy systems performance, startup, balancing, testing, troubleshooting, operation, and maintenance procedures, and energy systems automation control knowledge.
3. The CxA should clearly document and review the owner's project requirements (OPR) and the basis of design (BOD) for the building's energy-related systems (usually by A/E). Updates to these documents shall be made during design and construction by the design team. The commissioning process does not absolve or diminish the responsibility of the contractor to meet the contract documents requirements.

Design phase commissioning for both fundamental and enhanced commissioning is intended to achieve the following specific objectives:

- Owner's project requirements (OPR), i.e., the design and operational intent are clearly documented and fully understood. The OPR details the functional requirements of the different building systems from the owner's perspective, which should be fully measureable and verifiable, including facility use, occupant comfort, and project success. Should the owner lack sufficient experience to formally document these requirements, the CxA may conduct a workshop to facilitate the development of the OPR.
- Ensuring that the OPR recommendations are communicated to the design team during the design process in order to develop a basis of design (BOD) document that describes the system configurations and control sequences that will be put in place to meet the OPR and avoid later contract modifications.
- That the commissioning process for the construction phase is appropriately reflected in the construction documents. The CxA will conduct design reviews in the context of the BOD, and preferably perform an initial review prior to 50 percent completion of construction documents (CDs). The CxA is also required to develop specifications for the architect to incorporate into the CDs.

All tasks to be performed during commissioning are described in a commissioning plan developed by the CxA.

Before the design process is completed, the CxA will develop a construction phase commissioning plan. Construction phase commissioning for both fundamental and enhanced commissioning is intended to achieve the following objectives in line with the contract documents:

- Commissioning requirements and the OPR should be incorporated into the construction documents. During and immediately prior to the construction phase, a CxA may review contractor submittals related to the systems that will be commissioned.
- Develop and implement a proper commissioning plan. The CxA typically develops protocols for functional performance testing during this phase based on project specifics and the sequence of operations developed by the controls engineer and the CDs. Teamwork and accountability should be promoted.
- Hold a kickoff meeting with contractors and other stakeholders.
- Provide verification and documentation that the installation and performance of energy consuming equipment and systems meet the OPR and BOD. Upon completing equipment startup, the CxA conducts periodic pre-functional checks of installation progress to make sure contractor mounting of systems will allow straightforward and safe O&M access to ensure proper long-term maintenance.
- Provide verification and documentation that equipment and systems in place are installed according to the manufacturer's recommendations and to industry accepted minimum standards. Once equipment is fully installed, the CxA will conduct functional performance testing to evaluate the performance at all sequences of operation. As some functional testing can only be conducted in certain seasons, the commissioning process will usually extend beyond the completion of construction.
- Complete a final summary commissioning report. The report shall include, among other items, an executive summary, list of participants and roles, brief building description, outline of commissioning and testing scope, and an overall description of testing and verification methods used.
- Verify that O&M documentation left on-site is complete. Moreover, once the commissioning process is complete, the CxA prepares a final report and may also prepare an operations and maintenance (O&M) manual for the project.
- Verify that training of the owner's operating personnel is adequate and maintain a master Cx "issue log" throughout construction.

Without continuous maintenance, building systems tend to become less efficient over the years, mostly due to changing occupant needs, building renovations, and obsolete systems, which end up causing occupant discomfort and complaints. Unless these problems are appropriately addressed, such as investing in a commissioning process, a facility's operating costs will dramatically increase, making it less attractive to new and existing tenants. Commissioning will typically pay for itself in less than a year.

The following checked systems are to be commissioned:

HVAC Equipment and System

- ☐ Variable Speed Drives
- ☐ Hydronic Piping Systems
- ☐ HVAC Pumps
- ☐ Air Handling Units
- ☐ Underfloor Air Distribution
- ☐ Centrifugal Fans
- ☐ Ductwork
- ☐ Fire/Smoke Dampers
- ☐ Automatic Temperature Controls
- ☐ Laboratory Fume Hoods
- ☐ Testing, Adjusting, and Balancing
- ☐ Building/Space Pressurization
- ☐ Ceiling Radiant Heating
- ☐ Underfloor Radiant Heating

Electrical Equipment and System

- ☐ Power Distribution Systems
- ☐ Lighting Control Systems
- ☐ Lighting Control Programs
- ☐ Engine Generators
- ☐ Transfer Switches

- ☐ Boilers
- ☐ Chemical Treatment Systems
- ☐ Air Cooled Condensing Units
- ☐ Makeup Air Systems
- ☐ Switchboard
- ☐ Panelboards
- ☐ Grounding
- ☐ Fire Alarm and Interface Items with HVAC
- ☐ Renewable Energy Systems
- ☐ Security System

Plumbing System

- ☐ Domestic Water Heater
- ☐ Air Compressor and Dryer
- ☐ Storm Water Oil/Grit Separators

Building Envelope

- ☐ Building Insulation Installation
- ☐ Building Roof Installation Methods
- ☐ Door & Window Installation Methods
- ☐ Water Infiltration/Shell Drainage Plain

FIGURE 11.3 A detailed checklist of systems to be commissioned upon completion of the project. Although commissioning needs may differ from project to project, commissioning the building envelope systems, power distribution, domestic water heating, ductwork, and any hydronic piping systems is strongly recommended for any project. (Source: based on Building Green, Inc.-SECTION 01 91 00—General Commissioning Requirements.)

11.5.3 Enhanced Commissioning

For a LEED credit, enhanced commissioning is required in addition to the fundamental commissioning prerequisite. The intent of enhanced commissioning is to start the commissioning process early during the design process and execute additional activities upon completion of the systems performance verification. For the commercial interiors category the intent is to verify and ensure that the tenant space is designed, constructed, and calibrated to operate as intended. This requires the implementation of (or to have a contract in place requiring implementation of) further commissioning process activities in addition to the fundamental commissioning prerequisite requirements as stated in the relevant LEED Reference Guide (e.g., Green Building Design and Construction, 2009 Edition).

LEED states that the duties of the commissioning authority (CxA) include:

1. Prior to the end of design development and commencement of the construction documents phase, a commissioning authority (CxA), independent of the firms represented on the design and construction team, must be designated to lead, review, and oversee the completion of all commissioning process activities. Although it is preferable that the CxA be contracted with the owner, for enhanced commissioning the CxA may also be contracted through design firms or construction management firms not holding construction contracts. This person can be an employee or consultant of the owner, although this requirement has no deviation for project size. Furthermore, to meet LEED requirements this person must:
   - Have documented commissioning authority experience in at least two building projects
   - Be independent of the project's design and construction management
   - Not be an employee of the design firm, though the individual may be contracted through them
   - Not be an employee of, or contracted through, a contractor or construction manager of the construction project.
2. The CxA must report all results, findings, and recommendations directly to the owner.
3. The CxA must conduct a minimum of one commissioning design review of the owner's project requirements (OPR), basis of design (BOD), and design documents prior to the mid-construction documents phase, and back-check the review comments in the subsequent design submission.
4. The CxA must review contractor submittals and confirm that they comply with the owner's project requirements and basis of design

for systems being commissioned. This review must be conducted in parallel with the review of the architect or engineer of record and submitted to the design team and the owner.

5. The CxA or other members of the project team are required to develop a systems manual that provides future operating staff with the necessary information to understand and optimally operate the commissioned systems. For commercial interiors the manual must contain the information required for re-commissioning the tenant space energy related systems.

6. The task of verifying that the requirements for training operating personnel and building occupants have been completed may be performed by either the CxA or other members of the project team.

7. The CxA must be involved in reviewing the operation of the building with operations and maintenance (O&M) staff and occupants and have a plan in place for resolving outstanding commissioning-related issues within 10 months after substantial completion. For commercial interiors there must also be a contract in place to review tenant space operation for O&M staff and occupants.

11.5.4 Retro-Commissioning—Commissioning for Existing Buildings

Without doubt, building commissioning is an important aspect of new construction projects and is used mainly as a means of ensuring that all installed systems perform as intended. Yet most existing buildings have never encountered the commissioning or quality assurance process and, not unexpectedly, have been found to perform well below their intended design potential. This inefficiency fact is further confirmed by an LBNL study of 60 different types of buildings which showed that:

- Over 50 percent had control problems
- 40 percent had HVAC equipment problems
- 15 percent had missing equipment
- 25 percent had BAS with economizers, VFDs, and advanced applications that were simply not operating correctly.

(Source: Association of State Energy Research Technology Internships and U.S. Department of Energy.)

Retro-commissioning (RCx) simply refers to the commissioning of existing buildings (that have never been previously commissioned) and can be defined as a systematic process for investigating, analyzing, and optimizing the performance of building systems by improving their operation and maintenance to ensure their continued performance over time. A determination is first required on how the installed systems are intended to operate; then measure and monitor how they operate and prepare a prioritized list of the operating opportunities. The RCx process also reviews the functionality of equipment and systems installed, and optimizes how they work together to facilitate the reduction of energy waste, increase comfort, and improve building operation. Other advantages of retro-commissioning are to address issues such as modifications made to system components, function/space changes from original design intent, building systems that fail to operate according to designed benchmarks, complaints regarding IAQ, temperature, Building-Related Illness (BRI), Sick Building Syndrome (SBS), etc. Figure 11.4 is a photo of CEE hospital, a nine-story building that the owner decided to recommission. The Center for Energy and Environment declares, "Opportunities for recommissioning measures were identified through field measurements carried out by highly-skilled and experienced engineers to quickly zero in on sub-optimal central-system operating strategies that waste energy. Major recommissioning opportunities identified included:

1. Calibration of control system instrumentation.
2. Resetting supply air temperature set point.

FIGURE 11.4 A frontal view of Cee hospital, a nine-story, 600,000 square foot acute care facility in Minneapolis, MN, which recently underwent recommissioning. The first floor was built in 1981, and the rest of the floors were added in 1982. The owner's objective in recommissioning the building was to reduce operating costs while maintaining or improving indoor air quality and comfort. For this project, CEE partnered with the Energy Systems Lab at Texas A&M University, which is considered to be the most experienced and effective healthcare recommissioning provider in the U.S. (Source: Center for Energy and Environment.)

3. Resetting duct static pressure set point.
4. Replacing bad inlet guide vanes with VFDs.
5. Calibration of VAV terminal boxes.
6. Improving economizer operation.
7. Optimizing the chiller and chilled water pump operation.
8. Performing hot water and chilled water balance.
9. Optimizing heating water temperature reset schedule and on/off sequence.
10. Reducing outside airflow.
11. Calibration of thermostats.
12. Performing air balance.
13. Determining the minimum outside air damper position.
14. Repairing kinked flex ducts and leaky reheat control valves.

A key element of the recommissioning process is the diagnosis and correction of zone-level problems that might otherwise prevent key central-system measures from being fully implemented."

All operating improvements made should be recorded and the building operator trained on how to sustain efficient operation and implement capital improvements. RCx continues to witness increasing prominence as a cost-effective strategy for improving energy performance and helping to make the building's systems perform interactively in a manner that addresses the owner's current and anticipated facility requirements.

Recommissioning (ReCx) only applies to buildings that have previously been commissioned or retro-commissioned. The original commissioning process documentation shows that the building systems performed as intended at one point in time. The intent of recommissioning is to help ensure that the benefits of the initial commissioning or retro-Cx process remain valid. In some cases, ongoing-commissioning (ongoing Cx) may be necessary as an "ongoing process" to resolve operating problems, improve comfort, optimize energy use, and identify energy and operational retrofits for existing buildings.

11.5.5 Warranty Phase

Early occupancy of the building may continue through the warranty period and at least into the opposite season from when it was initially tested. During the warranty period, the CxA will provide warranty service and coordinate seasonal testing and other deferred testing requirements to ensure they are completed in line with the specifications. The contractor performs warranty service and corrects identified deficiencies witnessed by facilities staff and the CxA. The contractor updates and finalizes record documentation to reflect actual conditions at the end of the warranty period. Any final modifications to the O&M manuals and as-builts due to the testing are made. Operators work with the CxA and the design team to fine tune the facility to meet actual occupancy.

Finally, the high-performance building movement and the various energy rating prerequisites have brought commissioning well into the mainstream in recent years. And with the increasing complexity of mechanical systems and the continuous development of new technologies, the process of total building commissioning (TBCx) has taken on an increasingly important role. Cx entails not just commissioning typical systems like HVAC, but also includes many other elements, such as lighting systems and controls, as well as building envelope and fenestration to ensure total building performance.

Furthermore, since the 9/11 terrorist attacks, there has been a greater focus and urgency on providing occupant safety to visitors and workers in public facilities; this has created a need to deliver and commission facilities with enhanced building safety measures. This trend for increased security is likely to increase the standard of care required in the design and operation of all forms of public and corporate buildings.

Relevant Resources, Codes, and Standards

* AIA B211™—2004 Standard Form of Architect's Services: Commissioning—Fixed scope of services requires the architect to develop a commissioning plan, a design intent document, and commissioning specifications, based on owner's identification of systems to be commissioned.
* ASHRAE Guideline 0-2005: The Commissioning Process—The industry accepted model commissioning guide.
* ASHRAE Guideline 1.1—HVAC&R Technical Requirements for the Commissioning Process—2008.
* The Building Commissioning Guide, U.S. General Services Administration, 2005.
* BCA, Building Commissioning Association, www.bcxa.org/resources.
* Model Commissioning Plan and Guide Specifications, Version 2.05, PECI. Feb. 1998—Available from PECI, 921 SW Washington, Suite 312, Portland, Oregon 97205; Email peci@peci.org.
* No Operator Left Behind: Effective Methods of Training Building Operators. Proceedings of the 2007 National Conference on Building Commissioning. www.peci.org/ncbc/proceedings/2007/Brooks_NCBC2007.pdf.
* LEED Commissioning for New and Existing Buildings, HPAC Engineering (online). Wilkinson, Ronald, 2008. http://hpac.com/mag/leed_commisioning_new.

12.1 GENERAL

Being a "green" contractor is a marvelous occupation, and starting a successful green contracting company is a serious business that does not happen by accident. It requires careful planning to start a contracting company. It requires managing the business on a day-to-day basis with information about market analysis and planning, advertising, employee records and training, accounting and bookkeeping, determining your target market, analyzing the competition, and much more. In fact, it may be that the only businesses in building, renovation, and design that are currently succeeding are those that are embracing green building, using nontoxic building materials, making spaces more energy efficient, and using recycled building components.

With the downturn in the economy and the construction industry at a virtual standstill, many senior professionals are suddenly finding themselves on their own, seeking employment for the first time in years. These professionals have been forced either into retirement and/or to abandon the safety of an organization that regularly delivered their monthly paycheck. This has nudged many who happen to be at their peak to reevaluate their future prospects and employment strategy. Concerns over job satisfaction, location and stress, in addition to cash flow, health insurance, and retirement dominate this strategy.

After taking all this into consideration, the decision is made to incorporate and to search for new offices. Independence has many attractions and advantageous like being your own boss, having flexible hours, and taking control of your own future. As a green general contractor you bill clients for services you provide and which you are good at and enjoy doing. If you are new to contracting, you may initially decide to subcontract all or most of the work to other "specialty" contractors (who will typically bill you on a monthly basis). But in the final analysis, the foundation of a contracting business begins with the basics. Whether you are new in business or are an established contractor with solid gold clients, concentrating on the basic elements of your business is essential for survival in the near term and for healthy growth in the long term.

The freedom that comes with independence often comes with a heavy price tag, not the least of which is the potential loss of security. And while corporations respond rapidly to business cycles and job security can be tenuous at best, employment in any respectable organization still carries a degree of stability that many seek. By being independent, one suddenly breathes an illusion of freedom but the question that needs answering is, where is the next dollar coming from? Family members in particular need to be mentally prepared for such a reality.

12.2 THE OFFICE: HOME BASED OR BRICKS-AND-MORTAR?

The decision of whether to have a home-based business or an off site office is impacted by many factors such as available resources, whether foot traffic is necessary, the number of staff needed, whether the business is full-time or part-time, and whether it is Web-based or not, etc. Likewise, it helps if the business is to be located in a home that it has a separate walk-out basement with easy access.

There is no doubt that a legitimate and competitive business can be home-based, particularly with the new technology now available. Examples of this technology are the Internet, instant messaging, video conferencing, and other innovative workflow tools that make effective telecommuting a reality. Working from home will obviously save you the time you would spend traveling back and forth to an office.

Other upsides include less risk and startup cost, which allows you to test the waters without excessive expenditure. You can also outsource things like account management, public relations, Website management, etc. The main disadvantages are that working at home can be distracting, particularly when children are around. Moreover, meeting clients and subcontractors can be awkward and not provide a professional impression. You also need to ensure that a home office wouldn't conflict with zoning ordinances and that there is adequate parking space.

If the decision is made not to work from your own home, then an appropriate office space will be needed. The cost of rent is determined by the size and location of the office being rented. When a suitable office is found, have the lease prepared in the name of the corporation rather than your personal name to minimize liability exposure should the business not succeed. Also, while people can work in a tight space for short periods, particularly during the startup phase of an operation, over time, it will be difficult to support productivity and retain employees unless they are comfortable and appropriate space is allocated. It is also advisable to make the office space as "green" as possible to send the right message to visiting clients, etc.

The upside of having a bricks-and-mortar office is that a physical location causes fewer distractions and may even attract walk-in traffic (e.g., by noticing the sign, etc.). It also reflects more professionally on the firm and portrays an air of confidence and efficiency to clients and potential clients. The downsides are greater risks and more significant startup costs. It also requires a greater full-time commitment up-front to get the office ready for business. There will also be the need to hire some staff such as a secretary, etc.

When choosing a location for your business, ensure that it is not too far from your residence. Also having adequate parking space is always a plus. If you have any doubts about the success of the new venture, you should avoid taking out a long lease just in case the start-up is unsuccessful; it may even be possible to initially enter into a month-to-month rental agreement. If you find yourself tied into a long lease, check the agreement to see if there is a sub-lease clause that allows you to sub-lease the premises if you decide to close the business (although you will still typically need the landlord's approval).

Before preparing to view potential office space, decide on a budget. Likewise, make sure you are able to differentiate between "gross" square feet and "usable" square feet. Usable square feet consists of the area in square feet available for such things as workstations and consists basically of the total or "gross" square feet less areas occupied by lobbies, rest rooms, kitchens, etc. Thus, when inspecting a prospective office, make certain it is adequate to meet your current needs, but will also accommodate potential expansion in the future. It is important to check whether the lease is a net-net lease requiring tenants to pay all expenses including utilities, signs, lighting, taxes, insurance, garbage, maintenance, etc.

Depending on the size of the new business, space requirements will vary according to individual needs and the allocated budget. For typical startups, a space allocation of about 100 sq ft per workstation is recommended, plus about 30 sq ft per person for aisles and equipment and other shared areas. Placing professional looking signage at the front door entrance is necessary.

12.3 CREATING A BUSINESS PLAN

Prior to marketing your construction services, you will need to put together a business plan, which consists of a written document describing the business, its objectives, its strategies, the market it is in, and its financial forecasts. It also details precisely what services are being offered, who the proposed recipients of these services are, who the competition is, and methods used to advertise and promote your services during the first year of business and beyond. Having a business plan in place will help generate interest from potential lenders, prospective employees, and strategic partners. It is also an operating tool, which, when properly used, can help manage the business and effectively work toward its success.

When writing your business plan, keep it simple, concise, and neatly formatted, and preferably in a Microsoft Word document with attached or embedded spreadsheets in Microsoft Excel. Avoid unnecessary fancy graphics, "padding," and flowery language. If you prefer to use a business planning software package to prepare your business plan, that should not present a problem. One of the drawbacks of using business planning software is that it may not have the flexibility to accurately convey all of the features and potential of your new business. However, business planning software has the advantage of offering a logical step-by-step approach. It also generally formats your business plan for you and may also include sample business plans for specific types of businesses.

Proper planning is the real key to the success of any business and its importance cannot be overemphasized. The process of putting a business plan together, including the thought put into it before writing it, compels one to take a serious, objective, and unemotional look at the business project in its entirety. This will assist in identifying areas of strengths and weaknesses. To be effective and convincing, a plan must show the marketing strategy that is to be employed.

Business plans serve several functions, such as securing external funding and measuring success within the business. For new businesses it is used chiefly to ensure that the various aspects of running the business have been researched and thought out, thereby avoiding unexpected surprises. Additionally, it is required by lenders when applying for financing; it can help convince banks or potential investors that your firm is worthy of receiving financial assistance for the new venture. A business plan is therefore an important working document that should be used—not filed and forgotten.

The concept of a business plan is to communicate ideas to others while providing the basis for a financial proposal. Setting up a new business is fraught with difficulties and challenges; statistics show that over half of all new businesses fail within the first 10 years. The main cause for failures is lack of planning and adequate financing. Most new business owners will have difficulty in finding startup capital, which is why they will be expected to initially use their own funds or a bank loan linked to income or security other than the business, such as a home equity loan.

The main components of a business plan are:

Introduction

This consists of a brief but comprehensive summary of how the company was formed, what type of business it is, and the people linked to it.

Mission and Vision

This generally reflects the objectives, aspirations, and direction of the company's business, as well as its expected achievements. Both short-term and long-term goals are outlined and the factors to be focused on in the short and long term.

Management

Describe the management team with short biographies of principals and key personnel that will be instrumental to the business's success. Include each team member's role, background, position, and responsibilities and why they are specifically qualified for their role.

Activities and Services Offered

Outline in detail the type of services to be offered, e.g., green builder or sustainability consultant, the market for this service, and how you will fit into this market. Include drawings, specifications, previous projects executed, and anything else that would enhance your presentation. It is also important to point out any special skills, factors, and qualities that give your firm a competitive edge.

Financial Plan

This part of the business plan is critical and condenses the firm's strategies and assumptions into the cost of setting up the new business and the expected profits. This is the section that lenders and investors will be most interested in to evaluate the financial prospects of your business. The financial section should clearly show financial projections for the first few years of business (depending on the lender's requirements) and contain:

- Written narrative of key business assumptions
- A 12-month profit and loss projection
- Income statement
- Projected balance sheet and break-even point
- One year cash flow projection
- Cash management report.

Executive Summary

Although the executive summary appears as the first part of the plan, it is usually written last, upon completion of the whole document. It summarizes the most important information and aspects of your business plan and normally does not consist of more than two pages. It generally contains information relating to the key people, the idea and services offered, the market, the competition, and strategy to be employed. Investors will not normally spend more than a few minutes reviewing a business plan to determine whether they should read it in detail or move on to another plan. If an investor does decide to read any part of the business plan, it is typically the executive summary that is read, which is why it is imperative for it to be both appealing and convincing, and to be able to capture the investor's attention and imagination.

12.4 STARTUP COSTS, SETTING A BUDGET, AND CAPITALIZATION

Startup expenses are those expenses incurred before the business is up and running, and there are many. People often underestimate startup costs and start their business in a haphazard and unplanned way. Without adequate funding it would be almost impossible to establish, operate, and succeed in setting up a new business. Inadequate funding is one of the primary reasons many small businesses fail within their first year of operation. This is why it is important to accurately estimate these expenses and then determine where any required capital will come from. It has been shown that first-time business owners often greatly misjudge the amount of money needed to get their new business off the ground. New business owners also frequently fail to include a contingency amount to meet unforeseen expenses, and consequently, they fail to secure adequate financing to carry their business through the period before the business reaches a break-even status and starts to show a profit.

Many new startup businesses fail due to being "undercapitalized"— often the result of inadequate cost planning during the pre-launch phase. Likewise many new startups are likely to end up spending more money than planned, so it is better to err on the side of caution. The majority of experts recommend that startup funding should be adequate to cover operating expenses for six months to a year to allow the business to find customers and get established. But it is not possible to determine the amount of financing to seek without developing detailed cost projections. Some experts suggest a two-part process—developing an accurate estimate of your one-time startup costs, and putting together a projection of operating expenses for at least the first six months of operation. The performance of these two exercises will offer a clearer overview of the business and help identify potential problems that need rectification, thereby ensuring a business's success.

Startup Costs

These are expenses that are incurred prior to commencing with the plan, before the first month. For example, many new companies incur expenses for legal work, logo design, brochures, and other expenses. Using a startup worksheet to plan initial financing will help gather the necessary information to set up initial business balances and prepare an initial estimate of startup expenses (Figure 12.1). Needless to say, estimating the amount of funding needed to start a new business requires a careful analysis of a number of factors. A list would be needed of realistic expenses of one-time costs for opening your doors that would include all needed furniture, fixtures, and equipment. The list would also include the cost, down payment, or cash price of items, or if purchased on an installment plan, the amount of each monthly payment.

Expenses incurred during the first year should appear in the profit and loss statement of the first year, and expenses incurred before that must appear as startup expenses. When the approximate amount of cash needed to start your business is determined, an estimate can then be made of how much money is actually available or can be made available to help start the business. If this amount is inadequate, then where would the rest of the required money to set up the business come from?

When the approximate amount of cash needed to start is determined, one can then estimate how much money is actually available or can be made available to put into the business, and where the rest of the money needed to start the business will be coming from.

Employees and Required Forms

Always consult your accountant as to matters relating to whether you should hire yourself or others as full or part-time employees of the company, as you may then have to register with the appropriate state agencies or obtain workers' compensation insurance or unemployment insurance (or both). Many major firms now allow (or prefer) some of their employees to work from home and only come into the office say, once a week. This would be suitable for say, accountants or designers.

It takes many hours of hard work to prepare and file the various payroll reports and other necessary governmental forms; this would put a heavy burden on anyone trying to keep up with the whole enchilada on their own. When your business has grown sufficiently to allow the hiring of qualified employees and/or managers, don't hesitate—hire them. Qualified and well-trained personnel can significantly improve a company's performance and the bottom line.

Your accountant should be consulted when the decision is made to hire new employees to determine what type of personnel files will be needed for each person. Typically, the minimum forms needed would include an I-9 form, IRS form W-4, and the state equivalent form for

Start-up Costs Estimates: The first step is to put together a list of realistic expenses of one-time costs for opening your doors. Such a list would include what furniture, fixtures and equipment is needed, as well as the cost, cash price, down payment if purchased on an installment plan and the amount of each monthly or periodic payment. Record them in the costs table below:

Down Payment	$_____
Amount of each payment	$_____

The furniture, fixtures and equipment required may include such things as desks, moveable partitions, storage shelves, file cabinets, tables, safe, special lighting, and signs.

TYPICAL START-UP COSTS ITEMS TO BE PAID ONLY ONCE:

Furniture, Fixtures, & Equipment:

Interior decorating	$_____
Installation of fixtures and equipment	$_____
Starting inventory	$_____
Deposits with public utilities	$_____
Legal and other professional fees	$_____
Licenses and permits	$_____
Advertising and opening promotion	$_____
Advance on lease	$_____
Other miscellaneous cash requirement	$_____
TOTAL ESTIMATED CASH NEEDED TO START =	$_____

ESTIMATED MONTHLY EXPENSES:

Salary of owner-manager	$_____
All other salaries and wages	$_____
Payroll taxes and expense	$_____
Rent or lease	$_____
Advertising	$_____
Delivery expense	$_____
Office Supplies	$_____
Telephone	$_____
Other utilities	$_____
Insurance	$_____
Property taxes	$_____
Interest expense	$_____
Repairs and maintenance	$_____
Legal and accounting	$_____
Miscellaneous	$_____
TOTAL ESTIMATED MONTHLY EXPENSES =	$_____
Multiply by 4 (4 months)	$_____
Add: Total Cash needed to start above	$_____
TOTAL ESTIMATED CASH NEEDED	$_____

FIGURE 12.1 An example of a draft startup worksheet used to produce a preliminary startup costs estimate and to plan an initial financial strategy for a new business venture.

employee income tax withholding. If using independent subcontractors, they should sign IRS form W-9. Consult with your accountant as to whether state law requires subcontractors to be included on the firm's policy or not.

Utilities

Advance deposits, especially for new businesses, are often required when signing up for power, gas, water, and sewer. Also, once the decision is made to establish your own business and an office has been leased (if you are not working from home), ask when the next issue of the telephone directory is to be published and the deadline for getting listed. Ensure a display ad is placed in the yellow pages under the classification best describing the company's services—and before the deadline.

Expense Report

Many firms have developed standardized digitized expense report forms for their employees so that they can request reimbursement for their business expenses. Even with a new startup business, it is vital to monitor expenditures, and a standard form may be the best way to do so as it makes it easier for bookkeeping. The expense form, if not standardized, should be neatly typed and organized, identifying each location, project name and number, and applicable dates with all original receipts and supporting documentation attached in date order. It should then be given to accounting to process and record in a timely manner.

Office Equipment

Businesses will differ in the type of equipment they will need. It may also be prudent with a startup business to preserve cash for inventories or working capital and initially purchase good used fixtures and equipment at a much less expensive price. Obtain more than one quote on the equipment you will need. With the recent changes in the income tax laws you will have to do extra analysis to determine whether a lease program or direct purchase is the best way to proceed. Whether to buy or lease depends on several factors, so consult your accountant for advice. For new companies that want to keep their initial startup costs to a minimum, it may be smarter to initially lease as much as possible, especially electronic equipment, computers, copiers, printers, telephone systems, and certain other products due to the continuous advances taking place in these fields. Record any cash down payments for any equipment purchased on contract.

Furniture

Record the cost of all office furnishings (desks, credenzas, file cabinets, bookcases, chairs, end tables, lamps) so that you can deduct these expenses from your tax bill. If payment is by installments, note the down payment as a startup cost. When paying in cash, enter the full retail price.

Decorating/Remodeling

If you are moving into an office that needs some reconfiguration or redecoration, make an estimate of what the total cost will be. Try and negotiate with the landlord to pay for it or deduct it from the base rent. Talk to suppliers with whom you plan to purchase materials and other services, and record these expenses. It is unlikely that you will undertake major work unless you are contemplating the long term.

Internet Service

Apply for a phone number and domain name for the Internet Website for your new business (this is discussed later in the chapter). When you get the phone number, look into a yellow page advertisement (or at least listing), and consider whether to be listed in several headings, or just the most appropriate heading for your services.

Suppliers

Many suppliers are reluctant to ship their goods to new businesses. This is one of the reasons it is important to have a rapport with your banker as he can provide acceptable credit references to your suppliers. Your proposed suppliers may need convincing that you are honest, hard working, and in it for the long haul, and that your business is solid and has a good chance for success. Some suppliers may request you to pay C.O.D. while getting started; take this fact into consideration when preparing your financial planning and startup estimates. Once you become established with your suppliers, send your financial data to Dun and Bradstreet so your company becomes listed in their files. Dun and Bradstreet is the most well-known and reliable organization for obtaining correct credit information on businesses.

Accounting and Bookkeeping

It is necessary to set up a good accounting and recordkeeping system and to learn as much as possible regarding the taxes your new company is responsible for paying. Company documents and tax and corporate filings are generally required to be kept for three years, including: a list of all owners and addresses, copies of all formation documents, financial statements, annual reports, and amendments or changes to the company.

As a new business it is possible to do your own recordkeeping, but if you decide to go this route it is advisable to have an accountant help set up the books based on the simple method outlined below. If possible, let the accountant "keep the books" for the first few months while you learn how, then if you prefer, take them over yourself. After a short time you or an employee will probably be in a position to do the accounting on your own. Use a separate checkbook and bank account for your business—i.e., avoid comingling your private and business accounts. The use of the records of original entry plus a "general journal" to record extraneous transactions, as well as a "general ledger," to which accounts from the three records are posted at the end of each month, will provide the necessary data for a simple "cash" accounting system. This system can be readily converted to an accrual method of accounting by journalizing accounts receivable, payable, accruals, etc. Once these entries are posted, it is fairly simple to complete the balance sheet and income statement.

Upon preparing the financial statements, reverse the accruals and you will be equipped for the following month's entries. You can also now enter the gross payroll, payroll deductions and the net amount in your check register. It is expedient to give your employees a payroll slip itemizing all the facts while maintaining supplementary payroll sheets with all the information as to each employee. Having these individual payroll records and the control accounts in the general ledger provides you with all the information necessary to complete the various payroll tax reports and returns as they become due. At the end of each annual accounting period all the information for filing your income tax returns will be available for your accountant to go through and submit the final returns.

All files should preferably be stored on computers instead of in file cabinets (this also makes it easier to email and make off-site backup copies when traveling). Reviewing documents onscreen rather than printing them out also helps the environment, as does sending emails instead of paper letters. New software like Greenprint is also available and helps eliminate blank pages from documents before printing and can also convert to PDF for paperless document sharing.

Miscellaneous Issues

Other startup expenses that need to be considered when estimating the amount of cash you will need include both business and personal living expenses. If you are leaving a salaried job to start your business, you should include in your expense projection an estimate of the costs you and your household will incur for the months it will take to build up your business. At this point it probably makes sense to review certain categories, like equipment, office supplies, or advertising/promotions, with cost-control in mind. If it appears that your estimated startup costs are greater than originally anticipated, it may be time to review and reevaluate your list of projected expenses and decide for example if used office equipment or furnishings might make more sense. Likewise, instead of purchasing all new equipment and furniture, search the classified ads, which can lead you to bankruptcy auctions, house sales, and furniture resellers in addition to individual items in the classified section.

In the final analysis it may be wise to ask your attorney or accountant for referrals to business owners who have relevant experience in evaluating startup costs to ensure that you are on the right path.

You can get additional information and advice on startup costs from the U.S. Department of Commerce Minority Business Development Agency (www.mbda.gov). Articles can be found here that discuss the amount of money needed to start a new business. The Website includes helpful checklists and provides referrals to other information resources. The U.S. Small Business Administration (www.sba.gov) was created specifically to assist and counsel small businesses. Its publication, Small Business Start-up Kit, includes a checklist for calculating startup costs. The SBA has an Online Women's Business Center at www.onlinewbc.gov, which includes a helpful section on evaluating startup costs for new businesses.

12.5 CREATING A CORPORATE IMAGE

As discussed earlier in this chapter, starting a new business can be risky, but your chances of success significantly increase with proper planning and a clear understanding of the potential challenges you may face. Below are some of the steps needed to succeed in any new startup business venture, whether it is a green contracting business or a professional consultancy.

The Corporate Image

One of the first things to work on is to create an impressive corporate image and business identity that instills confidence, and creates loyalty and efficiency. This means hiring a professional designer to design a corporate logo, business card, letterhead, and promotional material for the business. The logo should be simple and not easy to forget; it represents the visual image of your company and will be used in a variety of applications. Moreover, an attractive and professionally created logo and letterhead can go a long way to giving clients the image of confidence and trustworthiness while reducing their perception of risk, making it easier to command a premium price. A logo also says who you are, how you're different from your competition, and why a client should do business with you.

Advertising and Promotion

The next step in the startup process relates to getting the word out about your business so that customers start coming through your door or your home page. To do that you'll need to study your target audience and develop a marketing message that will resonate with them. Some small new businesses start operating with a grand opening announcement and press releases to the local press and relevant business publications.

Circulars can be printed and distributed to potential clients or placed in the newspaper to be distributed to subscribers. The dollar cost of planned advertising and marketing announcing the launch of your new business should be recorded and should include the cost of all promotional items including flyers, sales letters, phone calls, signs, and brochures, etc.

Competitors' ads should be studied along with their Websites. It is also helpful to check out and record the promotional materials of similar businesses on the Internet and in specialized journals. Preferably choose ads and promotional stuff that fits within your promotional program or that can be modified to fit your needs. An ad should communicate what your company is and does as well as its distinct attributes and image. It should deliver a clear and compelling message.

Marketing Company Services

Before trying to market your services, adequate research is required to get all the facts, just as was done prior to writing the business plan. This will help you put together a successful marketing strategy that will be much more methodical and effective. Research will also help in the development of professionally designed brochures and other marketing materials by determining who your target audience is and what they want. As you are selling a specialized service, it is imperative to know how to market it. This requires choosing a marketing plan and strategy that will target your ideal customer. To do this you will need to answer several key questions such as:

- Who is the typical buyer?
- Is what you have to offer what they want?
- What is the budget of targeted customers and how much are they willing to pay?
- Why should potential customers prefer you over your competition?
- What media will best reach your target market?

Once these questions are satisfactorily answered, you are in a position to start developing a successful marketing strategy. Figure 12.2 is a typical letter to get the word out and let customers and potential clients know your doors are open for business.

Time Management

It goes without saying, the better organized you are, the more efficient you are and the less time you waste. This can be facilitated by the appointment of a secretary or office manager to deal with the operational aspect of the business and make it as automated and efficient as possible to allow you to concentrate on the business aspects. This will also free you from having to follow up on day-to-day issues like processing orders, paying utility and other bills, paying employees, maintaining your permits, etc.

Mr. Samuel Doe   November 24, 2009
President
XYZ Developers Inc.
1070 East Market St.,
Leesburg, VA 20176

(Tel)   (703) 777 1234
(Fax) (703) 777 2345
Email: sdoe@XYZ-developers.com

Re: Green Building Services

Dear Mr. Doe:

I am taking this opportunity to apprise you of green building services that we offer to property developers and investors.

ABC Green Building International has recently been formed to provide green construction services. Although ABC is a newly formed company, its principals have over twenty five years of experience in design, construction, and sustainable practices. Our specialty is green construction and we have a number of LEED Accredited Professionals on our staff to ensure improved occupant health, protection of ecosystems, and reducing energy consumption in our projects. For further information and an overview of our services, please view our website at: www.abc-greenbuildingt.com. We would be delighted to discuss our services with you and bid on any upcoming projects.

We are able to travel anywhere within the United States to provide services to meet your requirements.

I'll give you a few days to look things over, and then I'll call you to set-up a time to discuss your requirements.

Sincerely,

Sam Kubba, AIA, RIBA, Ph.D., LEED AP
Principal
ABC GREEN BUILDING INTERNATIONAL

SAK/bs
enclosures

cc: General Files
ABC/PROMO/LETTER/John Doe/Promo.doc

FIGURE 12.2 An example of a typical promotional letter offering green construction services that can be sent to potential clients to inform them that your company is open and ready for business. Promotional material should accompany the letter.

12.6 IDENTIFYING AND TRACKING SOURCES FOR LEADS

The approach used to identify potential sources and project leads will depend to some extent on whether your business is essentially a one-person organization or one that is well organized with several employees. These methods include:

- An excellent starting point would be sending out flyers, brochures, emails, etc. to potential clients and to advertise your services in the local press.
- Check specialized construction search engines like "bidclerk" (www.bidclerk.com). They can provide excellent construction leads for construction projects that are coming up for bid in your area.
- Scan the Internet, particularly real estate broker sites displaying vacant land. All major real estate firms typically have Websites and some of these firms have client lists to build up potential customer confidence on their Websites. These lists can be researched to see which, if any, names are worth following up on. Ask them if it is possible to get information on potential buyers so that you can send them promotional information.
- Contractors should make an effort to visit neighborhood commercial real estate agents to see what commercial properties are on the market, including vacant land. Some properties may require renovation. A list of all these possible leads should be made and followed up with letters and brochures offering the company's services.
- A general drive around the area can often highlight possible leads. Many of the clients will be lenders (e.g., banks, lending institutions) who will provide financing to property developers or individuals wanting to build a custom home or commercial building (Figure 12.3). It would be prudent to make a list from the yellow pages, Internet, and from research in the public library of these institutions and send them your promotion material.

12.7 SELLING YOURSELF

Dress

It is important for men and women to dress appropriately in corporate environments. For men, this implies dress slacks, a clean button up shirt with tie, and a blazer or waistcoat. A jacket may not be necessary in some situations, like if you are inspecting a building site in summer. And obviously, it is not necessary to wear a business suit everyday if you are working outdoors, etc. However, in the final analysis this will depend

FIGURE 12.3 An office building under construction in Arlington, Virginia. This building shows Wachovia Bank as the lending institution. The signs show that Clark Construction is the general contractor. It may be prudent to send promotional mail to the bank to have your firm's name placed on its contractor list and also a letter to Clark Construction to seek subcontracting work.

on your own individual situation, environment, and the audience you're dressing for. Another important aspect of proper business dress for a man is to be clean-shaven, as scruffiness is unprofessional and sloppy. To further enhance your image and complete your ensemble, hair should be clean and well-groomed and cologne or aftershave should be subtle.

Some dress consultants say that often companies have a business casual dress code for their employees to follow, without always producing guidelines for what this actually means. Business casual dress isn't as formal as wearing professional business clothes—suits are acceptable, but not necessary. It is dressing professionally, but comfortably, and outfits should create a relaxed, comfortable appearance, while still looking neat and smart. Examples of business casual are cotton trousers and khakis for men. Combine these with a collared shirt to create a professional but relaxed appearance. Loafers or other slip on shoes are acceptable. But even with a business casual dress code, you should still dress professionally, especially if you are meeting clients or customers face to face.

Women too sometimes get confused about what is appropriate, although they are not alone, as even women in executive positions sometimes admit to not knowing what styles suit their bodies. In fact, many women find themselves wearing clothing that is not the right size, being either too large or too small, something that can easily be avoided. Furthermore, clothing should not be too revealing. Great basics for most work environment include collared shirts, pencil skirts and good slacks. If you still remain unsure of what to wear to work, observe some of the professional, successful women in your industry as they can offer appropriate examples of what is acceptable in your particular environment (Figure 12.4).

A person's appearance is a very powerful form of communication and when used properly can be an effective tool for portraying confidence, trust, and ability. Although the importance of proper attire might seem logical and common sense for some, it does not come naturally to everyone. It is a well-known fact that first impressions can significantly impact how a person is ultimately perceived and this is why proper dress is so important. Also, when attending a business meeting with a client, bank manager, etc., start off with a firm handshake and follow up with eye contact.

Companies and industries have different norms in regard to business dress. For example, an accountant at an investment bank may not wear the same work attire as a site engineer at a small startup contracting company. Also, the dress code for a site meeting will differ from that for an important meeting with a client. And while a business suit is not necessarily the everyday work attire for an organization, potential clients do expect consultants and executives to look professional during a project meeting. When in doubt, it is normally better to err on the side of conservative dress. Clothing and accessories should not attract so much attention as to distract from a meeting's real purpose.

Introductions

Meeting new people and mingling can be uncomfortable experiences for some people, especially for those who are shy or introverted. When attending an event where you are likely to meet potential clients, e.g., a conference, be sure to carry business cards and perhaps some literature about the company. Also try and portray an air of confidence; this will give you the appearance of an accomplished professional to others. Be cool, calm, and collected, and, most importantly, *think* before you speak. Be organized and prepared and have the necessary knowledge to answer any questions you may be asked to show customers that you can execute the job successfully. The corporate world is rough at best, but, with your firm's specialty skills, you are in a position to achieve success and prosperity.

Correspondence

Correspondence is increasingly becoming an online affair which means less paper is being used. Also, business files are increasingly being kept on

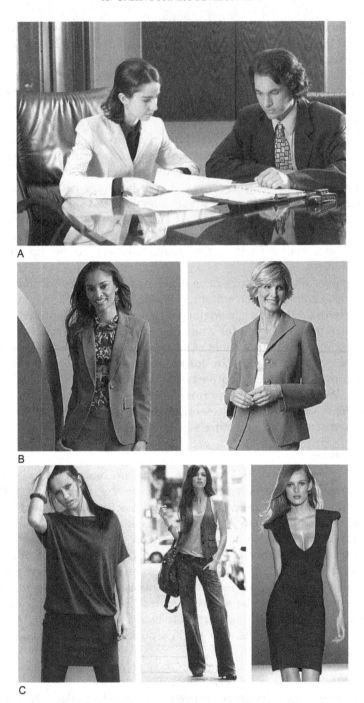

FIGURE 12.4 Examples of acceptable and unacceptable professional female attire in the office.

computers instead of in file cabinets (this also makes it easier to make off-site backup copies or take them with you when you move to a new office). Documents can be reviewed onscreen rather than having to print them out. Communicating by emails instead of paper letters is far more efficient and cost effective. Introverts tend to prefer email because it is efficient and avoids direct contact; extroverts on the other hand usually prefer direct face-to-face communication. Before sending an email or letter, make sure that it is sent to the correct person. Also, all correspondence should be reviewed for accuracy, and also use spell check before sending it. Some secretaries and executives are in the habit of selecting the email address last to avoid accidentally sending the email before it is complete.

There is also new software on the market like GreenPrint that eliminates blank or wasted pages from documents before printing, saving on both paper and ink. It can also convert files to PDF for paperless document sharing.

Meetings

The key to successful meetings largely depends upon good organization and adequate preparation. It is a time when you will meet people, clients, and executives to discuss relevant topics such as client projects, marketing strategies, financing, etc. Be prepared with questions and matters you want to cover, and anticipate in advance what you seek to achieve. During business meetings, be careful to stick to proper meeting etiquette, as this is an arena where poor etiquette can reflect negatively on you and your firm. Correct business meeting etiquette automatically improves your chances of success and communicates comfort and trust with everyone involved including colleagues, clients, and customers. In today's business world, it is these people who can influence your firm's ability to succeed.

Informal Meetings

Informal meetings are generally more relaxed affairs and may not necessarily take place in the office or meeting room. Even so, a sense of professionalism and good business etiquette is still required. In this respect, punctuality is always important. Also, the purpose of the meeting should be clearly outlined to the proposed attendees. Failing to relay the proper information is poor business etiquette and could cause embarrassment and prevent the meeting from succeeding and achieving its objectives. The person calling the meeting is usually the most senior person or the person with the most direct or urgent interest in the topic at hand. This person may also be responsible for determining (through consultation) the meeting's time, place, and agenda. Someone should be appointed (usually through the chair) to take minutes, which can later be typed and distributed to all attendees.

Formal Meetings

Although it is unlikely that the owner of a new startup firm will attend many formal meetings, nevertheless, it is important to have a clear understanding of required etiquette. Business etiquette for formal meetings such as management meetings, board meetings, and negotiations usually includes a set format and/or agenda. As a professional you should dress appropriately and be punctual. Mobile phones should be switched off during the meeting. It is imperative to be well prepared and any reports or other information to be used should be handed out prior to the meeting (with adequate time to review). If you are unsure about the seating pattern, you should ask. During the meeting you basically need to:

- Always address the chair when speaking, unless it is clear that no one else is doing so.
- When discussions are underway allow more senior figures to contribute first.
- Acknowledge opening remarks with a brief recognition of the chair and other participants.
- Do not interrupt a speaker—even if you strongly disagree. Note what has been said and come to it when appropriate with the chair's permission.
- It is very unethical and a serious breach to divulge to others confidential information about a meeting.

There are also many factors that will add to a new company's chances of achieving success and should be carefully considered, including:

1. Creating a Network

A good network is almost synonymous with business success. And while a lucrative contract may be the result of a single contact, it takes a strong network to generate a continuing stream of remunerative projects. Building a strong network requires access to potential clients, and what some might consider to be a well-developed network, successful consultants and contractors may see as little more than a starter list. Occasionally, a virtually unknown, but successful, executive is hired by one or two clients immediately after leaving an employer to begin his own business. And sometimes it is the former employer that immediately entices that person to return because of their valuable knowledge of the operations.

2. Communication Skills

Senior and executive-level professionals are expected to have excellent verbal skills, since this competency is a primary determinant for moving up the corporate ladder. However, writing skills can be a major challenge to those who have depended on others to put pen to paper, especially since consulting projects often require some form of written report. Publishing quality articles that attract the attention of potential clients in the industry is another cost-effective approach to spread the word. However face-to-face contact is the best form of communicating and potentially offers the best returns. The downside is that it is time-consuming and expensive, and few executives have the marketing budget or the time.

3. Hard Work is a Must

The probability of a startup business succeeding without putting in the hours and the effort is almost zero. There is obviously some flexibility in the work hours when you become your own boss, but this is no eight-to-five job and hard work and effort will definitely be needed to build the business. Some startup businesses may have been lucky and fallen immediately into client work and thus become complacent. Others may find themselves straddling the fence, not making a wholehearted commitment to the business and continuing to look for a suitable position. Not being fully committed prevents you from aggressively building a presence, aggressively marketing your firm's services, and aggressively obtaining a Web domain and building a Website. Even if you do not feel fully committed, there is no real justification for not projecting a professional image while seeking alternative employment, as doing so will ultimately reflect badly on you.

4. Marketing Skills

It is imperative to both identify your target market and to develop a detailed marketing strategy that provides a competitive edge and draws customers to you and your company rather than to the competition. To succeed you must be willing to engage in relentless self-promotion in order to bring in needed new business. Seek out specific target markets that will need your services and are willing to pay for it. Also develop a list of the principal competition in your field and provide an honest appraisal of their strengths and weaknesses and how you contemplate successfully competing against them with the available resources. In a saturated marketplace this may prove to be an uphill challenge. But one has to be flexible, as work can come from many unanticipated directions and sources.

Grassroots marketing is also an affordable type of marketing. It consists of taking advantage of available resources to spread the word about your service. It entails distributing your marketing material at local businesses, churches, chambers of commerce, and community centers. It also includes networking to connect with potential customers and strategic partners and spread the news about your business. Joining a chamber is very important for building your network and providing an ecosystem portal to members who may be looking for business, as well as sources of services.

5. Financial Security

To succeed at being your own boss in a new business requires financial stability and ability to survive the dry difficult periods that could easily last a year or more. If survival is difficult under such circumstances, it may be prudent to reconsider the decision to be an independent businessperson. One of many challenges confronting the small contractor is the need to be responsible for all aspect of a project. Persons that are exceptionally strong in one area, such as building systems, but very poor in say, marketing, may be wise to team up with others who can compensate for these weaknesses.

6. People Skills

Executives often find it easy to dictate orders to employees within their firm, but dealing with clients is much more complicated and takes skill. For example, consultants and contractors have to respond to a multitude of personalities with little or no background information on the people's likes and dislikes. Bullying techniques and intimidation that bosses can thrive on inside companies fail to get a welcoming response from clients and potential customers. Moreover, independent consultants or contractors may find themselves quickly dropped if their performance is less than top-notch. Possessing great people skills may bring in the work, but it will not necessarily help you retain it or get repeat business—only committed, persistent effort can do that.

7. Self-directed

Not all people are able to work on their own initiative; some experience great difficulty in performing without the umbrella of a structured environment. Independence can certainly be a freeing and exhilarating experience, but it can also be lonely without daily, face-to-face interaction. This is especially true of individuals who work out of a home office instead of a rented office space in some corporate office park. The bottom line, however, is about self-awareness, self-confidence, and the ability to go it alone while relying on your own abilities.

12.8 BUSINESS FORMS, LICENSES, PERMITS, AND INSURANCE

There are various bureaucratic and legal hurdles that an entrepreneur must overcome before being able to incorporate and register a new firm in addition to the time and cost involved in launching a new startup. These need to be fully examined before any attempt is made to launch such a venture.

Having made the decision to start a new business, it has become your responsibility to understand and comply with government laws and regulations that apply to your business. These laws are designed to protect you, your customers, and your employees. You may now be required to obtain a number of licenses and permits from federal, state, and local government before you can open your doors for business. Licensing and permit requirements for small businesses vary from one jurisdiction to another, so you will need to contact your state and local government to determine the permits, licenses, and other specific obligations required for your new business. Before doing so, however, a decision must be made on the proposed name for your new business and the legal structure your new business will be operating under.

12.8.1 Name and Legal Structure

There is no universally "right" structure for all businesses, and choosing the right one for you depends on the specific needs of you and your business. There are advantages and disadvantages to each type of business structure. It is important therefore to understand all the options available to you before setting up your company. Advantages and disadvantages of each business formation need to be evaluated, paying special attention to the tax implications and government formalities. It is always advisable to work closely with an attorney and possibly an accountant to ensure you make the right choice.

You basically have four forms of business ownership when selecting a legal structure: 1. Sole proprietorship, 2. Partnership, 3. Limited liability company (LLC), and 4. Corporation or S-corporation.

Sole proprietorships are a popular choice for the majority of new small business owners because they are the least complicated and simplest form of business organization to set up. The individual proprietorship business form is basically owned and operated by one person, and apart from local business licenses, there are minimal government fees and paperwork.

On the other hand, there are also considerable risks that need to be considered, such as the vulnerability to creditors of your personal assets and other liabilities such as lawsuits. In addition, certain tax breaks that are reserved for more formal business structures such as corporations or limited liability companies may not be available. Also, as a sole proprietorship, your company name is not protected. This means there is nothing to prevent another company from incorporating under your business name. Again this is why it is so important to work closely with an attorney when setting up a new business.

Partnerships are similar to sole proprietorships in that they are easy to set up and maintain, and require no government fees or annual state paperwork. It may also be the way to go if you need additional capital or expertise. A disadvantage with a partnership entity is that you and your partners are each held fully responsible for all company debts. Thus, if any of the partners default on a company loan, creditors can still go after you personally to satisfy the entire loan. This includes your personal bank accounts, property holdings, and other assets. Also, just as with sole proprietorships, your company name is not protected so that any new or existing business could incorporate using your company name.

Incorporating is the standard for many of today's businesses, largely because incorporating shields you and the company from personal liability. Thus, creditors are prevented from going after your personal assets to make up for any company shortfalls should your business hit hard times. In addition to protection from personal liability, the corporate business structure also offers significant tax savings, company name protection, and increased opportunities for raising capital. If you decide to incorporate you need to choose to set up your corporation as either a C-corp or an S-corp to take advantage of the various tax options. Also keep in mind that corporations do require some initial setup fees and a certain amount of regular maintenance.

A C corporation only makes sense if you have a significant amount of startup capital and you feel ready for the big time and you want to sell shares of stock in your business. This won't apply to the vast majority of new startups. A good alternative to this is the S corporation, which avoids the double taxation of a C corporation. This form provides a tax-efficient way to structure your business if you expect losses in the short term, allowing individual shareholders to report losses on their tax returns rather than pay the double taxation of the C corporation. Prior to making a final determination, consult with an attorney and check with your secretary of state (most states now are on-line).

Running your business as a corporation also comes with serious disadvantages, especially for the new small business, including strict laws and higher state income taxes in some states, in addition to increased legal work and heavier accounting and tax reporting requirements. Moreover, closing down a corporation is often more difficult than creating the initial incorporation.

For many new entrepreneurs, choosing a business structure basically comes down to liability protection, tax savings, and convenience. This is why many entrepreneurs today prefer forming a limited liability company or LLC since this type of entity requires fewer formalities and less on-going paperwork than corporations while maintaining the same personal liability protection and tax flexibility. Just as with a corporation, your company name is protected and the company is shielded from creditors and other company liabilities such as lawsuits. Likewise, with an LLC, minimal company records are required to be maintained. Many consider the LLC to combine the best aspects of incorporation with the tax advantages of partnership while omitting the red tape of both.

12.8.2 Federal Employer Identification Number (EIN)

A federal EIN is a unique nine-digit number assigned by the Internal Revenue Service (IRS) to business entities operating in the United States for the purposes of identification. The EIN is required for almost all types of businesses and acts as your business identifier on all types of regis-trations and documents, and most banks won't let you set up a business checking account or apply for a loan without this number. You must apply for a federal employer identification number (EIN) upon receiving your corporate charter back (if a corporation). You can do so by going to the IRS Website and downloading form SS-4. Once this is filled out, call toll-free (8 6 6) 816-2065 for your EIN. When the number is used for iden-tification rather than employment tax reporting, it is usually called a tax identification number (TIN).

Businesses that are considered proprietorships do not need an EIN and in this case the owner/operator's SSN is used on tax documents. Should you choose to form an LLC, you will need to decide how you pre-fer to be taxed (e.g., as a sole proprietorship, partnership, S-corporation, or C-corporation), and use IRS Form 8832 to make that election.

A fictitious business name permit, also called "dba" or "doing busi-ness as" permit, is required for almost all types of businesses. When choosing a business name it is generally good practice to choose a name that describes your product or service to make the public better aware of just what your firm has to offer. You should apply for a fictitious business name with your state or county offices if you plan on going into business under a name other than your own. The bank will also require a certifi-cate or resolution pertaining to your fictitious name at the time you apply for a bank account for your firm.

12.8.3 Licenses and Permits

Upon starting a new business, steps must be implemented to obtain a number of licenses and permits from federal, state, and local govern-ment. Since licensing and permit requirements for small businesses may vary from one jurisdiction to another, it is critical for the new business owner to contact state and local government to determine if any specific requirements are in place prior to setting up a new business. Keeping this in mind, below are some of the various federal, state, and local licenses and permits that may be required prior to opening for business.

Business Operation License

This is required from the city (for a fee) in which the business will be operating from, or from the local county if the business is located out-side the city limits. This license grants the company the authority to do business within that city/county. Most cities and counties require a business license, even when the business operates from a home. If the decision is to run the business from a home, you should carefully investigate zoning ordinances. Some residential neighborhoods have strict zoning regulations that prevent business use of the home.

When a license application is filed, the city planning or zoning department checks to ensure the location is zoned for the intended purpose and that there are enough parking spaces to meet the codes. If the area is not zoned for your type of business a variance or conditional-use permit will be needed before permission to operate is granted. This can be achieved by presenting your case before the city's planning commission. Getting a variance is usually quite straightforward as long as you can show the planning commission that your business in its proposed location won't have an adverse impact on the neighborhood. However, in many areas attitudes toward home-based businesses are gradually becoming more supportive, making it easier to obtain a variance for the home-based business.

Occupational Licenses

When you're overcome by the enthusiasm of starting a new business, it is easy to overlook the need for certain licenses and permits prior to opening for business. Certain types of new businesses will have to obtain an occupational license through the state or local licensing agencies. Such businesses include building contractors, real estate brokers, those in the engineering profession, electricians, plumbers, insurance agents, and many others. Moreover, in many states and jurisdictions, occupational licenses will not be granted to conduct business unless relevant state examinations are passed. Your state government offices can provide a list of occupations that require licensing and passing exams.

Signage Permit

Some cities and jurisdictions have sign ordinances that restrict the size, location, and sometimes the lighting and type of sign that can be installed outside a business. To avoid costly mistakes, the regulations will need to be checked to see if any signage restrictions are imposed by your city or county; written approval of the landlord (if renting a house or apartment) should be secured before going to the expense of having a sign designed and installed.

Other Licenses and Permits

Federal regulations control many kinds of interstate activities with license and permit requirements. In the majority of cases, this is not a cause for concern. However, a few types of businesses do require federal licensing, including investment advisory services. The Federal Trade Commission can tell you if your business requires a federal license.

County governments typically require the same types of permits and licenses as required by cities. If your business is outside the city or town's jurisdiction, these permits will apply to you. County regulations are often not as strict as those of adjoining cities. Localities may have individual variations or they may require additional permits or licenses (such as zoning, fire, or alarm permits), so both the city and county need to be contacted once you have your basic business information, tax ID number, and business address.

12.8.4 Insurance

The importance of insurance cannot be overstated, especially for general contractors and professionals. Premiums are usually high, especially for business liability, but no general contractor or consultant can operate with peace of mind without full coverage. There are many types of insurance on the market for businesses but these are usually packaged as "general business insurance" or a "business owner's policy." A good insurance agent can counsel you as to the types of insurance you will need and the available coverage, such as general liability insurance, health insurance, fire, property insurance, burglary, company vehicles, workers' compensation, business interruption, and malpractice insurance. It is advisable to have two or three agents submit estimates. It is also imperative to have adequate liability insurance, and anyone contemplating to offer contracting services is strongly advised to consult with an attorney. If you have employees and plan to offer them health insurance, talk to your agent about the up-front fee and record the premium payment you will need to make before opening your business. Health insurance costs are among the most important concerns of many small business owners.

12.8.5 Bank Account

Once you have received your tax ID number (also known as employer identification number), you can use that number to open a business checking account. Find a bank that is convenient and where you feel comfortable with the bank manager. He can be one of your best references. Ask his advice and get his help on financial matters. Even if you are not interested in qualifying for a loan yet, banks can provide numerous other financial services fundamental to your business. These can include business checking accounts, business credit cards, and even providing a credit reference. Added to this, banks have great contacts in the community and can be an excellent source of business referrals, which is why having good relations with your bank manager is very important. It may also be useful to develop a line of credit so it will be there should you need it further down the road. It is also imperative to maintain separate business and personal finances from the beginning to avoid complicating bookkeeping and tax returns.

You cannot establish a bank account without a federal ID number or social security number along with your certificate of assumed (fictitious) business name. If you are incorporated you will be required to provide a copy of the minutes and a corporate resolution authorizing the account. It is better to visit the bank you are considering and discuss with the bank manager the specific requirements for opening a business checking account and see if you feel comfortable with the bank manager. Requirements vary from bank to bank—from fairly simple to extremely complex. The more important issue is establishing a rapport and empathy early on with the bank manager.

12.9 TAXES STRATEGIES AND INCENTIVES

Tax planning is an important ingredient of any successful business, with the obvious objective of minimizing your business's tax bill. Whether it's capitalizing on business deductions, or finding tax-friendly ways to run your business, a good accountant can do wonders for reducing your tax obligations.

One of the first things to do when starting your new business is to submit applications for federal and state ID numbers and request "business startup" application forms from the Internal Revenue Service and from the state tax commission. After these are sent in you will be notified of your number and get a packet of information. After this you will periodically receive depository forms, quarterly report forms, W-2s, W-4-As, estimated tax forms, etc.

You will encounter various payroll expenses such as FICA taxes (social security), FUE taxes (federal unemployment), SUE taxes (state unemployment), or WC or SDI (workers' compensation). If you are a sole proprietorship business or a partner you will be required to file and pay federal estimated tax reports each quarter based on estimated annual income. Partnerships file an annual information return and each partner's share of profits is included in their individual personal income tax return. Corporations are also required to file for estimated taxes, which should be done by your accountant.

Tax Deductions and Write-offs

Maximize what you can deduct, and discover what you can write off by knowing what constitutes legitimate business expenses. A good accountant will be needed to prepare your accounts. Tax consultant David Wetzel says, "Proper planning will result in you getting all the deductions you deserve. Poor planning raises a red flag with the IRS. Sloppy looking returns and indications of poor recordkeeping will earn you a trip to see your friendly IRS agent." Below are some possible tax deductions:

- You can normally deduct the rent for a rented office as a business expense. With a home-based office the business must be located in a separate room within the home. Ideally, it would be located in a walk-out basement with a separate entrance. To claim home office expenses, calculate the square foot percentage of your home office in relation to your home. Then apply this percentage for deductions for utilities, mortgage or rent, insurance, Internet service, etc.
- Furniture: The amount that can be deducted for furniture purchase will vary, but you can reportedly now deduct 100 percent of all office furniture costs without having to depreciate it over several years. Check with your accountant.
- All supplies purchased for your office can be deducted. It is important to keep receipts.
- Utilities: These include water, electricity, and phones; they can normally be deducted in outside rented offices. In a home office setting it is better to install a second phone line for your business. This is the safest approach to take phone deductions on your business taxes.
- Website: This expense can be written off as a business expense.
- Computer equipment: A new or recently purchased computer is 100 percent deductible without having to depreciate it.
- Computer software: All software purchased for business use is 100 percent deductible.
- Business vehicles: Travel expenses can be claimed based on the actual mileage the vehicle is used for business. Check with your accountant to ascertain the mileage rate at the time.
- Insurance: Small business owners can deduct 100 percent of premiums, providing they do not exceed your business's net profits. Check with your accountant regarding requirements and stipulations.
- Entertainment: Legitimate entertainment expenses can be recouped. Check with your accountant.

12.10 THE BIDDING PROCESS AND TYPES OF BUILDING CONTRACTS

The essence of construction bidding consists basically of submitting a proposal to carry out a described residential or commercial construction project for an agreed price. Bidding for a project can occur at the construction manager, general contractor, or subcontractor level. Contractors can submit their bids for the total cost of a construction tender to the project owner, developer, or consultant. A decision can then be made to award the contract taking into account various factors such as price, contractor qualifications, time to complete the project, etc.

Construction contracts need to be carefully drafted and managed to avoid possible exposure to financial penalties and turning a potentially profitable project into an unprofitable one. In order for the contract to be lucrative, it must minimize or eliminate risk factors as much as possible.

The most common methods of construction project delivery are discussed in greater detail in Chapter 3. The three main delivery systems are:

- The traditional design-bid-build (DBB)
- The design-build (DB)
- The construction management (as advisor CM or at-risk CM).

Each of these methods has distinct advantages and disadvantages and all can be used to successfully plan, design, and undertake a given construction project.

Design-bid-build (DBB)

This is the traditional method of project delivery and has been the most widely used construction delivery method since ancient times. It commences with an owner selecting an architect to prepare construction documents. Most often the architect will release these construction documents publicly or to a select group of invited prequalified general contractors who will be asked to bid on the project, which reflects what they believe the total cost of construction will be. This bid includes various other bids from subcontractors for each specific trade. The general contractor's fee is generally built into the bid cost. The majority of government contracts are required to bid competitively using this method.

Design-build

This is a construction project delivery system, which differs from the traditional design-bid-build in that the design and construction aspects of the project are contracted for with a single entity. The system basically focuses on combining the design, permit, and construction schedules in a manner that allows the streamlining of the traditional design-bid-build environment. The design-builder is typically either the general contractor or design professional (architect or engineer). This system minimizes the project risk for an owner and reduces the delivery schedule by overlapping the design and construction phases of a project. Where the design-builder is the contractor, any required design professionals will typically be retained directly by the contractor.

As outlined in Chapter 3, there are potential problems associated with the design-build delivery system. As an example, cost estimating for a design-build project can be difficult when design documents are preliminary and liable to change over the course of the project. To address this situation, design-build contracts are often written to allow for unexpected situations without penalizing either the design-builder or the owner. Organizations like the Design/Build Institute of America can provide standardized form contracts for design-builders to use, although it is not unusual for the design-builder to provide its own contractual documents, particularly for well-established firms and firms that have previously constructed similar projects.

Construction Manager as Constructor

Under this delivery method, a construction manager is hired prior to completion of the design phase to act as a project coordinator and general contractor. As discussed in Chapter 3, hiring a construction manager during the design phase allows the construction manager to work directly with the architect and circumvent any potential design issues prior to completion of the construction documents. Upon completion of the tender documents the as advisor CM invites contractors or subcontractors to bid for the various divisions of work.

12.10.1 Bid Solicitation and Types of Building Contracts

This is the process of making published construction data (tender documents) readily available to interested parties, such as construction managers, contractors, and the public. Project owners release prepared project details including contract documents and specifications in the form of tender documents (normally for a fee) to interested general contractors, subcontractors, and other interested parties in an attempt to solicit bids. These services are usually subscription based or a flat fee is charged for a copy of the tender documents. It is difficult to structure a formula for quoting an exact fee, largely because it is highly uncommon for two properties to be exactly alike in terms of all the variables that need to be considered. Moreover, it becomes a guessing game trying to quote a fee on a cost per square foot basis without having all the information up front.

The owner is the person responsible for determining the type of contract to be used on the project. This will be partly determined by several factors such as the type of project and the amount of risk the owner is willing to accept. There are essentially four types of construction contracts currently in use, including:

- Lump sum
- Cost-plus
- Guaranteed maximum price (GMP)
- Unit price.

12.10.2 Lump Sum

A lump sum or fixed cost contract is one where the contractor agrees to supply all labor and materials for a fixed sum. Should the building contractor miscalculate, there is nowhere to go except to absorb the cost. It is the most widely used contract form in building construction, and is generally a more favorable type of building contract for the owner because the cost of the project is known from the outset. The cost of management is placed squarely on the shoulders of the contractor. It also adds responsibility in the pre-contract assessment carried out by the contractor to ensure that all potential contingencies are covered as agreed extras. Once the contract is signed, both parties (the owner and contractor) must live up to the terms of the contract and any modifications that follow will be considered "change orders," which will add to the cost and possibly cause delays. A lump sum agreement normally protects the project owner from unethical contractors hoping to take advantage of an owner.

Building contractors normally add 10 to 15 percent to the expected cost of the project to account for unforeseen contingencies. This contingency amount partly depends on the type of project and any special circumstances surrounding the project, and can either be agreed to be included in the total cost of the contract or agreed as a maximum contingency fund. This means that, provided no surprises eventuate, the owner will not be liable to pay this sum. However, if there are cost overruns associated with the work but which are not due to errors or omissions in the design, but instead are the result of the contractor's poor performance, or even the weather, then the contractor must absorb the loss with no additional requests for compensation from the owner. Similarly, if the contractor is able to achieve cost savings through superior performance, then these benefits will go solely to the contractor.

12.10.3 Cost-Plus

A cost-plus contract is one where the contractor agrees to supply labor and materials on a cost-plus basis, i.e., the owner reimburses the contractor for all costs associated with the contract in addition to a fee covering the contractor's profit and non-reimbursable overhead costs. Cost plus contracts can be either cost plus a percentage of costs, under which the fee is an agreed-upon percentage of the "costs," or costs plus a fixed fee where the fee is independent of the contractor's "costs." With this type of contract it is important to define exactly what is included in the "costs" (e.g., are soft costs such as supervision and overhead reimbursable?). Because of the discretionary and subjective nature of a cost-plus building contract—it ends up being the best for the contractor and the riskiest type of contract for the investor or building owner. The owner may suddenly discover that the project cost ends up being twice what was initially agreed to. Care must therefore be applied when using this type of agreement, as it is frequently open to abuse.

Contractors prefer cost-plus type bidding because it relieves them of sticking to a set price and guarantees them a profit regardless of project cost. However, this bid model must not be allowed to be used by building contractors as a haven for poor estimating. The various types of additional costs should be discussed with the contractor at the time the bid is being reviewed. Although surprises are not uncommon in construction, they can be kept to a minimum by a careful drafting of the contract. This type of contract is usually used for projects that are very specialized and where the scope is difficult to define, or when time to execute the project is the most important factor and construction is required to start before the completion of the contract documents.

12.10.4 Guaranteed Maximum Price (GMP)

This type of contract is essentially a variation of the cost-plus contract and is more suited for projects where the scope is well defined and is particularly suited to turn-key projects. With this type of contract, the contractor is reimbursed for actual costs incurred for materials, labor, equipment, subcontractors, overhead, and profit up to a maximum fixed price amount. Furthermore, the contractor warrants that the project will be constructed in accordance with the contract documents and that the project cost to the owner will not exceed the agreed total maximum price. All costs over the maximum price are to be borne by the contractor and any savings below the maximum price will revert to the owner. It is important for the owner to clearly state whether this maximum reflects the total costs (fee excluded) or the total costs the owner pays including the fee. Normally the owner pays the fee and the contractor pays for costs in excess of the maximum. Penalty and incentive clauses are often included in the agreement relating to costs, schedule, and quality performance. This type of contract is sometimes preferred when a design is less than 100 percent complete.

12.10.5 Unit Price Contract

Here the contractor is paid as the contract proceeds by requiring that the actual quantities of work completed one measured and these quantities are multiplied by a pre-agreed per-unit price. Tender estimates provided by contractors are based on specifications and estimated quantities supplied by the owner. However, during and after the work, the price is based on actual quantities completed and not estimated quantities. For the contractor this removes some of the risk in the bidding process because payment is based on actual quantities and not lump sum. The contractor's unit price must cover both direct and indirect costs, overheads, contingencies, and profit. For this reason, the owner usually provides fixed quantities for contractors to use as the basis of their unit-price costing. However, when additional work is required, a separate invoice should be presented (Figure 12.5).

BILLING FORM
ADDITIONAL WORK PERFORMED

EMPLOYEE NAME:

DATE:

PROJECT NO.

JOB NAME:

FOR PERIOD UP TO AND INCLUDING: _____

DESCRIPTION OF WORK PERFORMED, INCLUDING WHO WAS INVOLVED, DATES, ETC.

HOURS: _____

EXPENSES: _____

_____

_____

APPROVED: **SIGNATURE:**

FIGURE 12.5 An example of the type of billing form used for additional work completed.

This type of contract is typically suitable for projects where the quantities are ill defined and therefore cannot be accurately measured before the project starts, such as highway type projects. Thus, the owner could provide quantities for excavation, pipe laying, and backfill. The contractor would quote a dollar amount per cubic yard for soil excavation, a dollar amount per linear foot of piping laid, and a dollar amount per cubic yard of backfill installed and come up with a total bid based upon the quantities that the owner provided. The project's final price will not be known with certainty until the project is completed. Additionally, it is prudent for the owner or the owner's representative to "track" actual quantities by some method of measuring—counting truckloads of materials, weighing steel, etc.

Negotiated

This type of contract is not dissimilar to the design-bid-build method in that the project's design and construction are performed by different firms. In some cases a negotiated contract may be used in lieu of the tendering process, especially when an owner has had previous experience with a certain contractor. That contractor may be invited to submit a proposal or offer to the owner or the owner's representative based on the contract documents. This is followed by negotiations regarding price, scope of work, time to execute the project, and other contractual issues. If agreement is reached, a contract is signed for constructing the project. Negotiations may also form part of a tendering process. Upon evaluating the submitted tenders, a short list of top ranking firms is created. This is followed by negotiations regarding work content, risk, liability issues, and contract-related issues. If the owner decides on negotiating with the top ranking firms after the tenders are opened and analyzed, this must be made clear to all via the invitation to tender.

12.11 CREATING A WEBSITE AND INTERNET MARKETING

Having a Website for your new business was once considered a luxury, but today it is an absolute necessity, so much so that even so-called "brick and mortar" establishments have accepted the fact that their businesses cannot thrive without an online component. The reason is that the Internet has created enormous marketing opportunities to reach previously unimaginable numbers of people. The Internet also creates access to potential customers who you otherwise may not have access to. A business Website for your firm is now a high priority because not only is it a great marketing tool, but it also allows you to develop your services and launch successive marketing campaigns within a short time frame. Your Website is a specialized tool, one that enables you to reach countless new clients. This is why keeping up with what is happening in your field is vital in order to make possible modifications to your Website as needed.

Today the Internet is considered one of the best ways to generate high-qualified new business opportunities, largely because it is fast, easy to use, affordable and working for you 24/7. But to take full advantage of the Internet means more than just creating a Website and waiting for potential clients to find it. The Website should be but one part of an overall Internet marketing strategy. Property developers or persons in search of a green building contractor should be able to quickly find your photo and CV on your Website. Your CV should at a minimum provide background information and a list of relevant career highlights. A well thought-out blueprint will help guide all the other decisions you'll be making in the months and years to come.

Even before setting up a Website, it is imperative to have an email service. This is a high priority for communication and for sending promotional material to customers and potential clients. In fact, most clients today consider email availability vital and find it burdensome and unacceptable to have to communicate by posted mail. When choosing an email address, ensure that it is simple, yet professional. And once the Website is established and the domain name registered, the business email should be modified to reflect the domain name of the Website. The firm's email signature should also provide complete contact information and an active link to the Website. Many visitors prefer to make a first contact via email, either because they prefer keeping it impersonal or because it is easier for them to articulate what they are looking for by email, etc. This is why a "Contact Us" page and/or the footer on each Web page are good choices for listing your corporate mailing and email address. Usually the secretary/office manager will regularly monitor incoming email inquiries and respond quickly or transmit them to a principal for reply.

12.11.1 The Importance of a Website

Having a Website has many advantages, such as marketing your company's services to the world. It is important to consider what information you desire prospects to gather from visiting your Website. For example, a Website can inform consumers and end-users on the benefits of green contracting and how it adds to their bottom line. Well-designed Websites usually serve several functions, such as:

- Provide more information about the firm and its services
- Steer inquiries from potential customers to the Website
- Motivate users to visit the site and return
- Help resolve customers' outstanding issues
- Sell products (e.g., green products) and construction services online

- Provide downloadable files such as brochures, research, templates, and other information
- Help provide clients and customers with more efficient service
- Make it easy for the user/customer to contact you
- Receive customer feedback
- Staff recruitment.

A company Website can be a great asset and bring in great benefits to a firm by providing clients and users with better access to your company and its services. It can also facilitate resolving clients' issues speedily and satisfactorily. Happy and satisfied clients allow you to devote more of your time and energy to other pressing issues. The number of services that can be offered via a Website is enormous. Yet designers sometimes fail to understand the proper function of a Website, and therefore make wrong decisions about its form.

A "green" Website, for example, can include green construction projects that the firm has executed, green building costs and other facts and figures, including any awards that the firm may have received, contact information including company email, address, and telephone numbers, etc. The firm's phone number should be displayed prominently on the home page. The Website can also include information regarding the firm's structure and responsibilities. The site can be used to draw attention to upcoming events or other time-sensitive information. But whatever it is, the Website should be something that appeals to your target audience and not just yourself.

Many general contractors also use their Websites to communicate with subcontractors, consultants, clients, and other project team members to explain or ask questions regarding bidding guidelines, building schedules, variation orders, etc. They also provide downloadable forms, building fact sheets, and equipment procedures for field workers, subcontractors, or manufacturers who may be accessing the site remotely from an off-site location. Confidential information can be password-protected so that only authorized individuals can access the information.

12.11.2 Plan for Building a Website

Decide whether you want to hire a professional to build your Website or do it yourself because you are on a tight budget and have a modicum of computer savvy. If you are starting out on a tight budget, you should have no difficulty in creating a Website that is attractive and functional and will primarily serve as a source of information for your customers.

The Website is just another means of communication with your target audience. It basically updates and translates the results of the firm's day-to-day business to this new medium, which is why whenever a Website is being designed, the first consideration should always be the user. Determine the image you want the site to convey to visitors, and make sure that everything on the site is designed to contribute something toward that end. It is important therefore to try and contemplate what your potential site visitors will want to know or see featured on the site when they log on. The Website should be organized so that prospects can easily identify the firm's areas of expertise. It may be prudent for a new company starting from scratch to conduct a quick survey to help determine exactly what services are needed. Also, do periodic searches on your competitors, including other general contractors and green building Websites, to examine the types of services your peers are providing to customers and to see how they are marketing themselves on-line.

To get started on building a Website for your new business, a number of requirements need to be met, such as:

- Registering a domain name
- Signing up with a Web hosting service
- Purchasing Web authoring software or service to design the site.

12.11.3 Registering a Domain Name and Setting Up the Website

Before a commercial Website (as opposed to a personal home page) can be set up, a name for the company must be decided upon. Once this is determined, a search is conducted to see if that name is available and not already taken. There are many companies that let you register a domain name online by using the search tool to ensure that the name you are seeking is not reserved. If it is available, you can register the name immediately (for a small fee) and make it your Web address. If possible choose ".com," ".net," or ".us" extensions because far more users are familiar with these. A domain is essentially the name and address of your Website, all in one. For example, if your company name is ABC Green Contracting, your Web address may be www.ABCgreencontracting.com (or www.ABCgreencontracting.us, etc.). Domain registration is inexpensive and the domain name can be registered with the same hosting company that is to provide the Web space (i.e., space on a computer owned by a hosting company). Try and choose a domain name that is memorable and describes your business well. Furthermore, it should preferably be simple and easy to remember.

Web Hosting Service

Most startup companies don't wish to own or invest in a server (a powerful computer that's always online, and has the capacity to store the entire firm's Website files, as well as the content and operations of its network). It is therefore necessary to find and hire a reliable Web host for this purpose. A Web host will accept your site into its computers and securely store your files and data, while ensuring that it will be available to you and your customers every day, 24/7.

The space provided on the Web for your firm by the hosting company is set up so that when someone types your firm's domain name into their browser they will be connected to your Website. There are numerous hosting companies—some better than others. The prudent thing to do is to spend some time researching online for topics like "domain hosts" to find one that best meets your particular needs and budget. Below are a number of possible domain hosts to begin with:

- www.NetworkSolutions.com
- www.greengeeks.com
- www.justhost.com
- www.supergreenhosting.com
- www.GoDaddy.com.

Once an account has been established with a hosting service, it will provide you with instructions on how to upload your Website onto its server. This is usually achieved with FTP (file transfer protocol) utility software. Many hosting services also have file upload options within the control panel you use to help you manage your site.

You should also consider using a hosted service that allows you to build your site online using drag-and-drop, fill-in-the-blank templates, which are simple to apply and do not typically require much technical knowledge. Typical examples of this type of service include Yahoo! Small Business, StartLogic, and ValueWeb, all of which offer various packages that are affordable and include design, hosting, and site maintenance.

Although virtually anyone can put a Website together, especially if you can use a word processor, creating an efficient and successful Website is an entirely different matter, and requires considerable planning. Moreover, building a Website from scratch for your new business requires an investment of time and money. If you feel that you don't have the time, interest, or ability to personally design and go live with your business Website, it would be advisable to hire a professional designer or someone (such as a student designer) with sufficient knowledge and experience to build a functional site that meets all your needs and aspirations. If you decide to go this route, check out their references and possibly see examples of their work.

12.11.4 Website Components and Content

1. Basic Website Content

The beauty of the Internet is that it is constantly evolving, which is why it is necessary to regularly refresh, update, and add new content. This also helps increase visibility in the search engines, and gives your customers a reason to come back to your site. Blogs are tools that are increasingly being used to maintain current industry news and are discussed below.

The function of a Website should be attracting visitors and then converting them into clients by various means, including the articulation of services and/or products offered. This can be stated in a mission or vision statement. Since, with a few exceptions, the majority of Internet users won't actually know you, what you say on your Website must capture a visitor's attention, establish credibility, provoke their interest, and motivate them to action. This affords the opportunity to project the kind of image desired, in addition to being able to highlight any particular aspect of the organization that you wish to. Also keep in mind that the target visitor may have special needs that should be taken into consideration and included in your Website design. A customer should never have to work to get the information needed. Therefore, employ the minimum needed to enhance your central message and to tell it simply and clearly in an attractive setting.

Make sure that the corporate image you wish to portray on the Web pages matches that of the image displayed in other formats and media. Once a theme is decided upon, it is important to be consistent and stick to it. The firm's logo should be placed on all the site's pages in the same position, and in the same size. Also, unless you are considering hiring a professional designer to manage all the photos and other graphics on your Website, you'll need a graphics program to do it yourself (except of course if the hosting company has its own tools). However, basic digital photo and graphics editors can be found online for free, although sophisticated top-end programs like Adobe Photoshop can run into hundreds of dollars. The editing software program chosen should be capable of resizing and cropping images; resolving color and contrast issues; setting their resolution, which controls how sharp they are on the Web page; and saving them using appropriate color modes and formats.

A Website, whether for a small or large business, is a high priority; not only because it is a great marketing tool, but also because it allows development of the firm's numerous services along with multiple marketing campaigns. The Website should be looked at as a platform to promote company services, or just share commentary. It may also be prudent to include a list of bullet points on the home page citing the principal services offered. Each bullet could be expanded to link to a new page that describes your services in greater detail. A useful approach is that of exchanging links with complementary Websites. Search engines generally favor Websites with a high number of incoming links because it confirms a site's popularity and content value.

Creating a company Website has never been easier and a basic Website can be accomplished in a matter of hours. A simple method that can be used to start building your own Website is downloading a Website template (using the "Save As" option on your browser), and then editing it with a Web editing program. It is relatively easy now with a little research to find a suitable template so that you are not simply imitating another site. It should be emphasized here that it is illegal to copy another company's Web page, but locating and downloading a free template on-line is simple. Once this is accomplished, explore your options for upgrading and customization. Your favorite Website builder can be used to complete the design and customization, after which you can immediately publish it to the Web. The most popular examples of Web design software are Adobe Dreamweaver, Microsoft Expression Web (replaced FrontPage), and NetObjects Fusion 11. Before choosing what design software to get for designing your Website, make sure it has the ability to design in both HTML and drag-and-drop.

As you build the site, check for bugs in the system. This should be done one page at a time and in a thorough manner. Once a page is error-free and everything is working as intended, move on to the next page, repeating the process until the Website has been completely checked. Also, when checking for bugs, make sure to test for the following:

- Check for spelling and grammar mistakes. The most professional site will appear amateurish if it contains basic spelling or text errors.
- Ensure that all the links work, taking users where they expect to go.
- Keep page navigation consistent across the site.
- Ensure that all the site pages are printable.

Figure 12.6 illustrates two examples of designs that can be found on the Internet which have been edited. There are many Websites that offer free Web templates as well as design services. In Figure 12.6a the contracting buttons links are on the left side of the template, whereas in the architecture template they are on the top of the page. Both can easily navigate you to where you want to go (e.g., about us, projects in progress, services offered, firm principals and qualifications, etc.). The final layout depends largely on personal preference in how the page is composed and designed. However, for a professional looking Website, it is almost always best to use a professional for the design.

2. Defining Your Target Audience

This depends very much on the answers that you have already come up with. Here, your primary purpose is to know your potential clients so well that you answer any questions they might have before they ask, then make it easy for them to buy the concepts you are selling. Your audience includes existing customers and clients, potential clients, people interested in your area of specialty who may never have heard of you, organizations, individuals,

A

B

FIGURE 12.6   **A.** An example of a contracting home page template that has been edited, showing the link buttons on the upper part of the page, available free from www.freewebsitetemplates.com. **B.** An example of a typical architecture home page template from www. freewebsitetemplates.com, which has the link buttons on the left of the page.

and so on. If you are a green builder/contractor, your groups might look a little like this:

- Facility managers and property owners
- Lenders
- Investors
- Architects and engineers
- Subcontractors
- Attorneys.

Create a prioritized list. Here your potential and existing clients will learn why you are qualified to do what you do, and why your firm can offer a better service than the competition. In fact, your new Website will sell your company's services as no one else can. Thus, if you want to promote an image of a sustainable contractor that specializes in green building, then building owners and investors may take a higher priority than general users. This should be reflected in the structure of the site's pages, and the weighting that you give to each aspect in your guidance.

Likewise, let potential clients know you are to be trusted—The majority of viewers may not know your company, which is why it is so important to continuously hammer this message home. There are several ways to reinforce this message, such as the use of your company logo, university crest, etc. If you have projects that have won awards, make sure everyone who visits the site is aware of this. It is imperative that your viewers feel reassured that your firm is trustworthy, and that its information is reliable. Since there is no telling which page a user will visit first, it is advisable to place the company logo on every page and in a consistent location on the page.

3. Convert Site Visitors to Clients

The time available to make a serious impression on a site visitor is very limited before they're gone. And since the window of opportunity to impress and sell your services is very small, and huge flashy graphics take forever to load, most visitors will flee because they don't have the patience to wait for them to appear on their screen.

Clearly articulate what your site is for; if you don't know then don't expect your viewers to. Easy and unobstructed navigation through your site is absolutely essential to a successful design. If the path you lay out for your potential customers is twisted and difficult to follow, they'll get lost and you'll lose a potential client. A successful Website is not necessarily a very attractive one or one consisting of the latest Web technology. It doesn't even rely on the number of site visitors, but rather on how many visitors come back and how many visitors are converted into clients.

A successful Website needs to be continuously refreshed and periodically updated. If the site is not continuously updated, or carries out of date information, it suggests that you don't care and this reflects badly on the firm. Determine who is to be the Webmaster, who will have the authority to update the site, introduce new content, and so on. Although the hard part may be over, the real key to long-term business success on the Internet is continuous maintenance of your Website.

4. Company Blogs

These have become very popular and are usually used to enhance a business Website and drive more traffic to it by bolstering your credibility. For example, blogs are used to report company and industry events, comment on relevant news stories and let people know when you'll roll out new products or services. Blogs may also attract professionals inside your industry and possibly calls from the press. Blogs can be housed on or off-site with their own URL.

5. Attracting Traffic

Your business may offer the best products or services on the Web, but it doesn't mean a thing if potential customers can't find them. Potential customers not only need to know the Website exists, but they should also be able to access it without difficulty. Additionally, the Website and email address should be clearly mentioned on all letterheads, brochures, cards, and advertising material. Regular promotional campaigns and strategies are often necessary to drive traffic to your Website. This is limited only by your creativity, imagination, and what you have to offer.

An excellent way to get your Website noticed is by registering your site with as many search engines as you can, but especially with the leading search engines like Google, Yahoo, and MSN. This will help bring your company to the fore whenever users ask these search engines to scan the Internet for your kind of services. This is absolutely essential for the Website to succeed, even though it may appear at times a very challenging task. It typically takes at least a few months for a Website to generate responses and become recognized. This is the time it normally takes the big search engines to index a new site. But the popularity of a site and the speed at which it becomes popular really depends on what services are being offered and how it is promoted.

Registering with online directories such as GreenSpec® Directory to promote your firm is another excellent way to increase your online recognition and visibility and drive qualified prospects to your Website. Many contractor and green associations also maintain online service directories.

It is important to understand how search engines evaluate your site; they basically use what is known as a "Web crawler" that reads the "meta tags" in the header of your HTML pages. This is why your site should include a title and description tag, as well as appropriate keywords. One of the concepts employed for high search engine ranking is called "keyword density," which simply means that the Web pages should include key terms that drive searches. Useful buzzwords for green contractors may include "green," "green builder," "sustainable," "contractor," and other key industry terms. But being able to drive visitors to your Website is not the main objective of a Website. The real objective of a Website is to be able to convert site visitors to clients, e.g., by making it easy for new prospects to learn about your services and contact you.

6. Website Security

One issue that is of paramount importance to any Website is that of security. The Webmaster needs to be vigilant about the security of your Website and content, your network, and your customers' private information. The Webmaster needs to take this into account, creating firewalls, installing the latest Internet security software programs, etc. Examples of popular security software programs include: Security Shield, CA Internet Security, Norton 360, Norton Internet Security, and McAfee. It is also important that whoever is to be responsible for keeping the site up to date must be able to write and update the pages as needed, while keeping abreast of new technologies in order to maintain the site and keep it fresh and interesting. Websites may not be difficult to set up, but as we mentioned earlier, really good ones are.

7. Add new Content

As the firm grows and develops, your Website will need to keep pace and add new content. This will mean:

- Adding the profiles of new senior staff to the "About Us" page.
- Add new service lines, awards, etc. on the home page with the appropriate links to them.
- Whenever new content is added to the Website, the affected pages need to be tested for bugs and to ensure they are optimized for peak performance.
- Update all links and site map after changes.

Finally, the Website should adopt a "best practices" approach. Successful business Websites are based on simple things done well and each of the objectives set for the company Website needs to be clearly articulated. Following are some suggested objectives:

- Define your target audience and what you want them to do or obtain from your site.
- Determine the method to promote the firm's Website and contents to your target audience to keep visitors motivated and returning.
- Determine what the Website needs to attract prospects to the site in the first place and to ensure that they come back again and again. Good Website content and design are both vital for market success.

- Deliver the message and services offered to the target audience in an understandable and motivating manner.
- Decide what service must be provided on or off-line to customers and site visitors to ensure first class service wherever and whenever required.
- Integrate the Website into the business and marketing operations to achieve efficiency and good service levels and so that it becomes a normal part of business.
- Continuously monitor Website traffic and being prepared to develop and make modifications to the site and its content as the business and customer needs change.
- Prioritize your investment of time and money over the development of the Website design, content, and marketing aspects.

Acronyms & Abbreviations

Organizations and Agencies:

ACEEE	American Council for an Energy Efficient Economy
AFA	American Forestry Association
AIA	American Institute of Architects
ANSI	American National Standards Institute
APCA	Air Pollution Control Association
ASAE	American Society of Architectural Engineers
ASCE	American Society of Civil Engineers
ASHRAE	American Society of Heating, Refrigeration, and Air-conditioning Engineers
ASID	American Society of Interior Designers
ASME	American Society of Mechanical Engineers
ASNT	American Society for Non-destructive Testing
ASPE	American Society of Plumbing Engineers
ASTM	American Society for Testing Materials
AWEA	American Wind Energy Association
BBRS	Board for Building Regulations and Standards
BCDC	Bay Conservation and Development Commission
BIFMA	Business and Institutional Furniture Manufacturer's Association
BLM	Bureau of Land Management
BOCA	Building Officials and Code Administrators
BREEM	Building Research Establishment Environmental Assessment Method
CEC	California Energy Commission
CFR	Code Federal Regulation
CIBSE	Chartered Institution of Building Services Engineers

Doi: 10.1016/B978-1-85617-676-7.00016-6

CRI	Carpet and Rug Institute CRS—Center for Resource Solutions
CRS	Center for Resource Solutions
CSI	Construction Specifications Institute; Construction Standards Institute
CUWCC	California Urban Water Conservation Council
DoD	Department of Defense
DOE	U.S. Department of Energy
DPW	Directorate of Public Works
DWR	Department of Water Resources (CA)
EIA	Energy Information Administration
EPA	U.S. Environmental Protection Agency
ERDC	USACE Engineer Research and Development Center
ESI	European Standards Institute
FEMA	U.S. Federal Emergency Management Agency
FSC	Forest Stewardship Council
GBCI	Green Building Certification Institute
HUD	U.S. Department of Housing & Urban Development
IEA	International Energy Agency
IEEE	Institute of Electrical and Electronics Engineers, Inc.
IESNA	Illuminating Engineering Society of North America
IPMVP	International Performance Measurement & Verification Protocol
ISO	International Organization for Standardization
NAE	National Academy of Engineering
NAHB	National Association of Home Builders
NAS	National Academy of Sciences
NBI	New Building Institute
NCARB	National Council of Architectural Registration Boards
NFRC	National Fenestration Rating Council
NIST	National Institute of Standards and Technology
OEE	Office of Energy Efficiency
OSHA	Occupational Safety and Health Administration (or Act)
OSWER	U.S. EPA Office of Solid Waste & Emergency Response
SBIC	Sustainable Building Industry Council
SCAQMD	South Coast Air Quality Management District
SEC	Securities and Exchange Commission
SMACNA	Sheet Metal and Air Conditioning Contractors' National Association
UL	Underwriters Laboratories
USACE	U.S. Army Corps of Engineers
USDA	United States Department of Agriculture
USEPA	U.S. Environmental Protection Agency
USFS	U.S. Forest Service
USGBC	United States Green Building Council

Referenced Standards and Legislation:

ADA	Americans with Disabilities Act
ASHRAE 90.1	Building energy standard covering design, construction, operation, and maintenance.
ASHRAE 52.2	Standardized method of testing building ventilation filters for removal efficiency by particle size.
ASHRAE 55	Standard describing thermal and humidity conditions for human occupancy of buildings
ASHRAE 62	Standard that defines minimum levels of ventilation performance for acceptable indoor air quality
ASHRAE 192	Standard for measuring air-change effectiveness
ASTM E408	Standard of inspection-meter test methods for normal emittance of surfaces
ASTM E903	Standard of integrated-sphered test method for solar absorptance, reflectance, and transmittance
CAA	Clean Air Act; Compliance Assurance Agreement
CASBEE	Comprehensive Assessment for Building Environmental
CERCLA	Comprehensive Environmental Response, Compensation, and Liability Act
CERL	Construction Engineering Research Lab—part of USACE
CWA	Clean Water Act (aka FWPCA)
EISA	Energy Independence and Security Act of 2007
EPAct	Energy Policy Act of 2005
EPAct	U.S. Energy Policy Act of 1992
FCAA	Federal Clean Air Act
FFHA	Federal Fair Housing Act
FOIA	Freedom of Information Act and similar state statutes
GS	Green Seal
MERV	Minimum Efficiency Reporting Value
NFPA	National Fire Protection Association
PURPA	Federal Public Utilities Regulatory Policy Act of 1978
RCRA	Resource Conservation and Recovery Act
TSCA	Toxic Substances Control Act
UBC	Uniform Building Code: the International Conference of Building Officials model building code

Abbreviated General Terminology:

A	Area
A or AMP	Ampere
AAQS	Ambient Air Quality Standards
AC	Alternating current
A/C	Air Conditioning Unit

ACH	Air Change per Hour
ACM	Asbestos Containing Material
ACT AGE	Actual Age
ADAAG	ADA Architectural Guidelines
AE	Architect-Engineer firm (typically contracted for design services)
AEI	Advanced Energy Initiative
AEO	Annual Energy Outlook, DOE/EIA publication
AF	Acre-foot (of water)
AFC	Application for Certification
AFV	Alternative-Fueled Vehicle
AFY	Acre-feet per year
AGMBC	LEED-NC Application Guide for Multiple Buildings and On-Campus Building Projects (USGBC document)
AHM	Acutely Hazardous Materials
AHU	Air Handling Unit
AIB	Air Infiltration Barrier
AIRR	Adjusted Internal Rate of Return
AL	Aluminum
APPA	America Public Power Association
AQMD	Air Quality Management District
AQMP	Air Quality Management Plan
ARB	Air Resources Board (CA)
ATC	Acoustical Tile Ceiling
A/V	Audiovisual
BAS	Building Automation System
BAAQMD	Bay Area Air Quality Management District
BC	Building Code
Bcfd	Billion Cubic Feet per Day
BEA	U.S. Bureau of Economic Affairs
BEEP	BOMA Energy Efficiency Program
BEES	Building for Environmental and Economic Sustainability Support
BG	Biomass Gassification
BIM	Building Information Model
BIPV	Building Integrated Photovoltaics
BL	Building Line
BM	Benchmark
BMP	Best Management Practices
BOD	Basis of Design
BOD	Beneficial Occupancy Date
BRI	Building-Related Illness
BT	Building Technologies
BTU	British Thermal Unit

BTUH	British Thermal Unit per Hour
BUR	Built-up Roofing
CAA	U.S. Clean Air Act
CAAQS	California Ambient Air Quality Standards
CalEPA	California Environmental Protection Agency
CAPM	Capital Asset Pricing Model
CARB	California Air Resources Board
CBC	California Building Code
CBECS	Commercial Building Energy Consumption Survey
CD	Construction Division
CDVR	Corrected Design Ventilation Rate
CEERT	Coalition for Energy Efficiency and Renewable Technologies
CEU	Continuing Education Unit
CFCs	Chlorofluorocarbons
CFM	Cubic Feet per Minute
CFR	Code of Federal Regulations
CFS	Cubic Feet per Second
CHPS	Collaborative for High Performance Schools
CO_2	Carbon dioxide
CO	Carbon monoxide
COC	Chain-of-Custody: proper accounting of materials flows, as used by the FSC
COS	Center of Standardization
CMBS	Commercial Mortgage Backed Securities
CMU	Concrete Masonry Unit
CPG	Comprehensive Procurement Guidelines
CSA	Canadian Standards Association
CWA	Clean Water Act
CxA	Commissioning Authority
dB	Decibel
DB	Design-Build single contract
DBB	Design-Bid-Build
DC	Direct Current
DEC	Design Energy Cost
DHWH	Domestic Hot Water Heater/Water Heater
DOC	Determination of Compliance
DOR	Designer of Record
DPB	Discounted Payback
DPTN	Demountable Partitions
DS	Daylight Sensing Control; Disconnect Switch; Down-spout
DW	Drinking Water; Drywall
E	Each; East; Modulus of Elasticity

E&C	Engineering & Construction
ECB	Engineering and Construction Bulletin
ECBEMS	Energy Management System
ECMs	Energy Conservation Measures
EER	Energy Efficiency Ratio
EFF AGE	Effective Age
EIA	Energy Information Administration
EIFS	Exterior Insulation & Finish System
EIR	Environmental Impact Report
EIS	Environmental Impact Statement
EL	Easement Line; Elbow
EL, ELEV	Elevation
EMCS	Energy Monitoring and Control System
EMF	Electro Magnetic Fields
EMP	LEED Energy Modeling Protocol
EPCA	Energy Policy and Conservation Act
EPDM	Etheylic-Propalene Dian-Molomer
EQP, EQUIP	Equipment
ESA	Environmental Site Assessment
ESC	Erosion and Sedimentation Control Plan
ESP	Energy Service Providers
ETS	Environmental Tobacco Smoke
EUL	Expected Useful Life
FAU	Forced Air Unit
FEMP	Federal Energy Management Program
FERC	Federal Energy Regulatory Commission
FF&E	Finishes, Furniture (fixtures) & Equipment
FFL	Finished Floor Line
FIO	For Information Only
FIX	Fixture
FOIL	Freedom of Information Letter
FS	Full Scale; Full Size; Federal Specification
FSC	Forest Stewardship Council
FTC	Federal Trade Commission
FTE	Full Time Equivalent
FTE	Full Time Employee
FTG	Footing
FY	Fiscal Year
GBI	Green Building Initiative
GC	General Contractor
GDP	Gross Domestic Product
GEP	Good Engineering Practice
GF	Glazing Factor
GHG	Greenhouse Gases
GIS	Geographic Information System

GPD	Gallons per Day
GPF	Gallons per Flush
GPM	Gallons per Minute
GRD, GD, G	Grade
GW	Gigawatt
GW(h)	Gigawatt (hour) = 1 billion watts
GWP	Global Warming Potential
GYP	Gypsum
GYP BD	Gypsum Board
H$_2$S	Hydrogen Sulfide
HAZMAT	Hazardous Materials
HBD, HDB	Hardboard
HCFCs	Hydrochlorofluorocarbons
HFCs	Hydrofluorocarbons
HP	Horse Power
HRA	Health Risk Assessment
HT	Height
HV	High Voltage
HVAC	Heating, Ventilating, and Air Conditioning
HVAC&R	Heating, Ventilation, Air Conditioning, and Refrigerants
HWD	Hardwood
HZ	Hertz
IAQ	Indoor Air Quality
IDG	Installation Design Guide
IEPR	Integrated Energy Policy Report
IEQ	Indoor Environmental Quality—one of the six LEED credit categories
IFMA	International Facilities Management Association
IN	Inch (es)
INFO	Information
INSUL	Insulate (ion)
IPCC	Intergovernmental Panel on Climate Change
IPLV	Integrated Part Load Value
IRR	Internal Rate of Return
ISO	Independent System Operator
KG	Kilogram
KIT	Kitchen
KM	Kilometer
KW(h)	Kilowatt (hour) = 1000 watts
KV	Kilovolt
KVA	Kilovolt-Ampere (transformer size rating)
KVAR	Kilovolt-Ampere Reactive
KW	Kilowatt
LADWP	Los Angeles Department of Water and Power
LAN	Local Area Network

LAV	Lavatory
LBNL	Lawrence Berkeley National Laboratory
LBS	Pounds
LBS/HR	Pounds per Hour
LCA	Life-Cycle Assessment
LCC	Life-Cycle Cost
LCCA	Life-Cycle Cost Analysis
LCGWP	Life-Cycle Global Warming Potential
LCODP	Life-Cycle Ozone Depletion Potential
LD BRG	Load Bearing
LEED	Leadership in Energy and Environmental Design (USGBC)
LEED AP	LEED Accredited Professional
LEED-EB	LEED Tool for Existing Buildings
LEED Homes	LEED Tool for Homes
LEED-NC	LEED Tool for New Construction & Major Renovations
LEED ND	LEED Tool for Neighborhood Development
LE/FE	Low-Emission/Fuel-Efficient vehicle
LID	Low Impact Development
LL	Live Load
LPG	Liquified Petroleum Gas (propane and butane)
LQHC	Low Quality Hydrocarbons (i.e., tar sands and oil shale)
LR	Living Room
LTV	Loan-to-Value
LV	Low Voltage
LVL	Level
LW	Lightweight
LZ	Lighting Zone
M	Meter, Million, Mega, Milli or Thousand
MAIN	Maintenance
MAX	Maximum
MDF	Medium Density Fiberboard
MEP	Mechanical, Electrical, and Plumbing
MERV	Minimum Efficiency Reporting Value
M/F Ratio	Male/Female ratio
MMT	Million metric tons
MOU	Memorandum of Understanding
MR	Moisture Resistant
M/S	Meters per Second
MSDS	Material Safety Data Sheet
MV	Megavolt
MVA	Megavolt-amperes
MW	Megawatt (million watts)
MWD	Metropolitan Water District
MWh	Megawatt (hour) = 1 million watts
N	North

NAAQS	National Ambient Air Quality Standards
NC	New Construction
NCPA	Northern California Power Agency
NEMA	National Electrical Manufacturers Association
NEPA	National Environmental Policy Act (federal "equivalent" of CEQA) of 1969
NES	National Energy Savings
NG	Natural Gas
NO_2	Nitrogen Dioxide
NO	Nitrogen Oxide
NOx	Nitrogen Oxides
NOM	Nominal
NPDES	National Pollutant Discharge Elimination System
NPV	Net Present Value
NS	Net Savings
NTS	Not to Scale
O_3	Ozone
OC, O/C	On Center
O&M	Operation and Maintenance
ODP	Ozone Depleting Potential
OH, OVHD	Overhead
OPR	Owner's Project Requirements Document
OSA	Outside Air
OSB	Oversight Board
OZ	Ounce
PBD	Particleboard
PBP	Payback Period
PCA	Property Condition Assessment
PCC	Precast Concrete
PDT	Project Development Team
PFL	Pounds per Lineal Foot
PL	Property Line
PM	Project Manager
PML	Probable Maximum Loss
PMO	Project Management Oversight
POC	Point of Contact
PPM	Parts per Million
PPT	Parts per Thousand
PRM	Performance Rating Method
PSF	Pounds per Square Foot
PSI	Pounds per Square Inch
PTO	Permit to Operate
PU	Per Unit
PUC	Public Utilities Commission
PV	Solar Photovoltaics

PV	Present Value
PV	Photovoltaic
PVC	Polyvinylchloride
QA/QC	Quality Assurance/Quality Control
QTY	Quantity
R, RD	Radius
RA	Return Air
Rc	Refrigerant Charge
RD&D	Research, Development, and Demonstration
REC	Renewable Energy Certificate
REF	Reference
REINF	Reinforcement
REQ	Requirement, Required
REV	Revision
RFP	Request for Proposals
RFQ	Request for Qualifications
RH	Relative Humidity
RTU	Rooftop Package Unit
RUL	Remaining Useful Life
S, SW	Switch
SA	Supply Air
SAN	Sanitary
SBS	Sick Building Syndrome
SBTF	Sustainable Building Task Force (CA)
SCH, SCHED	Schedule
SD	Smoke Detector; Shop Drawings; Storm Drain; Supply Duct
SEER	Seasonal Energy Efficiency Ratio
SF, SQ FT	Square Foot (Feet)
SFTWD	Softwood
SHGC	Solar Heat Gain Coefficient
SIR	Savings-to-Investment Ratio
SOG	Slab-on-Grade
SOx	Oxides of Sulfur
SO$_2$	Sulfur Dioxide
SOx	Sulfur Oxides
SOW	Scope of Work
SPB	Simple Payback
SPIRIT	Army developed point/credit-based system for measuring sustainability of buildings/development (modified version of LEED version 2.0)
SRI	Solar Reflectance Index
STC	Sound Transmission Coefficient
STD	Standard

SYM	Symbol; Symmetry (ical)
SYS	System
T	Ton
TAC	Toxic Air Contaminant
THK	Thick (ness)
TL	Total Losses
TOG	Total Organic Gases
TOPO	Topography
TP	Total Phosphorous
TPD	Tons per Day
TPY	Tons per Year
TS	Tensile Strength
TSP	Total Suspended Particulate Matter
TSS	Total Suspended Solids
TVOC	Total Volatile Organic Compounds, see VOCs
U, UR	Urinal
UFGS	Unified Facilities Guide Specifications
UH	Unit Heater
UMCS	Utility Monitoring and Control System
USGS	United States Geological Survey
UST	Underground Storage Tank
UTIL	Utility
UV	Ultraviolet Radiation
V	Volt (s)
VAR, VAV	Variable Air Volume
VB	Vapor Barrier
VENT	Ventilation; Ventilator
VMT	Vehicle Miles Traveled
VOC	Volatile Organic Compound
VOL	Volume
VP	Vent Pipe
W	Watt; Width; Wide; West; Wire
WB	Wet Bulb; Wood Base
WBDG	Whole Building Design Guide
WC	Water Closet
WD	Wood
WH, DHWH	Water Heater; Domestic Hot Water Heater
WP	Waterproof; Weatherproof
WPM	Waterproof Membrane
WW, WTW	Wall-to-Wall
Y, YD	Yard
YR	Year
ZEV	Zero Emissions Vehicle (minimum energy star rating of 40)

2

Glossary

Abatement Reducing the degree or intensity of or eliminating pollution.

Absorption The process by which incident light energy is converted to another form of energy, usually heat.

Accessible Describes a site, building, facility, or portion thereof that complies with these guidelines.

Accessible route A continuous unobstructed path connecting all accessible elements and spaces of a building or facility. Interior accessible routes may include corridors, floors, ramps, elevators, lifts, and clear floor space at fixtures. Exterior accessible routes may include parking access aisles, curb ramps, crosswalks at vehicular ways, walks, ramps, and lifts.

Acrylics A family of plastics used for fibers, rigid sheets, and paints.

Adapted plants Plants that reliably grow well in a given habitat with minimal attention from humans in the form of winter protection, pest protection, water irrigation, or fertilization once root systems are established in the soil. Adapted plants are considered to be low maintenance but not invasive.

Addendum A written or graphic instruction issued by the architect prior to the execution of the contract which modifies or interprets the bidding documents by additions, deletions, clarifications, or corrections. An addendum becomes part of the contract documents when the contract is executed.

Adhesive A bonding material used to bond two materials together.

Adobe A heavy clay soil used in many southwestern states to make sun-dried bricks.

Aggregate Fine, lightweight, coarse, or heavyweight grades of sand, vermiculite, perlite, or gravel added to cement for concrete or plaster.

Air conditioning A process that simultaneously controls the temperature, moisture content, distribution, and quality of air.

Air filter A device designed to remove contaminants and pollutants from air passing through the device.

Air handling unit A mechanical unit used for air conditioning or movement of air as in direct supply or exhaust of air within a structure.

Air pollution The presence of contaminants or pollutant substances in the air that may be hazardous to human health or welfare or produce other harmful environmental effects.

Aligned section A section view in which some internal features are revolved into or out of the plane of the view.

Allergen A substance capable of causing an allergic reaction because of an individual's sensitivity to that substance.

Alligatoring A pattern of rough cracking on a coated surface, similar in appearance to alligator skin.

Alternating current (AC) Electrical current that continually reverses direction of flow. The frequency at which it reverses is measured in cycles-per-second, or Hertz (Hz). The magnitude of the current itself is measured in amps (A).

Alternator A device for producing alternating current (AC) electricity. Usually driven by a motor, but can also be driven by other means, including water and wind power.

Ambient lighting Lighting in an area from any source that produces general illumination, as opposed to task lighting.

Ambient temperature The temperature of the surroundings.

American bond A brickwork pattern consisting of five courses of stretchers followed by one bonding course of headers.

Ammeter A device used for measuring current flow at any point in an electrical circuit.

Ampere (A) or amp The unit for the electric current; the flow of electrons. One amp is 1 coulomb passing in one second. One amp is produced by an electric force of 1 volt acting across a resistance of 1 ohm.

Analog The processing of data by continuously variable values.

Anemometer A device used to measure wind speed.

Angle of incidence Angle between a surface and the direction of incident radiation; applies to the aperture plane of a solar panel. Only minor reductions in power output within plus/minus 15 degrees.

Animal dander Tiny scales of animal skin.

ANSI The American National Standards Institute. ANSI is an umbrella organization that administers and coordinates the national voluntary consensus standards system, http://www.ansi.org.

Appeal A formal written request to review the content of an exam question for accuracy, validity, or errors in content and grammar. Appeals must be specific to an exam question and must be submitted by the candidate to GBCI's accreditation department within 10 days of the exam appointment. The appeal must describe the content of the exam question and, if possible, the nature of the error. Exam scores are not modified under any conditions.

Arc A portion of the circumference of a circle.

Architect's scale The scale used when dimensions or measurements are to be expressed in feet and inches.

Array A number of solar modules connected together in a single structure.

As-built drawings Record drawings completed by the contractor and turned over to the owner at the completion of a project, identifying any change or adjustments made to the conditions and dimensions of the work relative to the original plans and specifications.

Asphalt shingles Shingles made of asphalt or tar-impregnated paper with a mineral material embedded; very fire resistant.

Assumed liability Liability which arises from an agreement between people, as opposed to liability which arises from common or statutory law. See also contractual liability.

ASTM International Formerly the American Society for Testing Materials. They develop and publish testing standards for materials and specifications used by industry, http://www.astm.org.

Authority having jurisdiction (AHJ) The governmental body responsible for the enforcement of any part of the standard codes or the official or agency designated to exercise such a function and/or the architect.

Axial load A weight that is distributed symmetrically to a supporting member, such as a column.

Axonometric projection A set of three or more views in which the object appears to be rotated at an angle, so that more than one side is seen.

Backfill Any deleterious material (sand, gravel, etc.) used to fill an excavation.

Baffle A single opaque or translucent element used to diffuse or shield a surface from direct or unwanted light.

Ballasts Electrical "starters" required by certain lamp types, especially fluorescents.

Balloon framing A system in wood framing in which the studs are continuous without an intermediate plate for the support of second-floor joists.

Baluster A vertical member that supports handrails or guardrails.

Balustrades A horizontal rail held up by a series of balusters.

Banister That part of the staircase which fits on top of the balusters.

Bar chart A calendar that graphically illustrates a projected time allotment to achieve a specific function.

Base A trim or molding piece found at the interior intersection of the floor and the wall.

Beam A weight-supporting horizontal member.

Base building The core (common areas) and shell of the building and its systems that typically are not subject to improvements to suit tenant requirements.

Base flashing Consists of flashing that covers the edges of a membrane.

Batten A narrow strip of wood used to cover a joint.

Batt insulation An insulating material formed into sheets or rolls with a foil or paper backing to be installed between framing members.

Bearing wall A wall that supports any vertical loads in addition to its own weight.

Benchmark A point of known elevation from which the surveyors can establish all their grades.

Bill of material A list of standard parts or raw materials needed to fabricate an item.

Bio-based Materials derived from natural renewable resources such as corn, rice, or beets.

Biodiversity The tendency in ecosystems, when undisturbed, to have a great variety of species forming a complex web of interactions. Human population pressure and resource consumption tend to reduce biodiversity dangerously; diverse communities are less subject to catastrophic disruption.

Black water Wastewater generated from toilet flushing. Black water has a higher nitrogen and fecal coliform level than gray water. Some jurisdictions include water from kitchen sinks or laundry facilities in the definition of black water.

Blistering The condition that paint presents when air or moisture is trapped underneath and makes bubbles that break into flaky particles and ragged edges.

Blocking The use of internal members to provide rigidity in floor and wall systems. Also used for fire draft stops.

Blueprints Documents containing all the instructions necessary to manufacture a part. The key sections of a blueprint are the drawing, dimensions, and notes. Although blueprints used to be blue, modern reproduction techniques now permit printing of black-on-white as well as colors.

Board foot A unit of lumber equaling 144 cubic inches; the base unit (B.F.) is 1 inch thick and 12 inches square or $1 \times 12 \times 12 = 144$ cubic inches.

Boiler Equipment designed to heat water or generate steam.

Bond In masonry, the interlocking system of brick or block to be installed.

Boundary survey A mathematically closed diagram of the complete peripheral boundary of a site, reflecting dimensions, compass bearings, and angles. It should bear a licensed land surveyor's signed certification, and may include a metes and bounds or other written description.

Breezeway A covered walkway with open sides between two different parts of a structure.

Brick pavers A term used to describe special brick to be used on the floor surface.

British thermal unit (BTU) The amount of heat energy required to raise 1 pound of water from a temperature of 60°F to 61°F at one atmosphere pressure. One watt hour equals 3413 BTU.

Building codes Rules and regulations adopted by the governmental authority having jurisdiction over commercial real estate, which govern the design, construction, alteration, and repair of such commercial real estate. In some jurisdictions trade or industry standards may have been incorporated into, and made a part of, such building codes by the governmental authority. Building codes are interpreted to include structural, HVAC, plumbing, electrical, life-safety, and vertical transportation codes.

Building density The total floor area of a building divided by the total area of the site (square feet per acre).

Building envelope The enclosure of the building that protects the building's interior from outside elements, namely the exterior walls, roof, and soffit areas.

Building inspector A representative of a governmental authority employed to inspect construction for compliance codes, regulations, and ordinances.

Building line An imaginary line determined by zoning departments to specify on which area of a lot a structure may be built (also known as a setback).

Building permit A permit issued by appropriate governmental authority allowing construction of a project in accordance with approved drawing and specifications.

Building-Related Illness A discrete, identifiable disease or illness that can be traced to a specific pollutant or source within a building. (Contrast with Sick Building Syndrome.)

Building systems Interacting or independent components or assemblies, which form single integrated units, that comprise a building and its site work, such as pavement and flatwork, structural frame, roofing, exterior walls, plumbing, HVAC, electrical, etc.

Build-out The interior construction and customization of a space (including services, space, and stuff) to meet the tenant's requirements; either new construction or renovation (also referred to as fit-out or fit-up).

Caisson A below-grade concrete column for the support of beams or columns.

Callback A request by a project owner to the contractor to return to the job site to correct or redo some item of work.

Candela A common unit of light output from a source.

Cantilever A horizontal structural condition where a member extends beyond a support, such as a roof overhang.

Capillary The action by which the surface of a liquid, where it is in contact with a solid, is elevated or depressed.

Carbon footprint A measure of an individual's, family's, community's, company's, industry's, product's, or service's overall contribution of carbon dioxide and other greenhouse gases into the atmosphere. It takes into account energy use, transportation methods, and other means of emitting carbon. A number of carbon calculators have been created to estimate carbon footprints, including one from the U.S. Environmental Protection Agency.

Casement A type of window hinged to swing outward.

Catch basin A complete drain box made in various depths and sizes; water drains into a pit, then from it through a pipe connected to the box.

Caulk Any type of material used to seal walls, windows, and doors to keep out the weather.

Cavity wall A masonry wall formed with an air space between each exterior face.

Cement plaster A plaster that is comprised of cement rather than gypsum.

Central HVAC system A system that produces a heating or cooling effect in a central location for subsequent distribution to satellite spaces that require conditioning. See also all-air, all-water, and air-water HVAC systems.

Centrifugal A particular type of fluid-moving device that imparts energy to the fluid by high velocity rotary motion through a channel. Fluids enter the device along one axis and exit along another axis.

Certificate for payment A statement from the architect to the owner confirming the amount of money due the contractor for work accomplished or materials and equipment suitably stored, or both.

Certificate of insurance A document issued by an authorized representative of an insurance company stating the types, amounts, and effective dates of insurance in force for a designated insured.

Certificate of occupancy A document issued by a governmental authority certifying that all or a designated portion of a building complies with the provisions of applicable statutes and regulations, and permitting occupancy for its designated use.

Certificate of substantial completion A certificate prepared by the architect on the basis of an inspection stating that the work or a designated portion thereof is substantially complete, which established the date of substantial completion; states the responsibilities of the owner and the contractor for security, maintenance, heat, utilities, damage to the work, and insurance; and taxes the time within which the contractor shall complete the items listed therein.

Certified wood Wood-based materials used in building construction that are supplied from sources that comply with sustainable forestry practices, protecting trees, wildlife habitat, streams, and soil as determined by the Forest Stewardship Council or other recognized certifiable organizations.

Cesspool An underground catch basin for the collection and dispersal of sewage.

Chain of custody A document that tracks the movement of a product from the point of harvest or extraction to the end user.

Change order A written and signed document between the owner and the contractor authorizing a change in the work or an adjustment in the contract sum or time. The contract sum and time may be changed only by change order. A change order may be in the form of additional compensation or time; or less compensation or time known as a "deduction."

Checklist A list of items used to check drawings.

Chiller Equipment designed to produce chilled water. See also vapor compression chiller (centrifugal, reciprocating) and absorption chiller.

Circuit A continuous system of conductors providing a path for electricity.

Circuit breaker A circuit breaker acts like an automatic switch that can shut the power off when it senses too much current.

Circumference The length of a line that forms a circle.

Clear floor space The minimum unobstructed floor or ground space required to accommodate a single, stationary wheelchair and occupant.

Clerestory A window or group of windows that are placed above the normal window height, often between two roof levels.

Coefficient of utilization (CU) The ratio of light energy (lumens) from a source, calculated as received on the workplane, to the light energy emitted by the source alone.

Column A vertical weight-supporting member.

Combustion An oxidation process that releases heat; on-site combustion is a common heat source for buildings.

Commissioning (Cx) A systematic process to verify that building components and systems function as intended and required; systems may need to be re-commissioned at intervals during a building's life-cycle.

Common use Refers to those interior and exterior rooms, spaces, or elements that are made available for the use of a restricted group of people (for example, occupants of a homeless shelter, the occupants of an office building, or the guests of such occupants).

Component A fully functional portion of a building system, piece of equipment, or building element.

Composite wood A product consisting of wood or plant particles or fibers bonded together by a synthetic resin or binder. Examples include plywood, particle-board, OSB, MDF, and composite door cores.

Composting toilet A dry plumbing fixture that contains and treats human waste via a microbiological process.

Compressor A device designed to compress (increase the density) a compressible fluid; a component used to compress refrigerant; a component used to compress air.

Computer-aided drafting (CAD) A method by which engineering drawings may be developed on a computer.

Computer-aided manufacturing (CAM) A method by which a computer uses a design to guide a machine that produces parts.

Concrete block A rectangular concrete form with cells in it.

Condensation The process by which moisture in the air becomes water or ice on a surface (such as a window) whose temperature is colder than the air's temperature.

Condenser A device designed to condense a refrigerant; an air-to-refrigerant or water-to-refrigerant heat exchanger; part of a vapor compression or absorption refrigeration cycle.

Conductor A material used to transfer, or conduct, electricity, often in the form of wires.

Conduit A pipe or elongated box used to house and protect electrical cables.

Conservation The act of preserving and renewing when possible, human and natural resources and the use, protection, and improvement of natural resources according to recognized principles that ensure their highest economic or social benefits.

Construction documents A term used to represent all drawings, specifications, addenda, and other pertinent construction information associated with the construction of a specific project.

Contingency allowance A sum included in the project budget designated to cover unpredictable or unforeseen items of work, or changes in the work subsequently required by the owner. See budget, project.

Contour line A line that represents the change in level from a given datum point.

Contract A legally enforceable promise or agreement between two or among several persons. See also agreement.

Convection The transfer of heat through the movement of a liquid or gas.

Cooling tower Equipment designed to reject heat from a refrigeration cycle to the outside environment through an open cycle evaporative process; an exterior heat rejection unit in a water-cooled refrigeration system.

Cornice The projecting or overhanging structural section of a roof.

Cost appraisal Evaluation or estimate (preferably by a qualified professional appraiser) of the market or other value, cost, utility, or other attribute of land or other facility.

Cost estimate A preliminary statement of approximate cost, determined by one of the following methods. 1. Area and volume method; cost per square foot or cubic foot of the building. 2. Unit cost method; cost of one unit multiplied by the number of units in the project; for example, in a hospital, the cost of one patient unit multiplied by the number of patient units in the project. 3. In-place unit method; cost in-place of a unit, such as doors, cubic yards of concrete, and squares of roofing.

Coving The curving of the floor material against the wall to eliminate the open seam between floor and wall.

Cradle-to-grave analysis Analysis of the impact of a product from the beginning of its source gathering processes, through the end of its useful life, to disposal of all waste products. Cradle-to-cradle is a related term signifying the recycling or reuse of materials at the end of their first useful life.

Crawl space The area under a floor that is not fully excavated; only excavated sufficiently to allow one to crawl under it to get at the electrical or plumbing devices.

Cross-section A slice through a portion of a building or member that depicts the various internal conditions of that area.

CSI Construction Specifications Institute. A membership organization for design professionals, construction professionals, product manufacturers, and building owners.

Develops and promotes industry communication standards and certification programs, http://www.csinet.org.

Current The flow of electric charge in a conductor between two points having a difference in electrical potential (voltage) and measured in amps.

Curtain wall An exterior wall that provides no structural support.

Cut-off voltage The voltage levels at which the charge controller (regulator) disconnects the PV array from the battery, or the load from the battery.

Damper A device designed to regulate the flow of air in a distribution system.

Dangerous or adverse conditions These are essentially conditions which may pose a threat or possible injury to the field observer, and which may require the use of special protective clothing, safety equipment, access equipment, or any other precautionary measures.

Date of agreement The date stated in the agreement. If no date is stated, it could be the date on which the agreement is actually signed, if this is recorded, or it may be the date established by the award.

Date of commencement of the work The date established in a notice to the contractor to proceed or, in the absence of such notice, the date of the owner/contractor agreement or such other date as may be established therein.

Date of substantial completion The date certified by the architect when the work or a designated portion thereof is sufficiently complete, in accordance with the contract documents, so the owner can occupy the work or designated portion thereof for the use for which it is intended.

Datum point Reference point.

Daylight factor (DF) The ratio of daylight illumination at a given point on a given plane, from an obstructed sky of assumed or known illuminance distribution, to the light received on a horizontal plane from an unobstructed hemisphere of this sky, expressed as a percentage. Direct sunlight is excluded for both values of illumination. The daylight factor is the sum of the sky component, the external reflected component, and the internal reflected component. The interior plane is usually a horizontal workplane. If the sky condition is the CIE standard overcast condition, then the DF will remain constant regardless of absolute exterior illuminance.

Daylighting The controlled admission of natural light into a space through glazing with the intent of reducing or eliminating electric lighting. By utilizing solar light, daylighting creates a stimulating and productive environment for building occupants.

Dead load The weight of a structure and all its fixed components.

Decibel (dB) Unit of sound level or sound-pressure level. It is ten times the logarithm of the square of the sound pressure divided by the square of reference pressure, 20 micropascals.

Defective work Work not conforming with the contract requirements.

Deferred maintenance Physical deficiencies that cannot be remedied with routine maintenance, normal operating maintenance, etc., excluding de minimis conditions that generally do not present a material physical deficiency to the subject property.

Design-build construction When an owner contract with a prime or main contractor to provide both design and construction services for the entire construction project. Use of the design-build project delivery system has grown from 5 percent of U.S. construction in 1985 to 33 percent in 1999, and is projected to surpass low-bid construction. If a design-build contract is extended further to include the selection, procurement, and installation of all furnishings, furniture, and equipment, it is called a "turnkey" contract.

Details An enlarged drawing to show a structural aspect, an aesthetic consideration, a solution to an environmental condition, or to express the relationship among materials or building components.

Diffuser A device designed to supply air to a space while providing good mixing of supply and room air and avoiding drafts; normally ceiling installed.

Digital The processing of data by numerical or discrete units.

Dimension line A thin unbroken line (except in the case of structural drafting) with each end terminating with an arrowhead; used to define the dimensions of an object. Dimensions are placed above the line, except in structural drawing where the line is broken and the dimension is placed in the break.

Direct costs (hard costs) The aggregate costs of all labor, materials, equipment and fixtures necessary for the completion of construction of the improvements.

Direct costs loan; indirect costs loan That portion of the loan amount applicable and equal to the sum of the loan budget amounts for direct costs and indirect costs, respectively, shown on the borrower's project cost statement.

Direct current (DC) Electrical current that flows only in one direction, although it may vary in magnitude. Contrasts with alternating current.

Discount factor The factor that translates expected benefits or costs in any given future year into present value terms. The discount factor is equal to $1/(1 + i)t$ where i is the interest rate and t is the number of years from the date of initiation for the program or policy until the given future year.

Discount rate The interest rate used in calculating the present value of expected yearly benefits and costs.

Dormer A structure that projects from a sloping roof to form another roofed area. This new area is typically used to provide a surface to install a window.

Downcycling Recycling a material in a manner that much of its inherent value is lost.

Downspouts Pipes connected to the gutter to conduct rainwater to the ground or sewer.

Drip irrigation system An irrigation system that slowly applies water to the root system of plants to maximize transpiration while minimizing wasted water and topsoil runoff. Drip irrigation usually involves a network of pipes and valves that rest on the soil or underground at the root zone.

Drywall An interior wall covering installed in large sheets made from gypsum board.

Duct Usually sheet metal forms used for the distribution of cool or warm air throughout a structure.

Due diligence The process of conducting a walkthrough survey and appropriate inquiries into the physical condition of a commercial real estate's improvements, usually in connection with a commercial real estate transaction. The degree and type of such survey or other inquiry may vary for different properties and different purposes.

Dwelling unit A single unit which provides a kitchen or food preparation area, in addition to rooms and spaces for living, bathing, sleeping, and the like. Dwelling units include a single family home or a townhouse used as a transient group home; an apartment building used as a shelter; guestrooms in a hotel that provide sleeping accommodations and food preparation areas; and other similar facilities used on a transient basis. For purposes of these guidelines, use of the term "dwelling unit" does not imply the unit is used as a residence.

Easement The right or privilege to have access to or through another piece of property such as a utility easement.

Eave That portion of the roof that extends beyond the outside wall.

Ecological/environmental sustainability Maintenance of ecosystem components and functions for future generations.

Ecological impact The impact that a human-caused or natural activity has on living organisms and their nonliving environment.

Ecosystem The interacting system of a biological community and its nonliving environmental surroundings. An ecological community, together with its environment, functioning as a unit.

Egress A continuous and unobstructed way of exit travel from any point in a building or facility to a public way. A means of egress comprises vertical and horizontal travel and may include intervening room spaces, doorways, hallways, corridors, passageways, balconies, ramps, stairs, enclosures, lobbies, horizontal exits, courts, and yards. An accessible means of egress is one that complies with these guidelines and does not include stairs, steps, or escalators. Areas of rescue assistance or evacuation elevators may be included as part of accessible means of egress.

Electric current The flow of electrons measured in amps.

Electrical grid A network for electricity distribution across a large area.

Electricity The movement of electrons (a sub-atomic particle), produced by a voltage, through a conductor.

Electrode An electrically conductive material, forming part of an electrical device, often used to lead current into or out of a liquid or gas. In a battery, the electrodes are also known as plates.

Element An architectural or mechanical component of a building, facility, space, or site, e.g., telephone, curb ramp, door, drinking fountain, seating, or water closet.

Embodied energy The total energy that a product may be said to "contain," including all energy used in growing, extracting, and manufacturing it and the energy used to transport it to the point of use. The embodied energy of a structure or system includes the embodied energy of its components plus the energy used in construction.

Energy Power consumed multiplied by the duration of use. For example, 1000 Watts used for four hours is 4000 Watt hours.

Energy Star rating A designation given by the EPA and the U.S. Department of Energy (DOE) to appliances and products that exceed federal energy efficiency standards. The label helps consumers identify products that are energy efficient and will save money.

Engineer's scale The scale used whenever dimensions are in feet and decimal parts of a foot, or when the scale ratio is a multiple of 10.

Environmental tobacco smoke (ETS) Mixture of smoke from the burning end of a cigarette, pipe, or cigar and smoke exhaled by the smoker (also secondhand smoke or passive smoking). See Smoke-free Homes Program at www.epa.gov/smokefree.

Environmentally friendly A term often used to refer to the degree to which a product may harm the environment including the biosphere, soil, water, and air.

Epicenter The point of the earth's surface directly above the focus or hypocenter of an earthquake.

Expansion joint A joint often installed in concrete construction to reduce cracking and to provide workable areas.

Expected useful life (EUL) The average amount of time in years that an item, component, or system is estimated to function when installed new and assuming routine maintenance is practiced.

Exploded view A pictorial view of a device in a state of disassembly, showing the appearance and interrelationship of parts.

Extension line A line used to visually connect the ends of a dimension line to the relevant feature on the part. Extension lines are solid and are drawn perpendicular to the dimension line.

Façade The exterior covering of a structure.

Face of stud (F.O.S.) Outside surface of the stud. Term used most often in dimensioning or as a point of reference.

Fascia A horizontal member located at the edge of a roof overhang.

Facility All or any portion of buildings, structures, site improvements, complexes, equipment, roads, walks, passageways, parking lots, or other real or personal property located on a site.

Felt A tar-impregnated paper used for water protection under roofing and siding materials. Sometimes used under concrete slabs for moisture resistance.

Fiber optics Optical, clear strands that transmit light without electrical current; sometimes used for outdoor lighting.

Fillet A concave internal corner in a metal component, usually a casting.

Filter A device designed to remove impurities from a fluid passing through the device. See also air filter.

Final completion Term denoting that the work has been completed in accordance with the terms and conditions of the contract documents.

Final inspection Final review of the project by the architect to determine final completion, prior to issuance of the final certificate for payment.

Final payment The payment made by the owner to the contractor, upon issuance by the architect of the final certificate for payment, of the entire unpaid balance of the contract sum as adjusted by change orders.

Finish grade The soil elevation in its final state upon completion of construction.

Fire barrier A continuous membrane such as a wall, ceiling, or floor assembly that is designed and constructed to a specified fire-resistant rating to hinder the spread of fire and smoke. This resistant rating is based on a time factor. Only fire-rated doors may be used in these barriers.

Fire compartment of fire zone An enclosed space in a building that is separated from all other parts of the building by the construction of fire separations having fire resistance ratings.

Fire door A door used between different types of construction that has been rated as being able to withstand fire for a certain amount of time.

Fire resistance rating Sometimes called fire rating, fire resistance classification, or hourly rating. A term defined in building codes, usually based on fire endurance required. Fire resistance ratings are assigned by building codes for various types of construction and occupancies, and are usually given in half hour increments.

Fire-stop Blocking placed between studs or other structural members to resist the spread of fire.

Firewall A type of fire separation of noncombustible construction which subdivides a building or separates adjoining buildings to resist the spread of fire and which has a fire-resistance rating as prescribed in the NBC and has the structural ability to remain intact under fire conditions for the required fire-rated time.

Flashing A thin, impervious sheet of material placed in construction to prevent water penetration or to direct the flow of water. Flashing is used especially at roof hips and valleys, roof penetrations, joints between a roof and a vertical wall, and in masonry walls to direct the flow of water and moisture.

Floor joist A structural member for the support of floor loads.

Floor plan A horizontal section taken at approximately eye level.

Flush Even, level, or aligned.

Flush-out The operation of mechanical systems for a minimum of two weeks using 100 percent outside air upon completion of construction and prior to building occupancy to ensure safe indoor air quality.

Fly ash The fine ash waste collected from the flue gases of coal combustion, smelting, or waste incineration.

Footcandle A common unit of illuminance used in the U.S. The metric unit is the lux.

Footings Weight-bearing concrete construction elements poured in place in the earth to support a structure.

Footlambert The U.S. unit for luminance. The metric unit is the nit.

Formaldehyde A colorless, pungent, and irritating gas mainly used as a disinfectant and preservative and in synthesizing other compounds such as resins.

Fossil fuels Fuel derived from ancient organic remains such as peat, coal, crude oil, and natural gas.

Foundation plan   A drawing that graphically illustrates the location of various foundation members and conditions that are required for the support of a specific structure.

Frieze   A decoration or ornament shaped to form a band around a structure.

Frost line   The depth at which frost penetrates the soil.

Fungi   Any of a group of parasitic lower plants that lack chlorophyll, including molds and mildews.

Fuse   A fuse is a device used to protect electrical equipment from short circuits. Fuses are made with metals that are designed to melt, when the current passing through them is high enough. When the fuse melts, the electrical connection is broken, interrupting power to the circuit or device.

Galvanized   Steel products that have had zinc applied to the exterior surface to provide protection from rusting.

Gauge   The thickness of metal or glass sheet material.

General conditions   (When used by contractors) Construction project activities and their associated costs that are not usually assignable to a specific material installation or subcontract. Example: temporary electrical power. (When used by everyone else) The contract document (often a standard form) that spells out the relationships between the parties to the contract. Example: the AIA Document A201.

General contractor   A prime contractor for general construction.

Generator   A mechanical device used to produce DC electricity. Power is produced by coils of wire passing through magnetic fields inside the generator. Most alternating current generating sets are also referred to as generators.

Gigawatt (GW)   A measurement of power equal to a thousand million watts.

Gigawatt-hour (GWh)   A measurement of energy. One gigawatt-hour is equal to one gigawatt being used for a period of one hour, or one megawatt being used for 1000 hours.

Girder   A horizontal structural beam for the support of secondary members such as floor joists.

Glare   The effect produced by luminance within one's field of vision that is sufficiently greater than the luminance to which one's eyes are adapted; it can cause annoyance, discomfort, or loss in visual performance and visibility.

Grading   The moving of soil to effect the elevation of land at a construction site.

Gray water   Wastewater that does not contain toilet wastes and can be reused for irrigation after simple filtration. Wastewater from kitchen sinks and dishwashers may not be considered gray water in all cases.

Greenfields   Land not previously developed beyond agriculture or forestry use.

Greenhouse gas   A gas in the atmosphere that traps some of the sun's heat and prevents it from escaping into space. Greenhouse gases are vital for making the Earth habitable, but increasing greenhouse gases contribute to climate change. Greenhouse gases include water vapor, carbon dioxide, methane, nitrous oxide, and ozone.

Green power   Electricity generated from renewable energy sources.

Grid   An electrical utility distribution network.

Grid-connected   An energy producing system connected to the utility transmission grid. Also called grid tied.

Grout   A mixture of cement, sand, and water used to fill joints in masonry and tile construction.

Guardrail   A horizontal protective railing used around stairwells, balconies, and changes of floor elevation greater than 30 inches.

Halogen lamp   A special type of incandescent globe made of quartz glass and a tungsten filament, enabling it to run at a much higher temperature than a conventional incandescent globe. Efficiency is better than a normal incandescent, but not as good as a fluorescent light.

Harmonic content Frequencies in the output waveform in addition to the primary frequency (usually 50 or 60 Hz). Energy in these harmonics is lost and can cause undue heating of the load.

Harvested rainwater Captured rainwater used for indoor needs, irrigation, or both.

Hazardous waste Byproducts of society that can pose a substantial or potential hazard to human health or the environment when improperly managed. Possesses at least one of four characteristics–ignitable, corrosive, reactive, or toxic, or appears on special EPA lists.

Head The top of a window or door frame.

Header A horizontal structural member spinning over openings, such as doors and windows, for the support of weight above the openings.

Header course In masonry, a horizontal masonry course of brick laid perpendicular to the wall face; used to tie a double wythe brick wall together.

Heat exchanger A device designed to efficiently transfer heat from one medium to another (for example, water-to-air, refrigerant-to-air, refrigerant-to-water, steam-to-water).

Heat island effect The incidence of higher air and surface temperatures caused by solar absorption and re-emission from roads, buildings, and other structures.

Heat pump A device that uses a reversible cycle vapor compression refrigeration circuit to provide cooling and heating from the same unit (at different times).

Heat recovery A process whereby heat is extracted from exhaust air before the air is dumped to the outside environment. The recovered heat is normally used to preheat incoming outside air; may be accomplished by heat recovery wheels or heat exchanger loops.

Hertz (Hz) Unit of measurement for frequency. A home's main power is normally 50 Hz in Europe and 60 Hz in the U.S. The magnitude of the current is measured in amps.

High-performance green building Buildings that create healthy indoor environments and include design features that conserve water and energy; efficiently use space, materials, and resources; and minimize construction waste.

Hydronic system A heating or cooling system that relies on the circulation of water as the heat-transfer medium. A typical example is a boiler with hot water circulated through radiators.

Illuminance The density of the luminous flux incident on a surface, expressed in footcandles or lux. This term should not be confused with illumination (i.e., the act of illuminating or state of being illuminated).

Incandescent light An electric lamp which is evacuated or filled with an inert gas and contains a filament (commonly tungsten). The filament emits visible light when heated to extreme temperatures by passage of electric current through it.

Incident light Light that shines on to the surface of a PV cell or module.

Indirect cost statement A statement by the borrower in a form approved by the lender of indirect costs incurred and to be incurred.

Indoor air pollution Chemical, physical, or biological contaminants contained in indoor air.

Indoor air quality (IAQ) According to the U.S. Environmental Protection Agency and National Institute of Occupational Safety and Health, the definition of good indoor air quality includes the introduction and distribution of adequate ventilation air, control of airborne contaminants, and maintenance of acceptable temperature and relative humidity. According to ASHRAE Standard 62 1989, indoor air quality is defined as "air in which there are no known contaminants at harmful concentrations as determined by cognizant authorities and with which a substantial majority (80 percent or more) of the people exposed do not express dissatisfaction."

Indoor environmental quality (IEQ) The evaluation of five primary elements—lighting, sound, thermal conditions, air pollutants, and surface pollutants, to provide an environment that is physically and psychologically healthy for its occupants.

Infill site A site that is largely located within an existing community. For the purposes of LEED for Homes credits, an infill site is defined as having at least 75 percent of its perimeter bordering land that has been previously developed.

Inscribed figure   A figure that is completely enclosed by another figure.

Insolation   The amount of sunlight reaching an area, usually expressed in watt hours per square meter per day.

Inspection   Examination of work completed or in progress to determine its conformance with the requirements of the contract documents. The architect ordinarily makes only two inspections of work, one to determine substantial completion and the other to determine final completion. These inspections should be distinguished from the more general observations made by the architect on visits to the site during the progress of the work. The term is also used to mean examination of the work by a public official, owner's representative, or others.

Insulation   Any material capable of resisting thermal, sound, or electrical transmission.

Integrated design team   The team of all individuals involved in a project from very early in the design process, including the design professionals, the owner's representatives, and the general contractor and subcontractors.

Internal rate of return   The discount rate that sets the net present value of the stream of net benefits equal to zero. The internal rate of return may have multiple values when the stream of net benefits alternates from negative to positive more than once.

Inverter   An inverter converts DC power from the PV array/battery to AC power. Used either for stand-alone systems or grid-connected systems.

Irradiance   The solar power incident on a surface, usually expressed in kilowatts per square meter. Irradiance multiplied by time gives insolation.

Isometric drawing   A form of a pictorial drawing in which the main lines are equal in dimension. Normally drawn using 30 degree or 90 degree angles.

Jamb   The side portion of a door, window, or any opening.

Joist   A horizontal beam used to support a ceiling.

Joule (J)   The energy conveyed by one watt of power for one second, unit of energy equal to 1/3600 kilowatt-hours.

Junction box   A PV junction box is a protective enclosure on a PV module where PV strings are electrically connected and where electrical protection devices such as diodes can be fitted.

Key Plan   A plan, reduced in scale used for orientation purposes.

Kilowatt (kW)   A unit of electrical power, one thousand Watts.

Kilowatt-hour (kWh)   The amount of energy that derives from a power of one thousand watts acting over a period of one hour. The kWh is a unit of energy. 1 kWh = 3600 kJ.

Lattice   A grille made by criss-crossing strips of material.

Ledger   A structural framing member used to support ceiling and roof joists at the perimeter walls.

LEED   Leadership in Energy and Environmental Design. A sustainable design building certification system promulgated by the United States Green Building Council. Also an accrediting program for professionals (LEED APs) who have mastered the certification system, http://www.usgbc.org.

LEED Accredited Professional (LEED AP)   The credential earned by candidates who passed the exam between 2001 and June 2009.

Legend   A description of any special or unusual marks, symbols, or line connections used in the drawing.

Lien   A monetary claim on a property.

Life-cycle cost   A sum of all costs of creation and operation of a facility over a period of time.

Life-cycle cost analysis   A technique used to evaluate the economic consequences over a period of time of mutually exclusive project alternatives.

Light shelf   A horizontal element positioned above eye level to reflect daylight onto the ceiling.

Limit of liability   The maximum amount which an insurance company agrees to pay in case of loss.

Lintel   A load-bearing structural member supported at its ends. Usually located over a door or window.

Live load   A temporary and changing load superimposed on structural components by the use and occupancy of the building, not including the wind load, earthquake load, or dead load.

Load   The electrical power being consumed at any given moment or averaged over a specified period. The load that an electric generating system supplies varies greatly with time of day and to some extent season of year. Also, in an electrical circuit, the load is any device or appliance that is using power.

Load-bearing wall   A support wall that holds floor or roof loads in addition to its own weight.

Lumen (lm)   The luminous flux emitted by a point source having a uniform luminous intensity of one candela.

Luminaire   A complete electric lighting unit, including housing, lamp, and focusing and/or diffusing elements; informally referred to as fixture.

Lux   The International System (SI) unit of illumination. It is the illumination on a surface 1 square meter in area on which there is a uniformly distributed flux of 1 lumen.

Manifold   A fitting that has several inlets or outlets to carry liquids or gases.

Masonry opening   The actual distance between masonry units where an opening occurs. It does not include the wood or steel framing around the opening.

Master specification   A resource specification section containing options for selection, usually created by a design professional firm, which once edited for a specific project becomes a contract specification.

Master format   The industry standard for organizing specifications and other construction information, published by CSI and Construction Specifications Canada. Formerly a five-digit number system with 16 divisions. Now a six- or eight-digit numbering system with 49 divisions.

Masterspec®   Subscription master guide specification library published by ARCOM and owned by the American Institute of Architects, http://www.specguy.com, http://www.masterspec.com.

Mastic   An adhesive used to hold tiles in place; also refers to adhesives used to glue many types of materials in the building process.

Mechanical drawing   Applies to scale drawings of mechanical objects.

Mechanics' lien   A lien on real property created by statute in all states in favor of person supplying labor or materials for a building or structure for the value of labor or materials supplied by them. In some jurisdictions a mechanic's lien also exists for the value of professional services. Clear title to the property cannot be obtained until the claim for the labor, materials, or professional services is settled.

Megawatt (MW)   A measurement of power equal to one million watts.

Megawatt-hour (MWh)   A measurement of power with respect to time (i.e., energy). One megawatt-hour is equal to one megawatt being used for a period of one hour, or one kilowatt being used for 1000 hours.

Mesh   A metal reinforcing material placed in concrete slabs and masonry walls to help resist cracking.

Mezzanine or mezzanine floor   That portion of a story which is an intermediate floor level placed within the story and having occupiable space above and below its floor.

Modules   A system based on a single unit of measure.

Modulus of elasticity (E)   The degree of stiffness of a beam.

Moisture barrier Typically a plastic material used to restrict moisture vapor from penetrating into a structure.

Mortar The mixture of cement, sand, lime, and water that provides a bond for the joining of masonry units.

Multi-zone HVAC system A central air-all HVAC system that utilizes an individual supply air stream for each zone; warm and cool air are mixed at the air handling unit to provide supply air appropriate to the needs of each zone; a multi-zone system requires the use of several separate supply air ducts.

Native vegetation A plant whose presence and survival in a specific region is not due to human intervention. Certain experts argue that plants imported to a region by prehistoric peoples should be considered native. The term for plants that are imported and then adapt to survive without human cultivation is naturalized.

Natural ventilation The natural exchange of air or movement of air through a building by thermal, wind, or diffusion effects through doors, windows, or other intentional openings in buildings.

NEC The U.S. National Electrical Code contains guidelines for all types of electrical installations and should be followed when installing a PV system.

Negligence Failure to exercise due care under normal circumstances. Legal liability for the consequences of an act or omission frequently depends upon whether or not there has been negligence.

Net metering The practice of exporting surplus solar power during the day (to actual power needs) to the electricity grid, which either causes the home owner electric meter to (physically) go backwards and/or simply creates a financial credit on the home owner's electricity bill. (At night, the homeowner draws from the electricity grid in the normal way).

Net size The actual size of an object.

Noise reduction coefficient (NRC) Average of the sound absorption coefficient of the four octave bands 250, 500, 1000, and 2000 hertz rounded to the nearest 0.05.

Nominal discount rate A discount rate that includes the rate of inflation.

Nominal size The call-out-size. May not be the actual size of the item.

Nonbearing wall A wall that supports no loads other than its own weight. Some building codes consider walls that support only ceiling loads as nonbearing.

Nonconforming work Implemented work that does not fulfill the requirements of the contract Documents.

Nonferrous metal Metals such as copper or brass that contain no iron.

Oblique drawing A type of pictorial drawing in which one view is an orthographic projection and the views of the sides have receding lines at an angle.

Occupiable A room or enclosed space designed for human occupancy in which individuals congregate for amusement, educational or similar purposes, or in which occupants are engaged at labor, and which is equipped with means of egress, light, and ventilation.

Off-gassing A process of evaporation or chemical decomposition by which vapors are released from materials.

Ohm The resistance between two points of a conductor when a constant potential difference of one volt applied between these points produces in the conductor a current of one amp.

Ohm's Law A simple mathematical formula that allows either voltage, current, or resistance to be calculated when the other two values are known. The formula is: $V = I \times R$, where V is the voltage, I is the current, and R is the resistance.

Opinions of probable costs Determination of probable costs, a preliminary budget, for a suggested remedy.

Operating cost Any cost of the daily function of a facility.

Organic compounds Chemicals that contain carbon. Volatile organic compounds vaporize at room temperature and pressure. They are found in many indoor sources, including many common household products and building materials.

Orientation Position with respect to the cardinal directions, N, S, E, W.

Orthographic projection A view produced when projectors are perpendicular to the plane of the object. It gives the effect of looking straight at one side.

Outlet An electrical receptacle that allows for current to be drawn from the system.

Packaged air-conditioner A self-contained unit designed to provide control of air temperature, humidity, distribution, and quality.

Parapet A portion of wall extending above the roof level.

Partial occupancy The occupancy by the owner of a portion of a project prior to final completion.

Partition An interior wall.

Party wall A wall dividing two adjoining spaces such as apartments or offices.

Passive solar home A house that utilizes part of the building as a solar collector, as opposed to active solar, such as PV.

Patent defect A defect in materials or equipment of completed work that reasonably careful observation could have discovered; distinguished from a latent defect, which could not be discovered by reasonable observation.

Performance specifications The written material containing the minimum acceptable standards and actions, as may be necessary to complete a project.

Phase An impulse of alternating current. The number of phases depends on the generator windings. Most large generators produce a three-phase current that must be carried on at least three wires.

Photometer An instrument for measuring light.

Photovoltaic (PV) Refers to any device which produces free electrons when exposed to light.

Photovoltaic (PV) panel A term often used interchangeably with PV module (especially in single module systems).

Photovoltaic system All the parts connected together that are required to produce solar electricity.

Pile A steel or wooden pole driven into the ground sufficiently to support the weight of a wall and building.

Pillar A pole or reinforced wall section used to support the floor and consequently the building.

Planking A term for wood members having a minimum rectangular section of 1 1/2 inch to 3 1/2 inch in thickness. Used for floor and roof systems.

Plans All final drawings, plans, and specifications prepared by the borrower, borrower's architects, the general contractor or major subcontractors, and approved by lender and the construction consultant, which describe and show the labor, materials, equipment, fixtures, and furnishings necessary for the construction of the improvements, including all amendments and modifications thereof made by approved change orders (and also showing minimum grade of finishes and furnishings for all areas of the improvements to be leased or sold in ready-for-occupancy conditions).

Plat A map or plan view of a lot showing principal features, boundaries, and location of structures.

Plenum An air space (above the ceiling) for transporting air from the HVAC system.

Plug load Refers to all equipment that is plugged into the electrical system, such as task lights, computers, printers, and electrical appliances.

Polarity The direction of magnetism or direction of flow of current.

Pollutant   Generally, any substance introduced into the environment that adversely affects the usefulness of a resource or the health of humans, animals, or ecosystems.

Poly-vinyl chloride (PVC)   A plastic material commonly used for pipe and plumbing fixtures and as an insulator on electrical cables. A toxic material, which is being replaced with alternatives made from more benign chemicals.

Post   A vertical wood structural member generally 4 × 4 (100 mm) or larger.

Post-and-beam construction   A type of wood frame construction using timber for the structural support.

Postconsumer materials/waste   Recovered materials that are diverted from municipal solid waste for the purposes of collection, recycling, and disposition.

Postconsumer recycling   Use of materials generated from resident or consumer waste for new or similar purposes such as converting wastepaper from offices into corrugated boxes or newsprint.

Potable water   Water that is suitable for drinking, generally supplied by the municipal water systems.

Power   Basic unit of electricity equal to the product of current and voltage (in DC circuits). The rate of doing work. Expressed as watts (W). For example, a generator rated at 800 watts can provide that amount of power continuously. 1 watt = 1 joule/sec.

Precast   A concrete component which has been cast in a location other than the one in which it will be used.

Preconsumer materials/waste   Materials generated in manufacturing and converting processes such as manufacturing scraps and trimmings and cuttings. Includes print overruns, over issue publications, and obsolete inventories. Sometimes referred to as "postindustrial."

Present value   The current value of a past or future sum of money as a function of an investor's time value of money.

Pressed wood products   A group of materials used in building and furniture construction that are made from wood veneers, particles, or fibers bonded together with an adhesive under heat and pressure.

Primer   The first coat of paint or glue when more than one coat will be applied.

Progress payment   Partial payment made during progress of the work on account of work completed and or materials suitably stored.

Progress schedule   A diagram, graph, or other pictorial or written schedule showing proposed and actual of starting and completion of the various elements of the work.

Project cost   Total cost of the project including construction costs, professional compensation, land costs, furnishings and equipment, financing, and other charges.

Projection   A technique for showing one or more sides of an object to give the impression of a drawing of a solid object.

Project manual   The volume(s) prepared by the architect for a project that may include the bidding requirements, sample forms and conditions of the contract and the specifications.

Purlin   A horizontal roof member that is laid perpendicular to rafters to help limit deflections.

Quarry tile   An unglazed, machine made tile.

Quick set   A fast-curing cement plaster.

Rafter   A sloping or horizontal beam used to support a roof.

Radius   A straight line from the center of a circle or sphere to its circumference or surface.

Radon (Rn) and radon decay products   Radon is a radioactive gas formed in the decay of uranium. The radon decay products (also called radon daughters or progeny) can be breathed into the lung where they continue to release radiation as they further decay.

Rainscreen A method of constructing walls in which the cladding is separated from a membrane by an air space that allows pressure equalization to prevent rain from being forced in. Often used for high-rise buildings or for buildings in windy locations.

Rainwater harvesting The practice of collecting, storing, and using precipitation from a catchment area such as a roof.

Rapidly renewable Materials that are not depleted when used. These materials are typically harvested from fast-growing sources and do not require unnecessary chemical support. Examples include bamboo, flax, wheat, wool, and certain types of wood.

RAPS (remote area power supply) A power generation system used to provide electricity to remote and rural homes, usually incorporating power generated from renewable sources such as solar panels and wind generators, as well as non-renewable sources such as petrol-powered generators.

Readily accessible Describes areas of the subject property that are promptly made available for observation by the field observer at the time of the walk-through survey and do not require the removal of materials or personal property, such as furniture, and that are safely accessible in the opinion of the field observer.

Record drawings Construction drawings revised to show significant changes made during the construction process, usually based on marked-up prints, drawings, and other data furnished by the Contracts to the Architect. Preferable to as-built drawings.

Rectifier A device that converts AC to DC, as in a battery charger or converter.

Recycled material Material that would otherwise be destined for disposal but is diverted or separated from the waste stream, reintroduced as material feed-stock, and processed into marketed end-products.

Reference numbers Numbers used on a drawing to refer the reader to another drawing for more detail or other information.

Reflectance The ratio of energy (light) bouncing away from a surface to the amount striking it, expressed as a percentage.

Refrigerant A heat transfer fluid employed by a refrigerating process, selected for its beneficial properties (stability, low viscosity, high thermal capacity, appropriate state change points).

Register An opening in a duct for the supply of heated or cooled air.

Regulator A device used to limit the current and voltage in a circuit, normally to allow the correct charging of batteries from power sources such as solar panels and wind generators.

Relative humidity The amount of water vapor in the atmosphere compared to the maximum possible amount at the same temperature.

Release of lien An instrument executed by a person or entity supplying labor, materials, or professional services on a project which releases that person's or entity's mechanic's lien against the project property.

Remaining useful life (RUL) A subjective estimate based upon observations, or average estimates of similar items, components, or systems, or a combination thereof, of the number of remaining years that an item, component, or system is estimated to be able to function in accordance with its intended purpose before warranting replacement. Such period of time is affected by the initial quality of an item, component, or system, the quality of the initial installation, the quality and amount of preventive maintenance exercised, climatic conditions, extent of use, etc.

Renewable energy Alternative energy that is produced from a renewable source.

Requisition A statement prepared by the borrower in a form approved by the lender setting forth the amount of the loan advance requested in each instance and including, if requested by the lender:

Resistance (R) The property of a material which resists the flow of electric current when a potential difference is applied across it, measured in ohms.

Resistor An electronic component used to restrict the flow of current in a circuit. Sometimes used specifically to produce heat, such as in a water heater element.

Retainage A sum withheld from progress payments to the contractor in accordance with the terms of the owner-contractor agreement.

Retaining wall A masonry wall supported at the top and bottom designed to resist soil loads.

R-factor A unit of thermal resistance applied to the insulating value of a specific building material.

Return air The air that has circulated through a building as supply air and has been returned to the HVAC system for additional conditioning or release from the building.

Residual value The value of a building or building system at the end of the study period.

Roof drain A receptacle for removal of roof water.

Roof pitch The ratio of total span to total rise expressed as a fraction.

Rotation A view in which the object is apparently rotated or turned to reveal a different plane or aspect, all shown within the view.

Rough in To prepare a room for plumbing or electrical additions by running wires or piping for a future fixture.

Rough opening A large opening made in a wall frame or roof frame to allow the insertion of a door or window.

R-value The unit that measures thermal resistance (the effectiveness of insulation); the higher the number, the better the insulation qualities.

Sanitary sewer A conduit or pipe carrying sanitary sewage.

Scale The relation between the measurement used on a drawing and the measurement of the object it represents. A measuring device, such as a ruler, with special graduations.

Schedule of values A statement furnished by the contractor to the architect reflecting the portions of the contract sum allocated to the work and used as the basis for reviewing the contractor's applications for payment.

Schematic diagram A diagram using graphic symbols to show how a circuit functions electrically.

Scratch coat The first coat of stucco, which is scratched to provide a good bond surface for the second coat.

Section A view showing internal features as if the viewed object has been cut or sectioned.

Seismicity The worldwide or local distribution of earthquakes in space and time; a general term for the number of earthquakes in a unit of time, or for relative earthquake activity.

Septic tank A tank in which sewage is decomposed by bacteria and dispersed by drain tiles.

Sheet steel Flat steel weighing less than 5 pounds per square foot.

Shear distribution The distribution of lateral forces along the height or width of a building.

Shear wall A wall construction designed to withstand shear pressure caused by wind or earthquake.

Shoring A temporary support made of metal or wood used to support other components.

Short-term costs Opinions of probable costs to remedy physical deficiencies, such as deferred maintenance, that may not warrant immediate attention, but require repairs or replacements that should be undertaken on a priority basis in addition to routine preventive maintenance. Such opinions of probable costs may include costs for testing, exploratory probing, and further analysis should this be deemed warranted by the consultant. The performance of such additional services is beyond this guide. Generally, the time frame for such repairs is within one to two years.

Sick Building Syndrome (SBS) A term that refers to a set of symptoms that affect some number of building occupants during the time they spend in the building and diminish or go away during periods when they leave the building. Cannot be traced to specific pollutants or sources within the building. (Contrast with Building Related Illness).

Sill A horizontal structural member supported by its ends.

Single-line diagram A diagram using single lines and graphic symbols to simplify a complex circuit or system.

Single prime contract This is the most common form of construction contracting. In this process, the bidding documents are prepared by the architect/engineer for the owner and made available to a number of qualified bidders. The winning contractor then enters into a series of subcontract agreements to complete the work. Increasingly, owners are opting for a design-build contract under which a single entity provides design and construction services.

Site A parcel of land bounded by a property line or a designated portion of a public right-of-way.

Site improvement Landscaping, paving for pedestrian and vehicular ways, outdoor lighting, recreational facilities, and the like, added to a site.

Skylight A relatively horizontal, glazed roof aperture for the admission of daylight.

Slab-on-grade The foundation construction for a structure with no basement or crawl space.

Smart growth Managing the growth of a community in such a way that land is developed according to ecological tenets that call for minimizing dependence on auto transportation, reducing air pollution and increasing infrastructure investment efficiency.

Solar energy Energy from the sun.

Solar heat gain coefficient Solar heat gain through the total window system relative to the incident solar radiation.

Solar module A device used to convert light from the sun directly into DC electricity by using the photovoltaic effect. Usually consists of multiple solar cells bonded between glass and a backing material. A typical solar module would be 100 watts of power output (but module powers can range from 1 watt to 300 watts) and have dimensions of 2 feet by 4 feet.

Solar panel A device that collects energy from the sun and converts it into electricity or heat.

Solar power Electricity generated by the conversion of sunlight, either directly through the use of photovoltaic panels, or indirectly through solar-thermal processes.

Special conditions A section of the conditions of the contract, other than general conditions and supplementary conditions, which may be prepared to describe conditions unique to a particular project.

Specifications A part of the contract documents contained in the project manual consisting of written requirements for material, equipment, construction systems, standards, and workmanship. Under the Uniform Construction Index, the specifications comprise 16 divisions.

Specific gravity The ratio of the weight of a solution to the weight of an equal volume of water at a specified temperature; used with reference to the sulfuric acid electrolyte solution in a lead acid battery as an indicator of battery state of charge. More recently called relative density.

Stakeholders All parties that might be affected by a company's policies and operations, including shareholders, customers, employees, suppliers, business partners, and surrounding communities.

Storm sewer A sewer used for conveying rain water, surface water condensate, cooling water, or similar liquid wastes exclusive of sewage.

Stucco A type of plaster made from Portland cement, sand, water, and a coloring agent that is applied to exterior walls.

Structural frame   The components or building systems that support the building's non-variable forces or weights (dead loads) and variable forces or weights (live loads).

Stud   A light vertical structure member, usually made of wood or light structural steel, used as part of a wall and for supporting moderate loads.

Subcontract   Agreement between a prime contractor and a subcontractor for a portion of the work at the site.

Subcontractor   A person or entity who has a direct or indirect contract with a subcontractor to perform any of the work at the site.

Substitution   A material, product or item of equipment offered in lieu of that specified.

Superintendent   A contractor's representative at the site who is responsible for continuous field supervision, coordination, completion of the work and, unless another person is designated in writing by the contractor to the owner and the architect, for the prevention of accidents.

Supervision   Direction of the work by the contractor's personnel. Supervision is neither a duty nor a responsibility of the architect as part of professional services.

Surety bond   A legal instrument under which one party agrees to answer to another party for the debt, default, or failure to perform of a third party.

Surge   An excessive amount of power drawn by an appliance when it is first switched on. An unexpected flow of excessive current, usually caused by excessive voltage, which can damage appliances and other electrical equipment.

Survey   Observations made by the field observer during a walk-through survey to obtain information concerning the subject property's readily accessible and easily visible components or systems.

Sustainable   The condition that meets the needs of present generations without compromising the needs of future generations. Achieving a balance among extraction and renewal and environmental inputs and outputs, as to cause no overall net environmental burden or deficit.

Symbol   Stylized graphical representation of commonly used component parts shown in a drawing.

Synergy   Action of two or more substances to achieve an effect of which each is individually incapable. As applied to toxicology, two exposures together (for example, asbestos and smoking) are far more risky than the combined individual risks.

System   (a process) A combination of interacting or interdependent components assembled to carry out one or more functions.

Task lighting   Light provided for a specific task, versus general or ambient lighting.

Tee   A fitting, either cast or wrought, that has one side outlet at right angles to the run.

Temper   To harden steel by heating and sudden cooling by immersion in oil, water, or other coolant.

Template   A piece of thin material used as a true-scale guide or as a model for reproducing various shapes.

Tensile strength   The maximum stretching of a piece of metal (rebar, etc.) before breaking; calculated in kps.

Termite shield   Sheet metal placed in or on a foundation wall to prevent intrusion.

Terrazzo   A mixture of concrete, crushed stone, calcium shells, and/or glass, polished to a tile-like finish.

Thermal comfort   The appropriate combination of temperature combined with airflow and humidity that allows one to be comfortable within the confines of a building. Individually, an expression of satisfaction with the thermal environment; statistically, an expression of such satisfaction from at least 80 percent of the occupants within a space.

Thermal resistance (R)   A unit used to measure a material's resistance to heat transfer. The formula for thermal resistance is: R_Thickness (in inches)/k.

Thermostat   An automatic device controlling the operation of HVAC equipment.

Third-party certification   An independent and objective assessment of an organization's practices or chain of custody system by an auditor who is independent of the party undergoing assessment.

Three-phase power   A combination of three alternating currents in a circuit with their voltages displaced at 120 degrees or one-third of a cycle.

Timely access   Entry provided to the consultant at the time of the site visit.

Timely completion   The completion of the work or designated portion thereof on or before the date required.

Title insurer   The issuer(s), approved by interim lender and permanent lenders, of the title insurance policy or policies insuring the mortgage.

Tolerance   The amount that a manufactured part may vary from its specified size.

Topographic survey   The configuration of a surface including its relief and the locations of its natural and man-made features, usually recorded on a drawing showing surface variations by means of contour lines indicating height above or below a fixed datum.

Toxicity   A reflection of a material's ability to release poisonous particulate.

Transformer   A transformer is a device that changes voltage from one level to another. A device used to transform voltage levels to facilitate the transfer of power from the generating plant to the customer.

Transient lodging   A building, facility, or portion thereof, excluding inpatient medical care facilities and residential facilities, that contains sleeping accommodations. Transient lodging may include, but is not limited to, resorts, group homes, hotels, motels, and dormitories.

Transistor   A semi-conductor device used to switch or otherwise control the flow of electricity.

Trap   A fitting designed to provide a liquid seal which will prevent the back passage of air without significantly affecting the flow of waste water through it.

Triangulation   A technique for making developments of complex sheet metal forms using geometrical constructions to translate dimensions from the drawing to the pattern.

Trimmer   A piece of lumber, usually a 2 × 4, that is shorter than the stud or rafter but is used to fill in where the longer piece would have been normally spaced except for the window or door opening or some other opening in the roof or floor or wall.

Truss   A prefabricated sloped roof system incorporating a top chord, bottom chord, and bracing.

Turbulence   Any deviation from parallel flow in a pipe due to rough inner wall surfaces, obstructions, etc.

UL   Underwriters Laboratories, Inc. A private testing and labeling organization that develops test standards for product compliance. UL standards appear throughout specifications, often in roofing requirements, and always in equipment utilizing or delivering electrical power, http://www.ul.com.

Unfaced insulation   Insulation that does not have a facing or plastic membrane over one side of it.

Union joint   A pipe coupling, usually threaded, that permits disconnection without disturbing other sections.

Urea formaldehyde   A combination of urea and formaldehyde that is used in some glues and may emit formaldehyde at room temperature.

Utility plan   A floor plan of a structure showing locations of heating, electrical, plumbing, and other service system components.

Vacuum   Any pressure less than that exerted by the atmosphere.

Valley   The area of a roof where two sections come together to form a depression.

Valve   A device designed to control water flow in a distribution system; common valve types include globe, gate, butterfly, and check.

Vapor barrier   The same as a moisture barrier.

Vapor compression chiller   Refrigeration equipment that generates chilled water via a mechanically driven process using a specialized heat transfer fluid as refrigerant; comprised of four major components: a compressor, condenser, expansion valve, and evaporator; operating energy is input as mechanical motion.

Variable air volume (VAV) HVAC system   A central air-all HVAC system that utilizes a single supply air stream and a terminal device at each zone to provide appropriate thermal conditions through control of the quantity of air supplied to the zone.

Vegetated roof   A roof that is partially or fully covered by vegetation. By creating roofs with a vegetated layer, the roof can counteract the heat island effect as well as provide additional insulation and cooling during the summer.

Vehicular way   A route intended for vehicular traffic, such as a street, driveway, or parking lot.

Veneer   A thin layer or sheet of wood.

Veneered wall   A single-thickness (one-wythe) masonry unit wall with a backup wall of frame or other masonry; tied but not bonded to the backup wall.

Ventilation   The exchange of air, or the movement of air through a building; may be done naturally through doors and windows or mechanically by motor-driven fans.

Vent   Usually a hole in the eaves or soffit to allow the circulation of air over an insulated ceiling; usually covered with a piece of metal or screen.

Ventilation rate   The rate at which indoor air enters and leaves a building. Expressed in one of two ways: the number of changes of outdoor air per unit of time (air changes per hour, or "ach") or the rate at which a volume of outdoor air enters per unit of time (cubic feet per minute, or "cfm").

Vent stack   A system of pipes used for air circulation and to prevent water from being suctioned from the traps in the waste disposal system.

Vertical pipe   Any pipe or fitting installed in a vertical position or which makes an angle of not more than 45 degrees with the vertical.

View   A drawing of a side or plane of an object as seen from one point.

Vision glazing   That portion of exterior windows above 2 feet 6 inches and below 7 feet 6 inches that permits a view to the exterior.

Volatile organic compound (VOC)   A highly evaporative, carbon-based chemical substance that produces noxious fumes; found in many paints, caulks, stains, and adhesives.

Volt (E) or (V)   The potential difference across a resistance of 1 ohm when a current of 1 amp is flowing. The amount of work done per unit charge in moving a charge from one place to another.

Voltage drop   The voltage lost along a length of wire or conductor due to the resistance of that conductor. This also applies to resistors. The voltage drop is calculated by using Ohm's Law.

Voltage protection   A sensing circuit on an inverter that will disconnect the unit from the battery if input voltage limits are exceeded.

Voltage regulator   A device that controls the operating voltage of a photovoltaic array.

Waiver of lien   An instrument by which a person or organization who has or may have a right of mechanic's lien against the property of another relinquishes such right.

Warranty   Legally enforceable assurance of quality or performance of a product or work, or of the duration of satisfactory performance. Warranty guarantee and guaranty are substantially identical in meaning; nevertheless, confusion frequently arises from supposed distinctions attributed to guarantee (or guaranty) being exclusively indicative of the duration of satisfactory performance or of a legally enforceable assurance furnished by a manufacturer or other third party. The Uniform Commercial Code provisions on sales (effective in all states except Louisiana) use warranty but recognize the continuation of the use of guarantee and guaranty.

Waste Pipe A discharge pipe from any fixture, appliance, or appurtenance in connection with a plumbing system which does not contain fecal matter.

Wastewater Spent or used water from a home, farm, community, or industry that contains dissolved or suspended matter.

Water-cement ratio The ratio between the weight of water to cement.

Water hammer The noise and vibration which develops in a piping system when a column of non-compressible liquid flowing through a pipe line at a given pressure and velocity is abruptly stopped.

Water main The water supply pipe for public or community use.

Waterproofing Materials used to protect below- and on-grade construction from moisture penetration.

Watt (W) The unit of electrical power commonly used to define the electricity consumption of an appliance. The power developed when a current of one ampere flows through a potential difference of one volt; 1/746 of a horsepower. 1 watt = 1 Joule/s.

Watt hour (Wh) A unit of energy equal to one watt of power being used for one hour.

Wetland In storm water management, a shallow, vegetated, ponded area that serves to improve water quality and provide wildlife habitat.

Wind lift (wind load) The force exerted by the wind against a structure.

Wiring (connection) diagram A diagram showing the individual connections within a unit and the physical arrangement of the components.

Working drawings A set of drawings that provide the necessary details and dimensions to construct the object. May include specifications.

Wythe A continuous masonry wall width.

Xeriscape™ A trademarked term referring to water-efficient choices in planting and irrigation design. It refers to seven basic principles for conserving water and protecting the environment. These include: (1) Planning and design; (2) Use of well-adapted plants; (3) Soil analysis; (4) Practical turf areas; (5) Use of mulches; (6) Appropriate maintenance; and (7) Efficient irrigation.

Zenith angle The angle between directly overhead and a line through the sun. The elevation angle of the sun above the horizon is 90 degrees minus the zenith angle.

Zinc Non-corrosive metal used for galvanizing other metals.

Zone numbers Numbers and letters on the border of a drawing to provide reference points to aid in indicating or locating specific points on the drawing.

Zoning The legal restriction that deems that parts of cities be for particular uses, such as residential, commercial, industrial, and so forth.

Zoning permit A permit issued by appropriate governmental authority authorizing land to be used for a specific purpose.

Bibliography

Apgar, M. (May/June 1998). The Alternative Workplace: Changing Where and How People Work. *Harvard Business Review*, 121–135.

Armer, G. S. T. (2001). *Monitoring and Assessment of Structures*. New York: Spon Press.

Arnold, C. (July 2, 2006). *Green Movement Sweeps U.S. Construction Industry*. NPR.

ASHRAE 2008 Proposed Standard 189P: *Standard for the Design of High-Performance Green Buildings*, second public review, July 2008, Atlanta, GA.

ASTM, *Standard Guide for Property Condition Assessments: Baseline Property Condition Assessment, Designation: E2018-01*.

Bady, S. (2008). Green Building Programs More About Bias Than Science, Expert Argues. *Professional Builder*.

Bennett, F. L. (2003). *The Management of Construction: A Project Life Cycle Approach*. Butterworth Heinemann.

Bezdek, R. H. (2008). "Green Building: Balancing Fact and Fiction," S. E. Cannon and U. K. Vyas, moderators, *Real Estate Issues*, (33)2, 2–5.

Bonda, P., & Sosnowchik, K. (2007). *Sustainable Commercial Interiors*. John Wiley & Sons.

Build it Green, *New Home Construction—Green Building Guidelines*, 2007 Edition.

Burr, A. C. (Jaunuary 30, 2008). *CoStar Green Report: The Big Skodowski*. CoStar Group.

Butters, F. (2008). "Green Building: Balancing Fact and Fiction," S. E. Cannon and U. K. Vyas, moderators. *Real Estate Issues*, (33)2, 10–11.

Calow, P. (1998). *Handbook of Environmental Risk Assessment and Management*. Oxford: Blackwell Science Ltd.

Coggan, D. A., *Intelligent Building Systems (Intelligent Buildings Simply Explained)*, Website: http://www.coggan.com/intelligent-building-systems.html, viewed June 1, 2009.

Craiger, Philip (Ed.) and Shenoi, Sujeet (Ed.), *Advances in Digital Forensics III (IFIP International Federation for Information Processing) (IFIP International Federation for Information Processing)*, Springer, 2007.

De Chiara, Joseph., & Julius, Panero. (2001). *Time-Saver Standards for Interior Design and Space Planning*. New York: McGraw-Hill.

Deasy, C. M. (1985). *Designing Places for People: A Handbook for Architects, Designers, and Facility Managers*. New York: Whitney Library of Design.

Del Percio, S., "What's Wrong with LEED?", *American City*, Issue 14, Green Building, Spring 2007, (http://www.americancity.org), accessed April 2007.

DiLouie, C. (March 2006). Why Do Daylight Harvesting Projects Succeed or Fail? *Lighting Controls Association*.

Dorgan, C., R., Cox, and C. Dorgan, *The Value of the Commissioning Process: Costs and Benefits*, Farnsworth Group, Madison WI, paper presented at the 2002 U.S. Green Building Council Conference, Austin, Texas.

Dworkin, J. F. (March 1990). Waterproofing Below Grade. *The Construction Specifier*.

Edwards, S. (1986). *Office Systems—Designs for the Contemporary Workspace*. New York: PBC International Inc.

EIA, "Annual Energy Outlook," Environmental Information Administration, 2008; "Assumptions to the Annual Energy Outlook," Energy Information Administration, 2008; cited in *Green Building Facts*, U.S. Green Building Council, November 2008.

Federal Emergency Management Agency. (2004). FEMA 426. *Reference Manual to Mitigate Potential Terrorist Attacks in High Occupancy Buildings*.

Fisk, W. J., & Rosenfeld, A. H. (1998). The Indoor Environment—Productivity and Health—and $$$. *Strategic Planning for Energy and the Environment, 17*(4), 53–57.

Global Green Building Trends. (2008). *SmartMarket Report*. McGraw-Hill Construction.

Goldsmith, S. (1999). *Designing for the Disabled—The New Paradigm*. Oxford: *Architectural Press*, An imprint of Butterworth-Heinemann.

Gore, A.l. (May 2006). The Future is Green. *Vanity Fair: Special Green Issue*.

Gottfried, D. (2000). *Sustainable Building Technical Manual*, U.S. Green Building Council, U.S. Department of Energy, and U.S. Environmental Protection Agency.

Green Building Certification Institute, *LEED® Green Associate Candidate Handbook*, May 2009.

Green Building Certification Institute, *LEED® Professional Accreditation Handbook*, revised May 2009.

Green Building Rating Systems, *Vermont Green Building Network*, www.vgbn.org/gbrs.php, viewed June 2009.

GSA, *LEED Cost Study—Final Report*, submitted to the U.S. General Services Administration, Steven Winter Associates, Inc., October 2004.

Gunn, R. A., & Burroughs, M. S. (March/April 1996). Work Spaces That Work: Designing High-Performance Offices. *The Futurist*, 19–24.

Haasl, T. P., & Claridge, D. (October 2003). The Cost Effectiveness of Commissioning. *HPAC Engineering*, 20–24.

Harmon, S. K., & Kennon, K. E. (2001). *The Codes Guidebook for Interiors* (2d ed). New York: John Wiley & Sons.

Hellier, C. J. (2001). *Handbook of Nondestructive Evaluation*. New York: McGraw-Hill.

Hess-Kosa, K. (1997). *Environmental Site Assessment Phase I* (2d ed). CRC Press.

Horman, M. J., Riley, D. R., Pulaski, M. H., Magent, C., Dahl, P., Lapinski, A. R., et al. (2006). Delivering Green Buildings: Process Improvements for Sustainable Construction. *Journal of Green Building, 1*(1), 123–140.

Houghton-Evans, R. W. (2005). *Well Built? A Forensic Approach to the Prevention, Diagnosis and Cure of Building Defects*. RIBA Enterprises.

Iano, E. A., & Iano, J. (2003). *Fundamentals of Building Construction: Materials and Methods* (4th ed). New York: John Wiley & Sons, Inc.

Jackson, B. J. (2004). *Construction Management JumpStart*. Wiley Publishing, Inc.

Kats, G. (2007). *The Costs and Benefits of Green*. Capital E Analytics.

Kennett, S., "Making BREEAM Robust," *Building*, (http://www.building.co.uk/story), accessed March 2008.

Kibert, C. J. (2005). *Sustainable Construction—Green Building Design and Delivery*. New Jersey: John Wiley.

Koomen-Harmon, S., & Kennon, K. E. (2001). *The Codes Guidebook for Interiors* (2d ed). New York: John Wiley & Sons.

Kubal, M. T. (2000). *Construction Waterproofing Handbook*. The McGraw-Hill Companies, Inc.

Kubba, S. A. A. (2003). *Space Planning for Commercial & Residential Interiors*. New York: McGraw-Hill.

Kubba, S. A. A. (2007). *Property Condition Assessments*. New York: McGraw-Hill.

Kubba, S. A. A. (2008). *Architectural Forensics*. New York: McGraw-Hill.

Kubba, S. A. A. (2010). *LEED^{TM} Practices, Certification, and Accreditation Handbook*. Butterworth–Heinemann (Elsevier).

Langdon, Davis, Report: *The Cost and Benefit of Achieving Green Buildings*, www.davislangdon.com, 2007.

Langdon, Davis, L. F. Matthiessen, and P. Morris, Study: *Costing Green: A Comprehensive Cost Database and Budgeting Methodology*, July 2004.

Langdon, Davis and P. Morris, 2009 *Market Update—A Guide to Working in a Recession*, 2009.

LEED®, *LEED®-NC Green Building Rating System for New Construction & Major Renovations*, Version 2.2, USGBC. www.usgbc.org (viewed in August 2008).

Levy, Matthys, Mario G. Salvadori, and Kevin Woest (Illustrator), *Why Buildings Fall Down: How Structures Fail*, W. W. Norton & Co Inc., 1994.

Lewis, B. T., & Payant, R. (2001). *Facility Inspection Field Manual: A Complete Condition Assessment Guide*. New York: McGraw-Hill.

Lippiatt, B. C. and G. A. and Norris, *Selecting Environmentally and Economically Balanced Building Materials*, Sponsored by: National Institute of Standards and Technology, 1998.

Loveland, J., "Daylight By Design," *LD + A*, October 2003.

Luhmann, T. (2007). *Close Range Photogrammetry: Principles, Techniques and Applications*. Wiley.

Lupton, M., & Croly, C. (February 2004). Designing for Daylight. *Building Sustainable Design*.

Macaluso, J. (May 2009). *An Overview of The LEED Rating System*. Empire State Development Green Construction Data.

Macdonald, S. (Ed.). (2002). *Concrete Building Pathology*. Blackwell Publishing Ltd.

Madsen, J. J. (March 2008). The Realization of Intelligent Buildings. *Buildings*.

Markovitz, M. (August 27, 2008). The Differences Between Green Globes and LEED. *PROSALES*.

Martín, C. and A. Foss, *All That Glitters Isn't Green and Other Thoughts on Sustainable Design*, PATH Partners, National Building Museum, fall 2006.

May, S. (2009). *Do Green Design Strategies Really Cost More?* DC&D Technologies, Inc DCD Construction.

Mendler, S., & William, O. (2000). *The HOK Guidebook to Sustainable Design*. New York: John Wiley & Sons.

Morris, P., L. F. Matthiessen, and Davis Langdon, *Cost of Green Revisited*, July 2007.

Muldavin, S., "A Strategic Response to Sustainable Property Investing," *PREA Quarterly*, Summer 2007, pp. 33–37, www.muldavin.com.

Murphy, B. L., & Morrison, R. D. (2007). *Introduction to Environmental Forensics* (2d ed). Academic Press.

Nadel, B. A. (2004). *Building Security—Handbook for Architectural Planning and Design*. New York: McGraw-Hill.

National Institute of Building Sciences, *The Whole Building Design Guide* published online by www.wbdg.org/sustainable.php.

Needy, K.L., R. Ries, N. M. Gokhan, M. Bilec, and B. Rettura, "Creating a Framework to Examine Benefits of Green Building Construction," *Proceedings from the American Society for Engineering Management Conference*, October 20–23, Alexandria, Virginia, pp. 719–724, 2004

Piper, J. E. (2004). *Handbook of Facility Assessment*. Fairmont Press.

Poynter, Dan. (1997). *Expert Witness Handbook: Tips and Techniques for the Litigation Consultant* (2d ed). Para Publishing.

Propst, Robert. (1968). *The office—A Facility Based on Change*. Elmhurst, Illinois: The Business Press.

Prowler, D., Donald Prowler & Associates, Revised and updated by S. Vierra and Steven Winter Associates, Inc., *Whole Building Design*, Last updated: August 7, 2008.

Pulaski, M. H., Horman, M. J., & Riley, D. R. (2006). Constructability Practices to Manage Sustainable Building Knowledge. *Journal of Architectural Engineering*, 12(2).

Ratay, R. T. (2005). *Structural Condition Assessment*. New York: John Wiley & Sons.

Reznikoff, S. C. (1989). *Specifications for Commercial Interiors*. New York: Whitney Library of Design.

Rogers, E., & Kostigen, T. M. (2007). *The Green Book*. New York: Three Rivers Press.

Samaras, C. (2004). *Sustainable Development and the Construction Industry: Status and Implications*. Carnegie Mellon University http://www.andrew.cmu.edu/user/csamaras.

Sampson, C. A. (2001). *Estimating for Interior Designers*. Watson-Guptill Publications.

Sara, M. N. (2003). *Site Assessment and Remediation Handbook* (2d ed). CRC.

Scheer, R., & Woods, R. (April 2007). *Is There Green in Going Green?* SBM.

Smith, W. D., & Smith, L. H. (2001). *McGraw-Hill On-Site Guide to Building Codes 2000: Commercial and Interiors*. New York: McGraw-Hill Professional Publishing.

Starr, J., & Nicolow, J. (October 2007). *How Water Works for LEED, BNET*. CBS Interactive Inc.

Steven Winter Associates. (1997). *Accessible Housing by Design*. New York: McGraw-Hill.

Stodghill, A. (August 27, 2008). LEED vs. Green Globes. *It's the Environment Stupid*.

Sullivan, P. J., Franklin J. A., & Richard, K. T. (2000). *Practical Environmental Forensics: Process and Case Histories*. Wiley.

Syal, M., *Impact of LEED® Projects on Constructors*, AGC Klinger Award proposal, Michigan State University, Construction Management Program, School of Planning Design and Construction, East Lansing, MI, 2005.

Tilton, R., Jackson, H. J., & Rigby, S. C. (1996). *The Electronic Office: Procedures & Administration* (11th Ed). Cincinnati, Ohio: South-Western Publishing Co.

Turner, C., & Frankel, M. (March 4, 2008). Energy Performance of LEED for New Construction Buildings. *New Buildings Institute*.

Turner Construction, Study: *Green Building Market Barometer*, 2008 (www.turnerconstruction.com/greenbuildings).

United Nations, *Report of the World Commission on Environment and Development*, General Assembly Resolution 42/187, December 11, 1987. Retrieved: November 2007.

USACE Army LEED Implementation Guide, Headquarters, U.S. Army Corps of Engineers, January 15, 2008.

USGBC, *United States Green building Council*, www.usgbc.org, viewed in 2008-2009.

U.S. Green Building Council, *LEED Reference Guide for Green Building Design and Construction, For the Design, Construction and Major Renovations of Commercial and Institutional Buildings, Including Core & Shell, and K-12 School Projects*, 2009 Edition.

U.S. Green building Council, *LEED 2009 for New Construction and Major Renovations*, November 2008.

Woods, J. E., "Green Building: Balancing Fact and Fiction," S. E. Cannon and U. K. Vyas, moderators, *Real Estate Issues*, Vol. 33, No. 2, p. 10, 2008.

Woods, J. E., "Expanding the Principles of Performance to Sustainable Buildings," Focus on Green Building, *Entrepreneur Media, Inc.*, Fall 2008

Yoders, J. (April 1, 2008). Integrated Project Delivery Using BIM. *Building Design and Construction*.

Yudelson, J. (2008). *The Green Building Revolution*. U.S. Green Building Council.

Yudelson, J. (March 2008). *The Business Case for Green*. Yudelson Associates.

Zelinsky, Marilyn. (1998). *New Workplaces for New Workstyles*. New York: McGraw-Hill.

Index

Printed in the United States
By Bookmasters